# 101
# Select
# DREAM HOUSES

ROUGH SAWN
VERTICAL SIDING
PLANTER

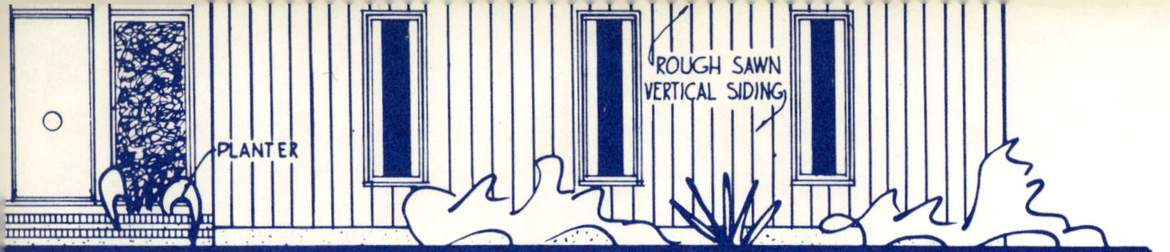

NT ELEVATION

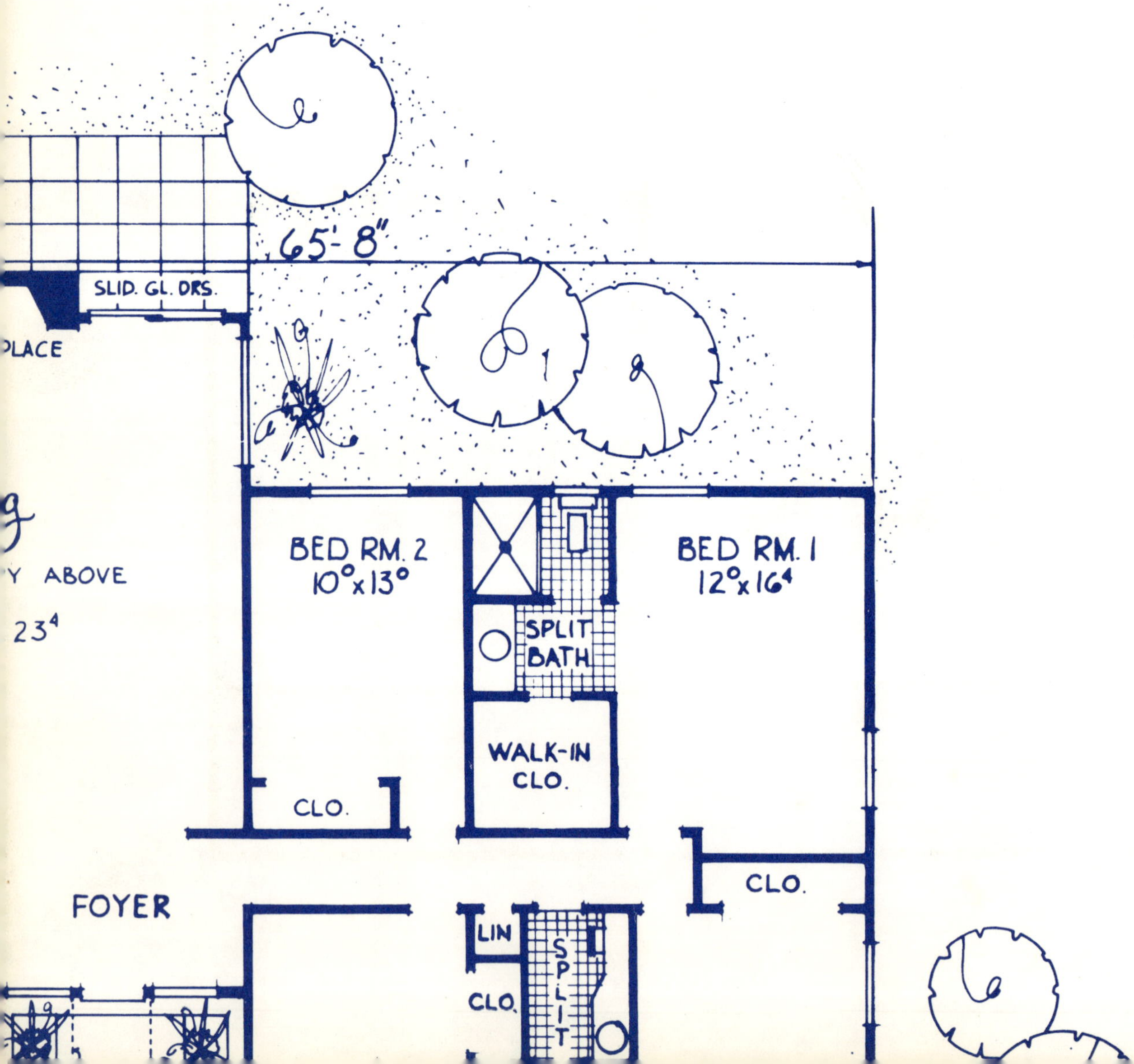
65'-8"
SLID. GL. DRS.
PLACE
Y ABOVE
23⁴
BED RM. 2
10⁰x13⁰
BED RM. 1
12⁰x16⁴
SPLIT
BATH
WALK-IN
CLO.
CLO.
CLO.
FOYER
LIN
CLO.
SPLIT

# 101 Select DREAM HOUSES

by Andy Lang

Library of Congress Cataloging in Publication Data

Lang, Andy, 1909-
101 select dream houses.

1. House construction — Amateurs' manuals.
I. Title.
TH4815.L36 643 72-6309
ISBN 0-8437-3240-7
ISBN 0-8437-3242-3 (pbk.)

# CONTENTS

# INTRODUCTION

The house designs shown in this book were carefully selected from among thousands in the popular House of the Week series, a creation of Associated Press Newsfeatures. The plans were designed for use in custom building and not merely redone from "development" plans intended for mass construction.

No matter what your age, your space requirements, your style preferences or the size of your pocketbook, you will find homes to meet your needs.

The houses have been arranged in the order of their square footage, with the smallest being shown first. Bear in mind, however, that it costs more to build horizontally than vertically.

We are indebted to the architects for permission to use their designs. Each is a professional with many years of experience in formulating plans for discriminating owners. A full set of each architect's plans is available at a surprisingly low cost. They meet and exceed the standards of the Federal Housing Administration and building codes.

Each house has a design number. When you have selected a plan you like and want the working drawings, consult the following list for the architect's name and address.

Samuel Paul,
107-40 Queens Blvd.,
Forest Hills, N.Y. 11375.
Z-82, Z-95, Z-98, S-11, S-25, S-30, S-44, S-48, S-53, S-74, S-79, S-87, S-94, L-12, L-16, L-19, L-22, L-28, L-33, L-36, L-40, L-46, L-47, L-50, L-55.

Herman H. York,
90-04 161st St.,
Jamaica, N.Y. 11432.
Z-94, Z-97, S-26, S-36, S-45, S-49, S-59, S-63, S-68, S-89, L-11, L-14, L-18, L-21, L-26, L-30, L-34, L-38, L-42, L-44, L-49, L-54, L-57.

Derick B. Kipp,
Architects' Home Plan Bureau,
Room 704, 48 West 48th St.,
New York, N.Y. 10036.
S-27, S-37, L-27, L-41, L-51.

Herbert C. Struppmann,
Architects' Home Plan Bureau,
Room 704, 48 West 48th St.,
New York, N.Y. 10036.
Z-92, S-51.

Rudolph A. Matern,
Master Plan Service,
89 East Jericho Turnpike,
Mineola, N.Y. 11501.
Z-84, Z-99, S-2, S-19, S-28, S-33, S-38, S-47, S-52, S-61, S-77, L-10, L-25, L-31, L-43, L-52.

William Chirgotis,
37 Mountain Ave.,
Springfield, N.J. 07081.
Z-83, S-15, S-24, S-29, S-34, S-39, S-50, S-55, S-73, S-78, L-9, L-15, L-24, L-29, L-35, L-39, L-48, L-53.

Lester Cohen,
Architects' Home Plan Bureau,
Room 704, 48 West 48th St.,
New York, N.Y. 10036.
S-64, S-86, S-99, L-8, L-13, L-17, L-20, L-23, L-32, L-37, L-45.

Fenick A. Vogel,
Architects' Home Plan Bureau,
Room 407, 48 West 48th St.,
New York, N.Y. 10036.
L-56.

# Chapter One
# THOSE IMPORTANT FIRST DECISIONS

## 1. How Much to Spend

Once you arrive at the decision to have a house built, you collide head-on with a second and more important determination. You must find the answer to the multi-thousand-dollar question: "How much can I afford to spend?"

Whatever you may have heard about a rough rule-of-thumb to help you decide on a price you can handle, including the one about two-and-one-half times your annual gross income, discard it! There is no way anyone in the world can even guess at your ability to buy a house costing $25,000 or $50,000 though he may have complete knowledge of the size of your annual gross income. In the days when families were on a pay-as-you-go basis, perhaps. Today, when the vital factor is not how much you make but how much you owe, a different method must be employed.

What you have left every month after you pay the bills is what counts most — if for no other reason than that's what your bank, savings and loan association or other lender will take into consideration before it gives you a mortgage. It knows that if you make $15,000 a year but are virtually debt-free, you probably can afford a costlier house than someone earning $25,000 but with heavy installment payments.

Entirely aside from the debt factor, there are other considerations affecting what you can afford. How large a down payment can you make? The larger the down payment, the smaller the monthly mortgage payments. How high are your present expenses? Which of them will be eliminated when you own your own house? What new expenses will you take on along with home ownership? What additional costs lie ahead because of your age and family situation? If you're young and have a family, will you be able to handle the inevitable heavy expenses of higher education for your children and still meet the mortgage payments? If you're getting along in years, have you thought about how you will handle such things as the mortgage and maintenance costs after retirement brings a reduced income?

If you already keep a family budget, you have a good idea how much you can lay out each month toward a mortgage payment, including taxes, interest and property insurance. Most monthly mortgage payments these days are made in this way.

If you presently are paying rent, you know approximately what monthly mortgage payment you can afford.

Even so, you should still put everything down on paper, taking into account certain economic changes that will occur when you become a home owner. The cost of your transportation may be different than it is now. You'll have to pay for your fuel. And you'll be responsible for maintenance costs.

Use the following list to determine how much of your income can be applied to monthly mortgage payments:

MONTHLY INCOME

Gross .................................................. $......................

Deductions .......................................... $......................

Take-Home Pay .................................. $......................

MONTHLY EXPENSES

Food .................................................... $......................

| | |
|---|---|
| Clothing | $ |
| Transportation | $ |
| Medical Care | $ |
| Life Insurance | $ |
| Savings | $ |
| Utilities and Fuel | $ |
| Contributions | $ |
| Dues | $ |
| Recreation | $ |
| Installment Payments | $ |
| Miscellaneous | $ |
| TOTAL EXPENSES | $ |
| TAKE-HOME PAY | $ |
| EXPENSES | $ |

Deduct your expenses from your take-home pay and you get the AMOUNT AVAILABLE FOR MONTHLY MORTGAGE PAYMENTS .............. $..................

## 2. Small or Large Down Payment?

You'll find, when you actually get down to having a house built, that the builder usually will have some understanding with a lending institution to handle mortgage loans for his customers. Examine the terms of the proferred mortgage agreement, but do not feel obligated to accept it. You may be able to do better on your own. More on mortgages later.

The larger the down payment you make, the smaller the size of the mortgage loan, which means that you will pay less interest over the years. You also will pay less total interest on a short-term mortgage of 15 years or less than on a long-term mortgage of 25 or 30 years.

Ideally, then, you should make a large down payment, obtain a low rate of interest and agree to repay the loan in 15 years or less.

But, most of the time, it doesn't work out quite that way. Most people can not afford to make a large down payment. Others, even if they have the money, should not do so. In this latter category are those who would be stripping themselves of all "rainy day" cash reserves if they made a large down payment. Also included are those who might be able to use the extra money in an investment which would return an attractive yield.

In any case, remember that you will need cash for a number of things connected with your purchase. These include the closing costs, which often run to several hundred dollars. It will cost you money to move and to do any redecorating or landscaping. It may be necessary to buy some new furnishings. And there will be legal fees.

A large down payment, therefore, is desirable but not always practical. To help you make your decision, after evaluating your financial status (many persons find they have more assets than they might have supposed in making a quick estimate), you should have some knowledge of how the size of your down payment affects the amount you will have to pay every month over the years.

The examples that follow do not include property insurance and taxes.

EXAMPLE 1 — The price of the house is $30,000. You make a down payment of $8,000 and obtain a $22,000 mortgage at 7 per cent for 20 years. Your monthly payments will be $170, which takes care of both amortization and interest.

EXAMPLE 2 — Assume the same purchase price of $30,000 at the same interest rate of 7 per cent over the same 20-year period. But this time you make a down payment of $12,000. Your monthly payments will be $139.

EXAMPLE 3 — Again the same purchase price, the same interest rate and the same mortgage period — only this time you can afford a $15,000 down payment. Your monthly payments drop to $116.

No doubt about it. The savings are considerable over the period of the mortgage. However, should you be one of those able to make either a large or a small down payment, don't jump to a quick conclusion. Money NOT used for the down payment will earn interest in a savings institution or can be used for investment purposes.

## 3. Selection of Neighborhood

Whether you are placing your new house on a lot of your own or one which the builder has made available, you should exercise considerable care in determining that the neighborhood will be satisfactory. It is surprising how many persons are in a neighborhood they like even though the house doesn't quite meet their needs. And others move out of a satisfactory home because they are dissatisfied with the neighborhood.

The selection of the right neighborhood for you and your family should be guided by a number of factors. It is not enough that someone you know — a relative, a friend or a fellow worker — lives in a particular section. While this has some advantage, especially if similar economic and social backgrounds exist, there may be a wide disparity of family requirements. You must decide what your own needs are and whether the community will meet them. Far more important than whether you have to go several blocks or several miles to visit your relative or friend will be the distances between your home and the facilities you use regularly.

A neighborhood is made up of things, including the general design of the houses, the location of the schools and churches, the shopping conveniences, the types of utilities and other material matters on which you can place a figurative finger and say: "This is the way it is."

But it is the character of the neighborhood that will play the most important role in determining whether the area will wear well. If you live near the neighborhood, the chances are you've already formed an opinion about it. You know whether the houses and lawns are kept in good condition. You know the type of people who live there. You know what the reputation of the section is. In short, you know ahead of time whether it's the kind of place you want to live in.

In many cases, however, you have no such advance information. You look at a neighborhood and decide it has a nice appearance. You investigate and find there's a school and a place of worship nearby. You can see that the shopping center is convenient. The transportation doesn't appear to be too bad. But, you wonder, what about the people who will be your neighbors? Will you get along with them?

If those last two questions give you concern, investigate further. Talk to the local minister or priest or rabbi. Ask questions of a real estate broker. Try to find someone who knows somebody who lives there and can arrange an introduction for you. Visit the neighborhood several times, at night as well as during the day. Attend some community function if possible. Go shopping. Observe and ask questions.

Some persons, who make friends easily and generally are well liked, may not be too concerned about getting along with their neighbors. They're confident that things will work out all right. For this type of person, things usually do. But even they should not neglect visits to the neighborhood at different times of the day. Such visits can be sharply revealing.

Do you have small children? Then you will want to find out whether many of the families in the neighborhood also have them. Entirely aside from the matter of companionship for your youngsters is the attitude the neighborhood will have toward them. Adults whose children are grown and have moved away are sometimes impatient with the noise and antics of children enjoying themselves.

Remember that neighborhoods change over the years. Older neighborhoods usually have settled down to a certain level. New ones have not, which makes the checking-up process even more important. The following list will help you to make your overall estimate:

| | GOOD | FAIR | POOR |
|---|---|---|---|
| Attractiveness | ☐ | ☐ | ☐ |
| Congeniality | ☐ | ☐ | ☐ |
| Reputation | ☐ | ☐ | ☐ |
| Privacy | ☐ | ☐ | ☐ |
| Safety | ☐ | ☐ | ☐ |
| Probable Stability | ☐ | ☐ | ☐ |
| Transportation | ☐ | ☐ | ☐ |
| Shopping | ☐ | ☐ | ☐ |

| | | | |
|---|---|---|---|
| Schools | ☐ | ☐ | ☐ |
| Churches | ☐ | ☐ | ☐ |
| Play Areas | ☐ | ☐ | ☐ |
| Recreation Facilities | ☐ | ☐ | ☐ |
| Utilities | ☐ | ☐ | ☐ |
| Zoning Restrictions | ☐ | ☐ | ☐ |

Most of the time you will have to make a compromise somewhere. Rarely does a neighborhood have everything you want. Only you can determine where you should give a little in order to get the benefit of the neighborhood's other advantages.

## 4. What Kind of House?

Should you choose a ranch, a split-level, a raised ranch, a two-story or a $1\frac{1}{2}$-story house? If you already own a home and have decided to sell it and buy another, you probably have already come to one of two conclusions: you want a similar type, larger or smaller, in the same or a different neighborhood, but the same kind; or you definitely don't want a similar one because you have learned that it has certain disadvantages which you would like to avoid in the new dwelling.

As an aid to making a proper choice, here is a capsule summary of the things that should be taken into account:

RANCH — Generally, easier to construct. Stair climbing is eliminated. Maintenance is easier. More opportunity to create a relationship between the indoors and the outdoors. Easier to sound-control. Little inside space is wasted. Usually requires more land. Bedrooms not as private as some persons prefer. Horizontal walking distances may be greater. Requires more foundation, roofing and insulation. Cost, on the basis of amount per square foot, is usually higher.

TWO-STORY — Permits more interior space on less land. Cost, on the basis of amount per square foot, is usually lower. Often has an impressive exterior appearance. Bedrooms usually can be larger. Bedrooms are more private. Upstairs need not be meticulously groomed when casual company drops in. Maintenance of second-floor exterior more difficult. Requires more stair climbing. Wasted space in area of stairway. Needs at least one extra bathroom.

SPLIT-LEVEL — Can be built on rolling terrain. Blends better with two-story houses than ranch houses. More livable space for the money than a ranch, but requires more land than a two-story. Lends itself to attractive exterior design. Some space wasted in stairway areas. Less stair climbing when going from one level to another, but overall stair climbing may be as much or more than in a two-story. Requires zoned controls to provide even temperatures at various levels.

ONE AND ONE-HALF STORY — Can be built on small lot. Second floor need not be finished at time of original construction. Permits master bedroom to be on first floor, children's bedrooms upstairs. Provides extra storage space under eaves. Needs second bathroom on upper floor. Has knee walls and sloping ceilings upstairs.

RAISED RANCH — Looks larger than it is, but must be carefully designed to avoid awkward appearance. Blends with different types of houses. Can be built on flat or rolling terrain. Wasted stairway space. More livable space for the money, but more difficult to maintain than a regular ranch. Utilizes what ordinarily would be the basement.

These kinds or types of houses are built in many architectural styles. There was a time when the style of a house automatically indicated which type it was. Thus, a Cape Cod was a $1\frac{1}{2}$-story, a Dutch Colonial a two-story, a Spanish hacienda of the Southwest a single-story or ranch, and so on. That is no longer true. While some excellent examples of traditional architecture are still being built, most houses are hybrids, mixtures of two or more styles.

Whereas the kind of house you select will be dictated by your needs, the style will be a matter of personal taste. And if you don't have a feeling for the characteristics of a traditional house, you may want a contemporary, an all-encompassing term for a home with styling of the present period. Contemporary styling can be exciting or dull, but should not be confused with modernistic styling, which is esoteric and likely to please only the owner and a tiny minority of those who see it.

## 5. Room Arrangements

Speaking of style, it is the life style of your family that must be considered when deciding on a suitable room arrangement. What is perfectly satisfactory for a young couple with one or two small children or an elderly couple with no children may be all wrong for a husband and wife with one or more teen-agers.

Because this is so, you must carefully consider the matter of traffic flow or circulation from room to room and area to area; in fact, the entire walk pattern of a house.

The requirements for some families call for a distinct separation between the living and the sleeping quarters. This is especially true when one part of the family does entertaining at a time when another part usually has retired for the night. If you plan on making one room into a nursery, be sure it is convenient to the master bedroom. To get a better idea of whether the traffic pattern in this particular house is what you want, try to visualize a busy day's activities to determine whether the room arrangement will be suitable.

Do you need a small or large kitchen and does the one in the house you are considering fit your needs? Is it located so that it will serve as a kitchen and not as a walk-through in going from one room to another? Is the dining area large enough for your family? Will you have enough space for a dining table when you have company? Are there enough cabinets, drawers and shelves for dishes, silverware and food containers? How about counter space?

In the one-floor house, mother can do her work, keep an eye on young children or aged grandparents living in the household, and move horizontally through all areas without the effort and time-consuming feat of going upstairs to see what needs to be done. But many mothers, and older householders as well, feel much more secure if they go upstairs to bed, less exposed to the threat of a housebreaker when they are on an upper floor.

Privacy is another major item: for the parents where there are very young children, and for the juniors as they grow up and need the all-important assurance that there is a welcoming, private area for them in their own home. This can be achieved in a one-story house by use of a wing for bedrooms, separated from the rest of the house by good halls. The family room is popular with most householders for this reason, too, offering an informal area for the children to use, with the major entertaining rooms of the house kept for all-together or company activities. Particularly important in the informal, space-for-every-generation scheme in a ranch house is the larger-sized basement, offering excellent areas for teen-agers. Because the full basement under the sweep of the ranch plan is so large, and the basement in a moderate two-story plan is smaller, the former gives widest scope for play, hobbies and other activities in a space all their own. The split-level usually has a very good recreation area on the ground level, and access is particularly good. Interestingly, the bi-level provides the largest basement area at grade level of any house style, and perhaps this accounts for the growing popularity of this style. In the two-story home, the upper floor offers complete privacy, and mother's admonition "take your friends up to your room" gives her the chance to see her own friends in peace and quiet that is usually harder to achieve in a one-floor house.

The split-level plan offers a compromise, and a good one for many families. Where the two-story house has 14 steps from first to second floor, the split-level breaks this pattern and still provides the "other floor" aspects for privacy and a lot of extra space potential as well. However, we repeat what we said earlier: a split-level is not necessarily the answer.

# Chapter Two
# ARRANGING THE DETAILS

## 1. Do You Need an Architect?

You are a prospect for building your own house rather than buying a ready-made one *if* you own or can acquire a plot of ground in a neighborhood of your choice — or *if* you've toured the housing developments and inspected individual new homes for sale without finding anything that satisfies you — or *if* your desires are highly individual and you have a fair idea of the kind of plan that will suit you — and *if* your budget is flexible enough to meet the costs, since a house made-to-order obviously commands a higher price than one which has been produced on a mass basis.

The advantages of building new as opposed to buying new are many. They include:

LOCATION — You can put the house where you want it.

STYLE — Any style can be yours if you're willing and able to pay.

SPACE — If you've owned or rented a house previously, you have a good idea of your minimum space requirements and can provide for them.

TRAFFIC PATTERN — You know the habits of your family best and can work out the house's traffic pattern.

BUILDING MATERIALS — You can make your own selection of materials which give the best service and will look best.

Other advantages of custom construction include rooms in sizes you like, work and play areas where you want them, and less waste space generally. Careful planning is the most important factor in getting started.

You need house plans and specifications in order to have a house built. There are two ways to get these from an architect. One is to select an architect, talk it over and have him draw up plans according to your ideas and suggestions. The other is to look over the drawings and floor plans of houses that appear in publications, such as those selected for this book. Each house bears a number. At the beginning of the book is a list of the architects, their addresses and the numbers of the houses they have designed. You may directly communicate with the one whose house you have chosen. You will find that you can obtain four sets of plans (at least three and usually four are needed) for between $40 and $50 total.

Since some of you may be searching for ideas to merge with your own with the idea of hiring an architect to draw up the plans, it is well to know something about architects and what they do and charge.

There is a general misconception that an architect is a person whose sole job is to design. But an architect offers many supervisory services which transfer all the problems of home building from your shoulders to his. He can analyze the bids of contractors and select or assist you in selecting the proper one. He can supervise day-to-day construction to see that the work is proceeding according to the plans. He can tell you when and when not to honor requests from contractors for payments. He can be sure that no liens exist. He can take care of the many details necessary to be certain you are completely satisfied before you move in. He can, in short, act as your agent from start to finish.

The architect's fee is determined by whether you want his full services, or whether you merely want him to draw up the plans or, in less usual circumstances, whether you want only his supervision, having obtained the plans elsewhere. Fees for an architect who handles everything usually run about 10 per cent of the total cost of the house but can be more or less depending on the arrangements you work out with him. Some architects will set a flat fee if you so desire.

Since there can be no doubt that a good architect often can save an owner a sum equal to or larger than his fee, it is important that you obtain the services of a good one. Nothing is better than the recommendation of a satisfied client. Lacking this, look at the houses in the neighborhood where you intend to build. If you see a house you admire, find out who designed it, since many architects are specialists. Banks and other lending institutions can make recommendations and advise you where to check examples of work.

When you have made a selection and wish to double-check your choice, get in touch with the local office of the American Institute of Architects, whose members must conform to high standards of professional practice. Non-membership in A.I.A. does not reflect in any way on an architect's competence and integrity, but the organization has mandatory standards for its own members. Membership or not, an architect is right for you if he is reliably recommended, if you like the kind of work he does and if you can reach an agreement with him on price, services and other details. It is usually a good idea to choose an architect whose office is not too far from the area where you are building, especially if he will have supervisory duties.

## 2. Choosing a Builder

Since the workmanship that goes into a house is all-important, the quality of the finished product depends on how well it is put together.

If you have hired an architect, leave the selection of the contractor or builder up to him. He knows who the competent and responsible builders are. He usually will choose one through competitive bidding, issuing bid invitations only to those builders who can do a good job. The lowest bidder usually will be given the assignment.

If you are building with the use of an architect's plans but have not personally engaged an architect, you must choose a contractor carefully. Don't shop around to see who can do the job at the lowest price. You then may wind up with a builder to whom an architect would not even have submitted the plans for bidding.

Just as in the choice of an architect, the best recommendation is that of a satisfied client. If you know someone who had a house built and was satisfied with the result, ask for the name of his builder. If not, a local bank or other institution that issues mortgage loans will be glad to make a recommendation in the hope of obtaining your business.

## 3. Choosing a Lot

Nearly everyone who hires an architect or who deals directly with a builder is already a lot owner. He knows where he wants the house to be; now he wants to see it come to life. But if you are one of the minority who is ready to build and has not yet bought a lot, make your purchase with care.

Follow our previous recommendations regarding selection of a neighborhood, then exercise additional care in the choice of the lot itself. It is especially important that you investigate every aspect of the zoning laws or you may find out too late that you can't build the type of house you want on the land you own. Visualize your house on the lot. Will the sun be where you want it when you want it? Will there be enough room for a lawn and a backyard? Will you need a garage and where will be the best place for it? And—this is vital—is the lot located on a street and in a neighborhood where your house will not be out of tune with the others structurally, or out of line on cost?

## 4. Where Not to Skimp

In having a house built to order, you are automatically evincing a desire to own a house particularly suited to your needs. Make sure you get one. Don't insist on costly frills that add to the price when you actually need plenty of closet space or an extra bath. Individually is fine within bounds and good taste. But don't overdo it. Some day you may want to sell the house, and you may discover that your prospective purchasers are interested in that closet space and extra bath, and in whether there is sufficient insulation as well as a dozen other practical things that make family living comfortable.

The matter of which frills to avoid is often individual, depending on needs,.taste and the amount of money available. It is wiser, for instance, to spend money to be certain a house is protected against termites than to have expensive hardware. It is more important to be sure there is sufficient electricity and there are are enough convenient outlets than to have ornate electrical fixtures. But if money is no object, you can have anything you want. In all cases, avoid excessive ornamentation, which is often a bar toward the sale of an otherwise attractive house.

## 5. Extra Costs

It often happens that a family which has closed a deal with a contractor eventually pays $35,000 for a house even though the contract called for a payment of $30,000.

How does this come about, and how can you avoid it? The first and foremost way is not to ask for seemingly unimportant structural changes *after* the contract is signed. Deciding, for example, on larger windows than those called for in the plans will cost you money if the decision comes too late. Builders usually will agree to certain plan changes without extra charges if the decision is made ahead of time.

If you think too much stress is being placed on this aspect of having a house built — probably because you think *you* are not the type to change his mind — these are the words of a prominent builder of custom houses: "I have rarely built a house without having the buyer seek one or more costly change before construction was finished."

## 6. Additions--Now or Later?

Leaving some part of the house unfinished so that you can do it yourself later will save you money only if you have the ability — and, more important, the time to complete the job. Otherwise it usually is cheaper to have the work done during the original construction. It is fairly common for a man to call in a remodeling contractor to finish an attic, for instance, because he discovered six months after buying the house that, for some reason, he could not tackle it himself. On the other hand, the attic might well be left unfinished if the owner had no immediate use for it as living quarters and knew he could make the transformation if the need arose.

The houses shown in this book have different kinds of foundations, but in some cases the builders can make changes to suit your requirements.

There are three major types of foundations: those which provide for full or partial basements; those which are low and have crawl spaces separating the houses from the earth; and those of the concrete slab type set right into the ground surface. In most cases, you should be guided by the advice of your architect or builder, since weather conditions affect the suitability of the kind of foundation most desirable. Where temperatures are very low in the winter, the foundation must be placed well below the frost line of that section. Since the foundation already will be very deep, it is the better part of economy to spend a little more and get the extra space of a basement. But houses are sometimes built on concrete slabs in the northern part of the country, and houses with basements often are built in the south.

# Chapter Three
# PROTECTING YOUR INVESTMENT

## 1. Drawing up the Contract

An architect hired to supervise the construction of your home — whether or not he drew up the original plans — has the responsibility of providing the necessary technical information for the contract between you and the builder and seeing to it that the work is being done in accordance with the specifications.

But whether you are acting through an architect or dealing directly with the contractor, you must protect yourself at all times. As with any contract, it is not only what goes into it that is important but what doesn't. It is safe to say that most disputes occur over some point that wasn't included in the contract rather than one that was. If you have an architect, don't hesitate to call his attention to an omission of statement in the contract even though there may have been a verbal agreement about that aspect of the construction. If you don't have one, you are wise to obtain the services of an attorney who specializes in real estate transactions. He may uncover some "double talk" in the terms that might have escaped your attention.

Once in a while an owner may decide to do the sub-contracting on his own; that is, make a series of arrangements with various companies doing different kinds of work — carpenters, masonry workers, plumbers, electricians, roofers, etc. Unless the owner has extensive experience in building, he is likely to regret it. The details and complexities are so many and so varied that he is almost certain to wind up with a king-sized headache. It is far better, and sometimes not any more expensive, to hire a single general contractor. He chooses the sub-contractors. They are responsible to him. They must satisfy him or risk losing his business.

## 2. Kinds of Mortgages

We have already told you what a difference the size of the down payment can make in the amount of your monthly payments and the total cost of your house over the duration of the mortgage.

Also affecting the costs—monthly and long-range—is the rate of interest you are charged. Because this is so important to you you should shop for the best mortgage you can get.

To someone who has never owned a house, the term mortgage loan may seem a bit mysterious. It shouldn't. It is merely a loan for which property is pledged.

When you buy a house, you have to borrow enough money to make up the difference between your down payment and the total price. You agree to pay off the loan at so much per month, including interest, for a specified period of time. By the terms of the mortgage, you pledge the house as a guaranty that you will make those payments as scheduled. If you fail to do so, the lender, under certain legal conditions, may take over your house and sell it in order to get back the money you still owe.

It is a truism that most home buyers pay less attention to the terms of the mortgage than to anything else connected with the purchase. Yet a hasty, careless acceptance of a mortgage, without knowing its terms, can cost a home owner thousands of dollars over the years.

Mortgage loans are made by various kinds of financial institutions, the most common of which are banks and savings and loan associations.

Many persons, having heard about Federal Housing Administration (FHA) and Veterans Administration (VA) loans, assume that it is the government agencies which lend the money for home purchases. Except in one special case, which will be explained shortly, they do not.

The loans are made by private lending institutions. With an FHA loan, the Federal Housing Administration insures that the lender will not lose on the loan. To pay expenses and cover possible losses, FHA charges an insurance premium of ½ per cent per year on the unpaid loan. With a VA loan, made to eligible veterans, the Veterans Administration guarantees the lender a major portion of the loan in the event of default, with the borrower in this case paying no insurance premium. The only instance in which the VA makes a direct loan is the case of a shortage of credit in a rural area, and the veteran finds it impossible to obtain a loan in that section.

The third type of mortgage loan, in which the government plays no part, is generally called a conventional loan. It is the most common type in use today. If the borrower defaults, the lender must recover his money by taking over the selling of the house.

With an FHA or a VA loan, you usually can get by with a smaller down payment and have a longer period of time to repay the loan. This is an advantage if it dovetails with your financial situation. But the total cost of your house will be considerably more because you will be paying interest on a larger loan over a longer period of time.

Also, if you hope to get an FHA or VA loan, you must usually allow a longer period of time between the start of negotiations and the granting of the mortgage, since it takes longer to process a government-guaranteed loan. Many sellers do not want to wait too long to find out whether you can obtain a satisfactory mortgage, because if something goes wrong, they will have to begin all over again to try to sell the house. Do not put down any money unless you get a specification in writing that your deposit will be returned if you are unable to get a mortgage on certain terms in a reasonable, specified period of time. And be sure there is an immediate understanding, also in writing, whether such things as appliances, rugs, drapes, storm windows, screens, etc., are included in the sale price.

It is false economy not to use the services of a lawyer in buying a house. He not only will give you valuable advice; he will take care of the seemingly countless details which must be handled from the moment you make your selection until you move in. Home purchasers who attempt to get by without a lawyer stumble over numerous pitfalls. Try to find a lawyer who specializes in housing transactions. The fee is probably less than you think, usually about one per cent of the purchase price of the house. You and your lawyer can agree on a set price. The set fee is more common in new house transactions, since there are fewer details for the attorney to handle.

Interest rates vary according to general business conditions. When business is very good and demands for money are high, interest rates go up. When business is bad and borrowing demands slacken, interest rates go down.

On loans insured by the FHA and the VA, the government sets a maximum interest rate that can be charged to the borrower. This rate has changed several times in recent years and undoubtedly will move up or down in the future as conditions warrant. In order to prevent lending institutions from charging "what the traffic will bear," many states have laws limiting the amount of interest that can be charged on conventional mortgage loans. The important thing to remember is that, even when maximum rates are set by law, different lending establishments in the same section may charge different interest rates, at or below the maximum.

The uninitiated person is usually surprised to discover what a difference there is between two mortgages which vary only one-half of one per cent. Here is the effect of interest rates on the cost of a $20,000 loan over a 25-year period:

| Interest rate (per cent) | Monthly payment (principal and interest) | Total interest (over 25 years) |
|---|---|---|
| 6 | $129 | $18,600 |
| 6½ | 135 | 20,440 |
| 7 | 141 | 22,390 |
| 7½ | 148 | 24,330 |
| 8 | 154 | 26,280 |
| 8½ | 161 | 28,200 |
| 9 | 168 | 30,220 |
| 9½ | 175 | 32,370 |
| 10 | 182 | 34,460 |

Now you can see why the rate of interest, as well as the size of the down payment and the duration of the mortgage, is so important to you. But the fact remains that, once he has selected a house, the average buyer can't wait until he puts his signature on a mortgage, giving only fleeting attention to its terms.

Those terms, by the way, include a lot of other things not previously mentioned. You will want to know whether there is a pre-payment clause. This enables you to make extra payments ahead of time without suffering a penalty. Don't forget that if you pay your loan ahead of time, the lending institution loses a certain amount of interest. Perhaps you think the situation would never arise in your case. But this has happened more than once: a home owner, in moderate circumstances, pays regularly for a few years. He then comes into a sizable sum of money, perhaps by inheritance. "I will pay off my mortgage," he thinks, "and save all that interest I would have to pay for the next 15 years." But he discovers in some cases, the mortgage terms will not permit him to do this or, if they do, there is a financial penalty for doing so. This could have been avoided if he had seen to it that his mortgage included pre-payment terms satisfactory to him.

Another thing you will want to know is whether your mortgage contains some kind of "grace" period, so that if you delay one or more payments, no action will be taken until after a certain period of time has elapsed. With some mortgages, such "grace" periods automatically go into effect after the loan has been reduced to a certain level.

In some cases, you can get an open-end mortgage. This enables you to borrow money at a future time for improving your home or for other purposes—at the same interest rates and terms as provided in the original mortgage.

If the wage earner dies before the loan is paid off, will the survivors be forced to give up the house because they can't meet the monthly payments? If that is the case, then you may want to make some sort of insurance arrangement to cover that eventuality.

Sometimes a mortgage application asks whether sufficient life insurance is carried on the wage earner to take care of the loan. Some mortgages carry what is called a "loan modification" agreement. This allows adjustment of the original terms to cut down the size of the monthly payments if there should be a drop of income or a serious illness.

Be sure you know whether your monthly payments include taxes and hazard insurance as well as interest and reduction of principal. This won't mean more or less money to you, but it's usually more convenient to have the lending institution take care of the taxes and hazard insurance.

These are some of the important things that will or will not be in that fine print so many home owners neglect to read. The time to find out what will be in your mortgage is *before* it is drawn up. Later, you and your lawyer can check to see that it includes the terms you desired.

Many mortgages include all or most of the matters we have discussed. Many do not. It is your job to get the kind of mortgage that will best fit your needs. How? The same way you buy a suit or a dress.

You shop around until you find what you

want. A real estate broker can make a recommendation. The bank or savings and loan association where you have an account will be more than glad to talk it over with you. In fact, any lending establishment will take the time to explain everything in detail and answer all your questions. When you seek a mortgage, you're a potential customer in a competitive business. Unless you are a poor credit risk, you will have no trouble getting a satisfactory mortgage.

## 3. Present and Future Taxes

The property taxes on your new home will be determined by its assessed value and the local tax rate.

In most cases, the assessed value is set by authorities at a figure far below the actual market value. Thus, a $30,000 house might be assessed at $10,000 or $15,000, depending on what percentage of the market value is being used in that community.

The tax rate is set after the local budget has been approved. If the rate is $9, it means $9 for each $100 of the assessed value.

While it is possible to appeal the assessment that has been placed on your house — if you feel that it is unfair — most requests of this nature are rejected. After such a rejection, the matter can be taken to court if an attorney feels that you have good grounds for your position.

You can generally figure that your taxes will increase at a faster rate in a developing neighborhood than in an established one, especially where there is a need for additional schools. Both your taxes and the mortgage interest can be deducted from your gross income at tax time providing that you use the long form rather than the short form where there is a set deduction.

Some states have homestead exemptions which decrease the assessed value of your house or lower the tax rate. In some cases, they apply to all home owners, in some only to special categories of persons. If you are a senior citizen, a veteran, a widow or disabled, be sure to check with local authorities to determine whether you are eligible for some sort of tax break. If the exemption applies to everybody, you will get it automatically.

# Chapter Four

# 101 HOUSE DESIGNS

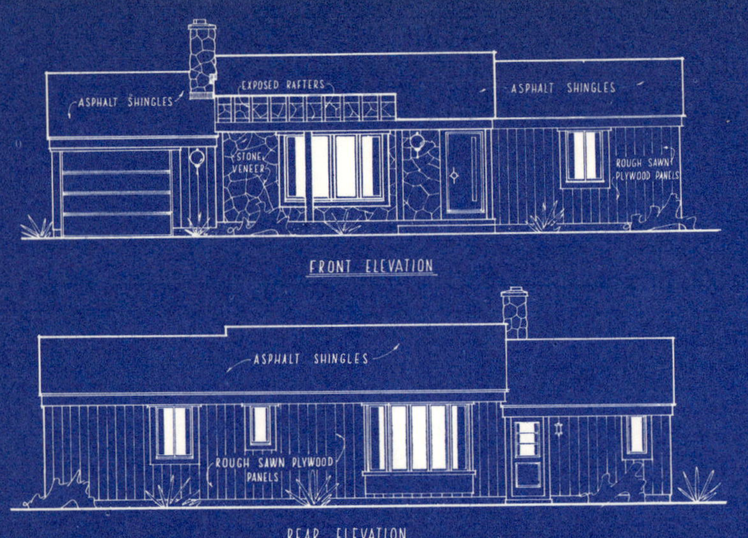

THE CONTEMPORARY

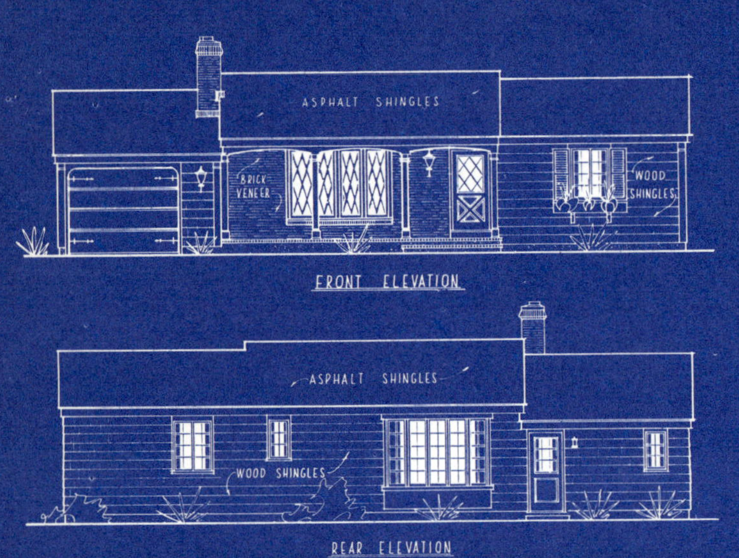

THE COLONIAL

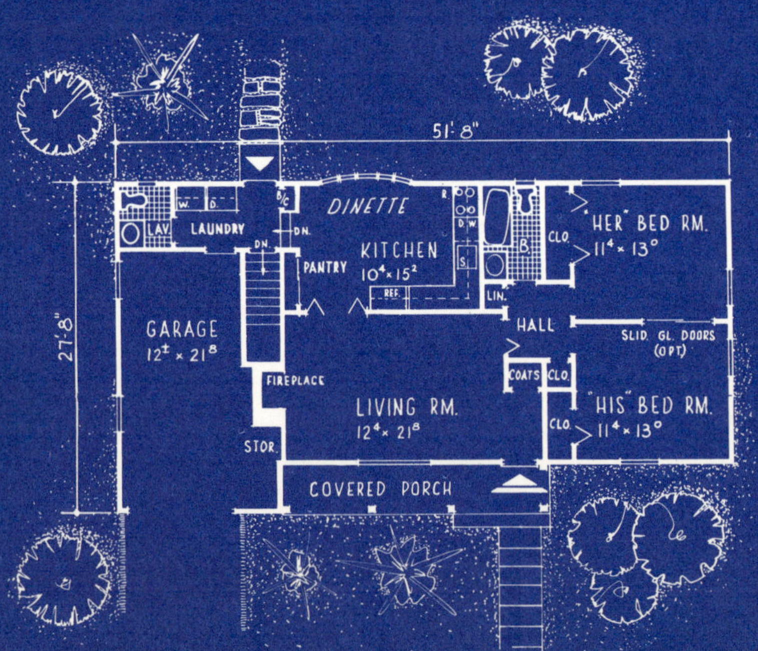


FLOOR PLAN

## HOUSE DATA

| | | |
|---|---|---|
| HABITABLE AREA | 904 | SQ. FT. |
| LAUNDRY AREA & LAV. | 78 | SQ. FT. |
| GARAGE & STORAGE | 249 | SQ. FT. |

DESIGN L-49

# Retirement House with Choice of Exteriors

Builders of retirement housing have noted that some buyers have begun to ask questions about contemporary styling.

This comparatively new trend influenced the design of this retirement home, which gives the owner the choice of a Colonial or contemporary design.

It is a small house — only 904 square feet of habitable area — yet it has a full-sized kitchen, a long living room and two bedrooms arranged in an unusual and practical manner. The same floor layout is used for both the traditional and contemporary models with changes only in the sidewall materials, window types and a few details, among them open rafters and wide overhangs.

The two bedrooms are identical in size and have been marked "his" and "hers" on the floor plan. There is a sliding glass door between the two rooms, creating "togetherness" yet retaining separate sleeping areas. Actually, one of the rooms can be used for guests by closing the door and drawing the drapes. The bathroom is convenient to both bedrooms as well as to the rest of the house.

A pair of bi-folding doors separates the living and dining areas. In retirement housing, there is need for casual, less formal living and dining. In this plan the dining room is incorporated as part of a large kitchen with the table set in front of a bow window overlooking the rear garden.

## Material List

**CONCRETE WORK**

| | | |
|---|---|---|
| Concrete Walls | | 697 cu. ft. |
| Slabs | | 426 cu. ft. |
| Foundation Footings | | 216 cu. ft. |
| Misc. Concrete | | 112 cu. ft. |

**STRUCTURAL STEEL**

| | | |
|---|---|---|
| Lally Columns | 3½" diam. | 5 pcs. |
| Girder | S—6 12.5# | 38 lin. ft. |

**BRICK WORK**

| | | |
|---|---|---|
| Chimney | Brick or Stone | 160 cu. ft. |
| Flue Lining | T. C. | 36 lin. ft. |
| Veneer | Brick or Stone | 158 sq. ft. |

**CARPENTRY**

| | |
|---|---|
| Framing Lumber | 5349 B.F. |
| Studs | 2000 B.F. |
| Plates | 600 B.F. |
| Roof Sheathing | 1850 sq. ft. |
| Sub Flooring | 892 sq. ft. |
| Side Wall Sheathing | 1430 sq. ft. |
| Insulation Walls | 1070 sq. ft. |
| Insulation Ceilings | 960 sq. ft. |
| Wood Flooring | 695 sq. ft. |
| Kitchen Plywood | 157 sq. ft. |

**MILLWORK**

| | |
|---|---|
| Exterior Doors & Frames Compl. | 2 |
| Garage Door Complete Set | 1 unit |
| Interior Doors & Frames Complete | 10 |
| Bi Fold Door Units | 4 units |
| Sliding Glass Door Unit | 1 unit |
| Sliding Doors | 2 |
| Window Units | 14 units |
| Fascia | 232 lin. ft. |

**KITCHEN CABINETS**

| | |
|---|---|
| Base Cabinets | 20′-9″ long |
| Wall Cabinets | 9′-9″ long |

**SHEET METAL**

Saddle & Counter Flashing 16 oz. Copper

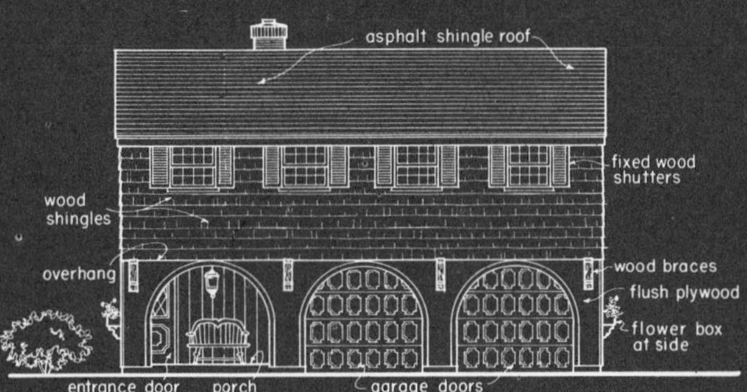


front elevation

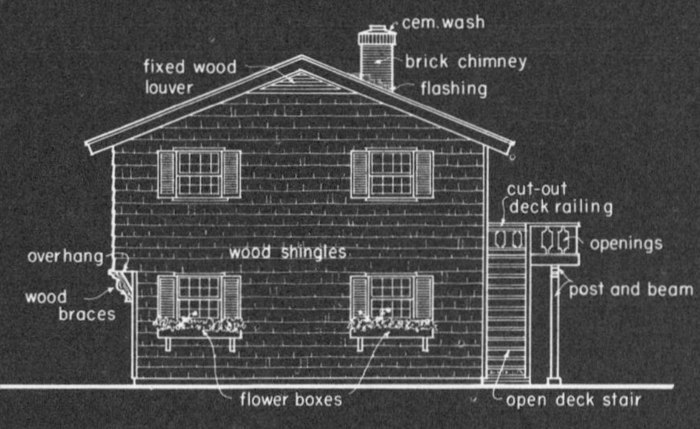


right side elevation

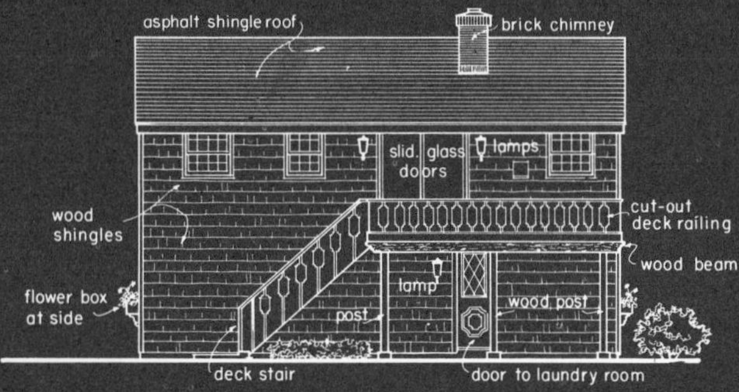


rear elevation

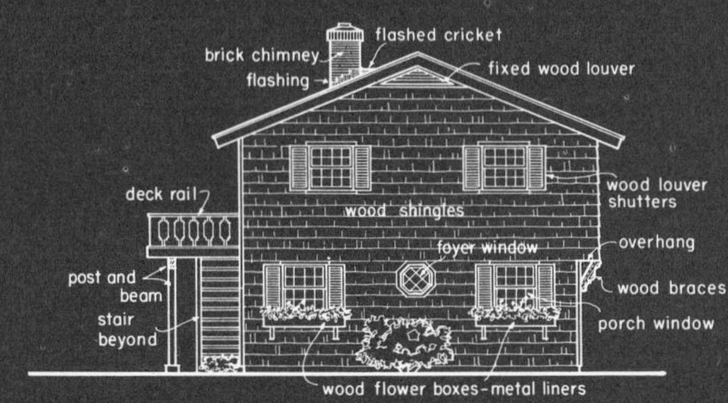


left side elevation

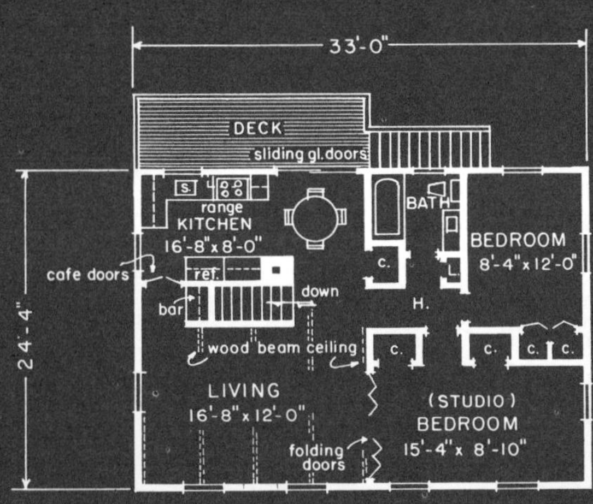


upper level

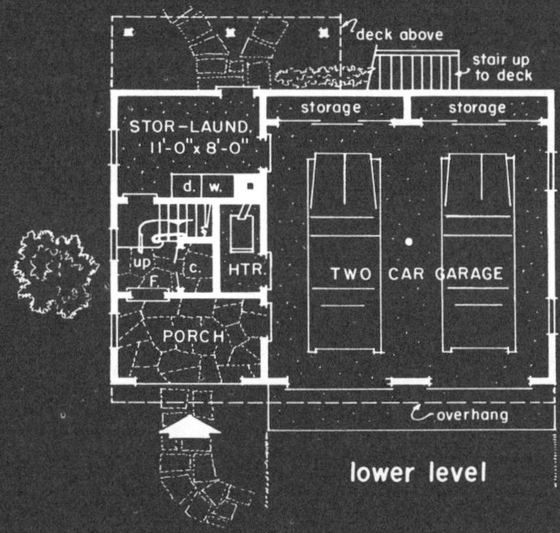


kitchen elevations

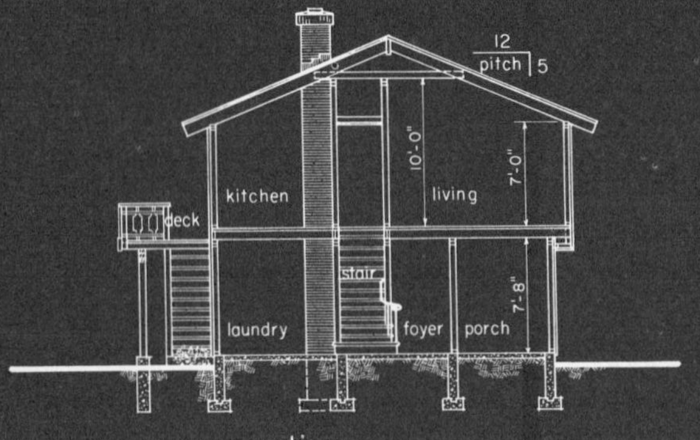


**size data**

803 square feet on upper level excluding deck.

142 square feet on lower level excluding, garage, porch and heater room.

DESIGN L-52

# Apartment-Garage for All Seasons

Not a complete house, but definitely a home! It's an apartment-garage — a structure with as many varied uses as people have varying conditions in their life patterns. Small in size and simple in construction, it can be built at a modest cost. That in itself is a big plus.

It could be built on residential property on which there already is a house with or without garage, so that it might be used for rental above and the owner's garage use below — or by renting both levels for a total income producer — or by having it occupied by a newly married couple of the same family for modest living quarters, with the garage rented.

It could be occupied by in-laws for their retirement with an arrangement that enables them to retain their independence by having their own apartment yet be close by. The garage could be rented or used by the owners and the in-laws, since it has a two-car unit.

It could be built as a vacation home, with the garage used for car or boat storage or as a rainy day play area.

It could be built on its own piece of property as a retirement home, with one or both of the garage units rented to supplement a fixed income.

It could be built by an individual to rent out as an income producer. Three rooms, a kitchen and a bath make up the efficiency apartment. A sloping beamed ceiling runs through the living room and kitchen.

## Material List

**CONCRETE**

| | |
|---|---|
| All Walls, Footings, Piers, Floors | 930 cu. ft. |
| Flagstone Porch & Walls | 85 sq. ft. |
| Waterproofing | 800 sq. ft. |
| 16 Lin. Iron Rail | |
| Brick Chimney 2'0 x 2'0 x 24'0 High | |

**CARPENTRY**

| | |
|---|---|
| Sub Floor Sheathing — Roof | 3740 sq. ft. |
| All Studs, Headers, Joists, etc. | 7160 BM |
| Vertical Siding | 150 BM |
| 24" Wood Shingles | 15 sq. |
| 1/4 Hardboard Underlay | 800 sq. ft. |
| Cornice & Trim | 180 BM |
| 3/8 Ext. Plywood Good I Side | 320 sq. ft. |
| 3/4 Int. Plywood & Half Round | 400 sq. ft. |
| Deck Posts, Beams Joists & Floor, Rails & Stairs | 550 BM |

**INSULATION**

| | |
|---|---|
| Batt Type | 2700 sq. ft. |

**ROOFING**

14 Sq. 235 lb. Shingles
10 Rolls 15 lb. Sat. Felt

**PLASTERBOARD**

| | |
|---|---|
| 1/2" Avg. Plasterboard Taper Joint | 6340 sq. ft. |

**MILLWORK**

| | |
|---|---|
| 3 O.S. Doors & Frames 3'0 x 6'8 | Trim 1 Side |
| 1 Alum. Sliding Door & Frame | Trim 1 Side |
| 11 Hollow Core Doors | Trim 2 Sides |
| 24 - 2 Panel Doors Avg. 1'0 x 6'8 | Trim 2 Sides |
| 1 Pair Cafe Doors 2'6 x 3'0 | Trim 2 Sides |
| 2 Garage Doors & Frames | Trim 1 Side |
| 15 DH Windows 3'0 x 3'2 | Trim 1 Side |
| 1 Octagon Window 2'0 | Trim 1 Side |
| 650 Lin. Ft. Moldings | |
| 24 Shutters 1'4 x 3'3 | |
| 4 Flower Boxes 6'0 x 10" x 10" | |
| 1 Flight Box Stairs 3'0 Wide | |
| 2 Gable Louvers 7'0 Base | |

ASPHALT SHINGLES
STONE
Vert
FRONT ELEVATION
vent
vent
SIDING
RIGHT ELEVATION
BRICK
flashing
glass doors
SIDING
REAR ELEVATION
vent
SIDING
Vert
LEFT ELEVATION
47' 0"
PATIO
glass doors
MASTER BEDROOM 13' x 12'5"
BATH
R F R G
S
DW
11'6" x8'6"
KIT
DINING ROOM 11'6" x10'
W D DN
laund
mud rm
C C C C
BEDROOM 2 13' x 10'
C
UP
LIVING ROOM 22' x 13'4"
GARAGE
32' 8"
PORCH
FIRST FLOOR
D W
storage and mud rm
HTR
ALTERNATE BASEMENTLESS VERSION
GUTTER
INSULATION 2 x 10 - 16" o.c.
2 x 4 STUD CONST
2 x 10 = 16" o.c.
air space basementless version
APPROX GRADE
2" CONC
CB FOUND
FULL BASEMENT
CONC FTG
4" CONC
CROSS SECTION
dropped
LIN REF
OV
RANGE
OV
dropped
SINK
dropped
Snack Counter
dropped
DW
KITCHEN CABINETS
ROOF
storage
C
B 2
BEDROOM 4 13'x13'
HALL
BEDROOM 3 14'x13'
DOWN
L
walk in clos
eave storage
ROOF
SECOND FLOOR

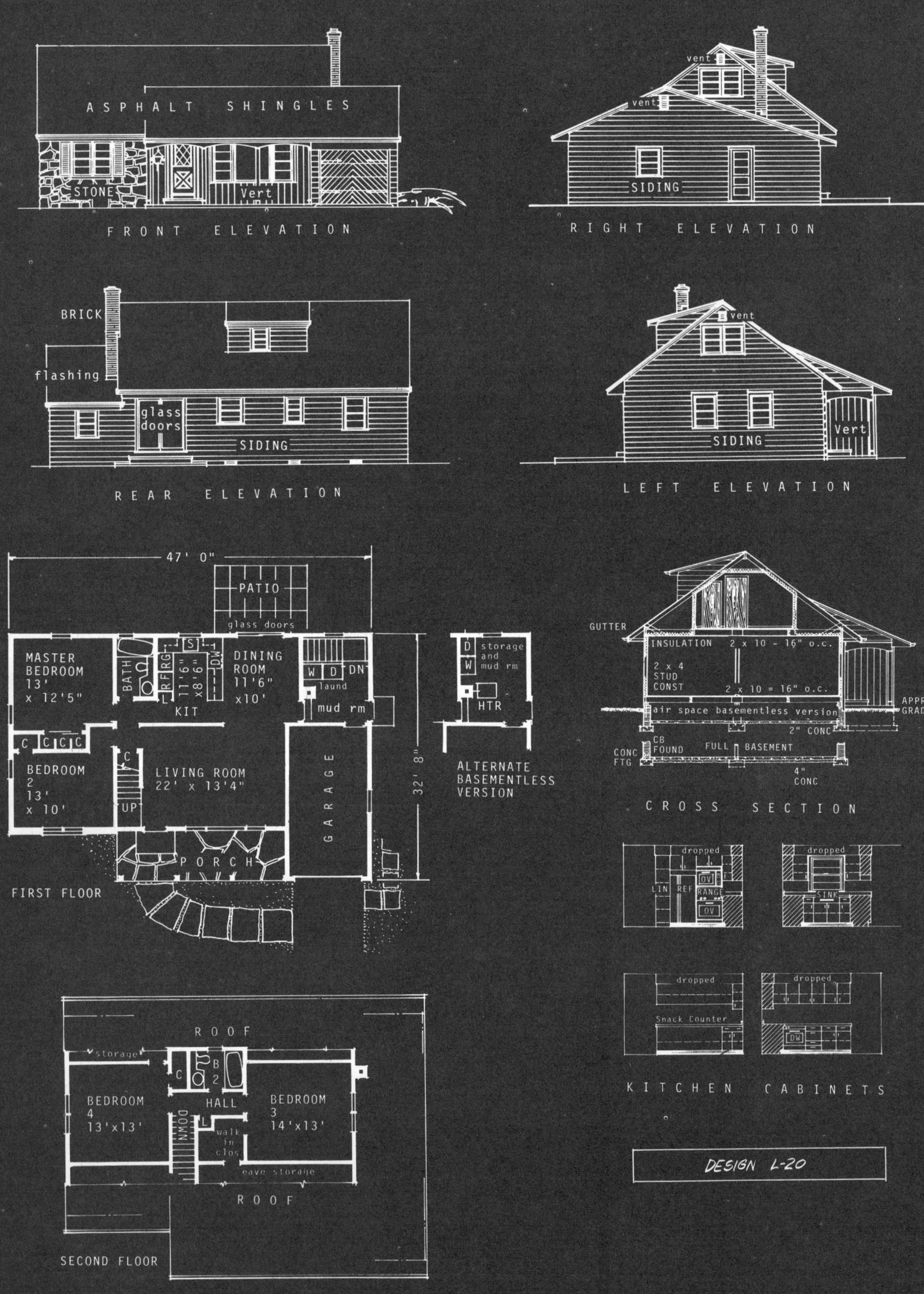

DESIGN L-20

# Choose Between Two or Four Bedrooms

Consideration can be given to this plan whether you're a small family looking for a two-bedroom house or a large family requiring four bedrooms.

In either case, it's a house that will fit on a small lot, since its overall dimensions are only 47′ by 32′8″, including a one-car garage. Even if a two-car garage should be needed, that would not take it out of the small-lot category. This is a truly flexible design.

By keeping the upstairs area unfinished and using it strictly as an attic, the owners have a two-bedroom, five-room house, all on one floor and with a habitable area of 961 square feet. In addition, there is a basementless version which eliminates the down stairway and places the utility equipment in the mud room.

The living room is spacious. A wide arch at the back gives a view of the dining room. From the living room this offers a traditional, off-center L-shaped layout. The kitchen layout, in the convenient U-shape, has a bright window overlooking the backyard. While the dining room is not large, it is at the family's disposal for company dinners.

On the main floor, the two bedrooms share a closet wall and each has two exposures. If the second floor is finished as shown in the plans, there are two bedrooms and a bathroom, along with spacious closets. A storage plus is the readily accessible eave storage area, easily reached from each room.

## Material List

|  |  |
|---|---|
| **CONCRETE WORK** | |
| Footings | 9½ cu. yds. |
| Floors (basement & platforms) | 15 cu. yds. |
| **MASONRY** | |
| 4 x 8 x 18 CB | 125 lin. ft. |
| 8 x 8 x 18 CB | 50 sq. ft. |
| 10 x 8 x 18 CB | 1020 sq. ft. |
| 12 x 8 x 18 CB | 100 sq. ft. |
| Chimney brick | 480 sq. ft. |
| Flagstone - Stone veneer 110 sq. ft. | 145 sq. ft. |
| **CARPENTRY** | |
| 4″ Lally Columns | 4 pieces |
| Framing Lumber | 6230 B.F. |
| Stud & Plates | 3035 B.F. |
| Roof sheathing 1 x 6 | 2620 sq. ft. |
| Plywood | 2160 sq. ft. |
| Fascia #1 pine 1 x 6 | 105 lin. ft. |
| Soffit ⅜″ x 2′ plywood | 100 lin. ft. |
| Shingle mould | 105 lin. ft. |
| Wall sheathing 1 x 6 | 2690 sq. ft. |
| Plywood | 2220 sq. ft. |
| Siding | 2175 sq. ft. |
| Roofing 210# asphalt | 22 squares |
| Insulation walls | 1865 sq. ft. |
| Insulation ceiling | 1600 sq. ft. |
| Finish wood flooring | 1740 sq. ft. |
| **DRY WALL** | |
| Ceilings ⅜″ | 4865 sq. ft. |
| Walls ½″ | 1830 sq. ft. |
| **MILLWORK** | |
| Base | 500 lin. ft. |
| Closet Pole | 25 lin. ft. |
| Shelving 12″ | 40 lin. ft. |
| Hook Strip | 85 lin. ft. |
| Windows | 16 pieces |
| Exterior doors | 4 pieces |
| Interior doors | 17 pieces |
| Louvers | 3 pieces |

Front elevation

Right side elevation (Front elevation for narrow lot)

Rear elevation

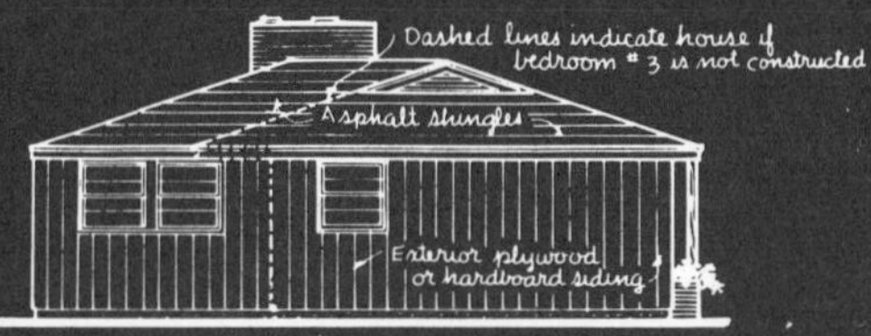


Left side elevation (For left elev of 2 bedroom house - see cross section)

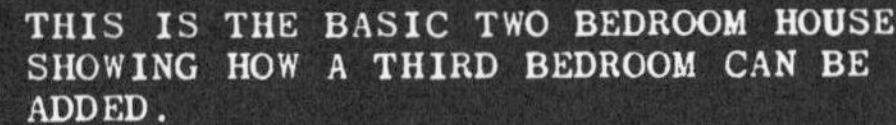

THIS IS THE BASIC TWO BEDROOM HOUSE SHOWING HOW A THIRD BEDROOM CAN BE ADDED.

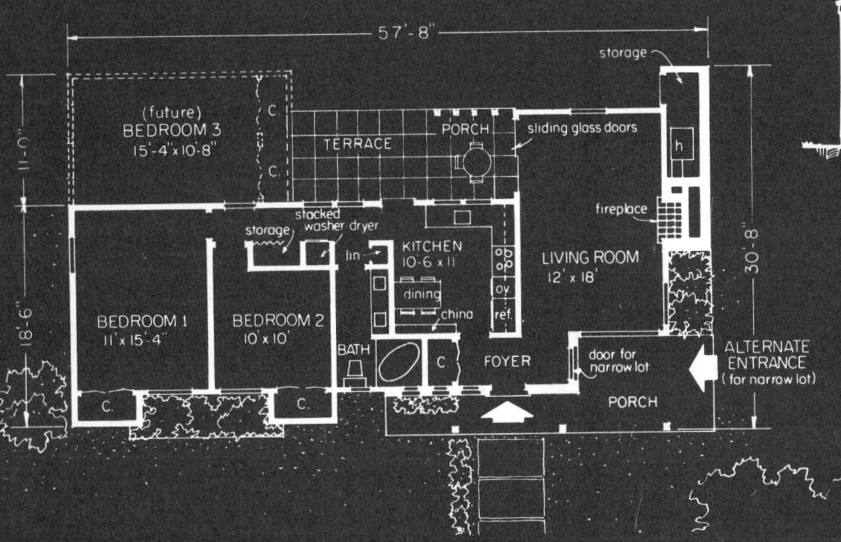


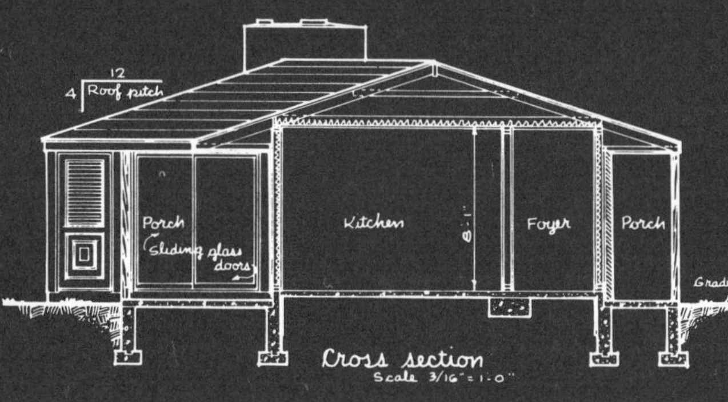


Cross section
Scale 3/16" = 1'-0"

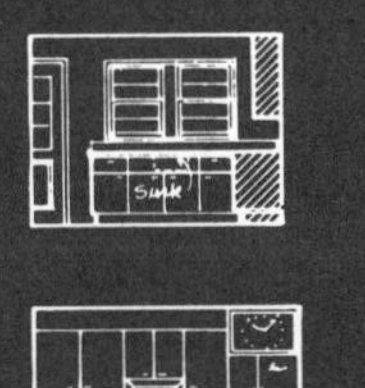


Kitchen elevations

BOTH HOUSES CAN BE TURNED SO RIGHT ELEVATION BECOMES THE FRONT FOR NARROW LOTS (FITS ON 50' LOT)

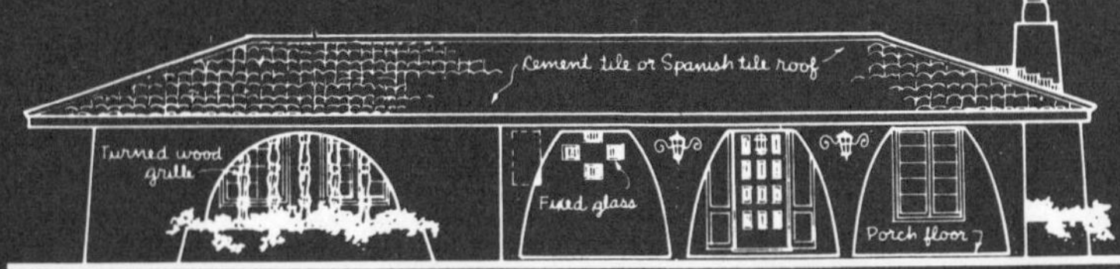


Front elevation

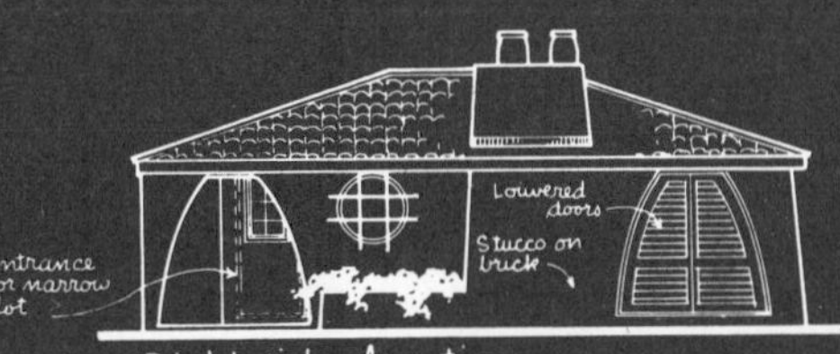


Right side elevation

THIS IS AN ALTERNATE EXTERIOR (SPANISH) OF THE BASIC HOUSE.

DESIGN **S-61**

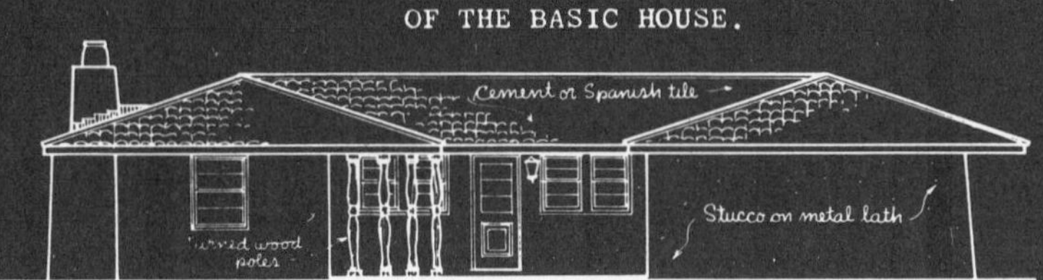


Rear elevation

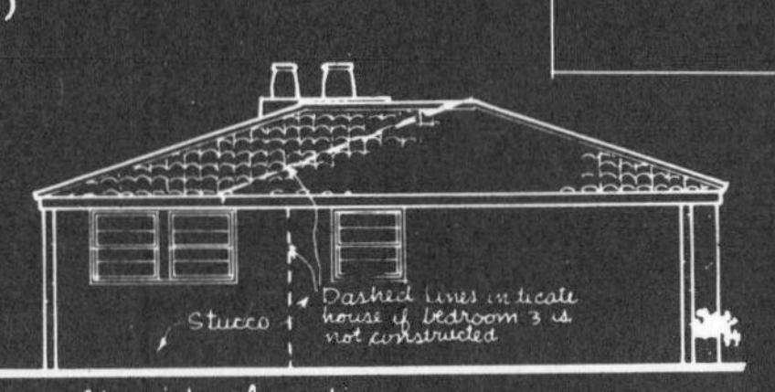


Left side elevation

# Small House with Choice of Exteriors

Modest in both size and cost, here's a house full of pleasant surprises.

It's a two- or three-bedroom house. It has a contemporary or a Spanish-style exterior. It can sit on a lot covering 57′8″ of the frontage — or a narrow lot, using only 31′ of frontage.

To hold costs down this unusual dwelling is built on a concrete slab and without a garage. Part of the savings has been used to create an interesting exterior, using wall breaks.

Design S-61 is a two-bedroom contemporary with 970 square feet of habitable area plus front and rear porches. A most attractive pattern of opaque glass lights on the exterior winds up inside as foyer decorations, as a source of light in a closet and over the tub in the bathroom.

The kitchen, beyond the foyer, has sufficient space for a dining table. To the left rear of the kitchen is the hall to the bedrooms.

The two bedrooms are at the left side of the house, sound buffered from the kitchen by the bathroom.

On the right side of the house is the living room, profusely glazed and with a fireplace.

On a narrow lot, the living room would be at the front. The floor plans show the alternate entrance for such an arrangement.

Exterior materials for the contemporary version are vertical board siding and brick. In the Spanish version, masonry becomes the predominating material.

## Material List

**CONCRETE**

| | |
|---|---|
| Poured concrete walls, footing & slabs | 1250 cu. ft. |
| Exterior reinforcing mesh | 280 sq. ft. |
| 2″ rigid insulation | 300 sq. ft. |
| Plastic vapor barrier | 900 sq. ft. |

**MASONRY**

| | |
|---|---|
| 4′ brick veneer — planters, etc. | 504 sq. ft. |
| 1 4′0 angle iron | |
| 4″ cement block back up | 140 sq. ft. |
| Brick back up 1900 brick | |
| 24 Lin. ft. flue liner & (1) fireplace damper | |

**CARPENTRY**

| | |
|---|---|
| All studs - plates headers | 5590 BM |
| rafters etc. | 3180 |
| Plywood sheathing roof & walls | 100 BM |
| Finish lumber | 230 BM |
| Ext. plywood eaves & ceils | 570 sq. ft. |
| 3/8 exterior scored-siding | 730 sq. ft. |

**ROOFING**

20 sq. — 240 lb. self sealing asphalt shingles

**PLASTERBOARD**

| | |
|---|---|
| For walls & ceilings | 3620 sq. ft. |

CONDENSED LIST — For future bedroom #3 (Plan A)

**CONCRETE**

| | |
|---|---|
| Walls & footing & floor slab | 170 cu. ft. |
| 2″ rigid insul. | 50 sq. ft. |
| Plastic vapor barrier | 200 sq. ft. |

**CARPENTRY**

| | |
|---|---|
| Studs — header ceiling, joists rafters, etc. | 1230 BM |
| Plyscore sheathing wall & roof | 870 sq. ft. |
| 3/8 exterior scored siding | 280 sq. ft. |
| 2 Rolls — 15# Felt | |

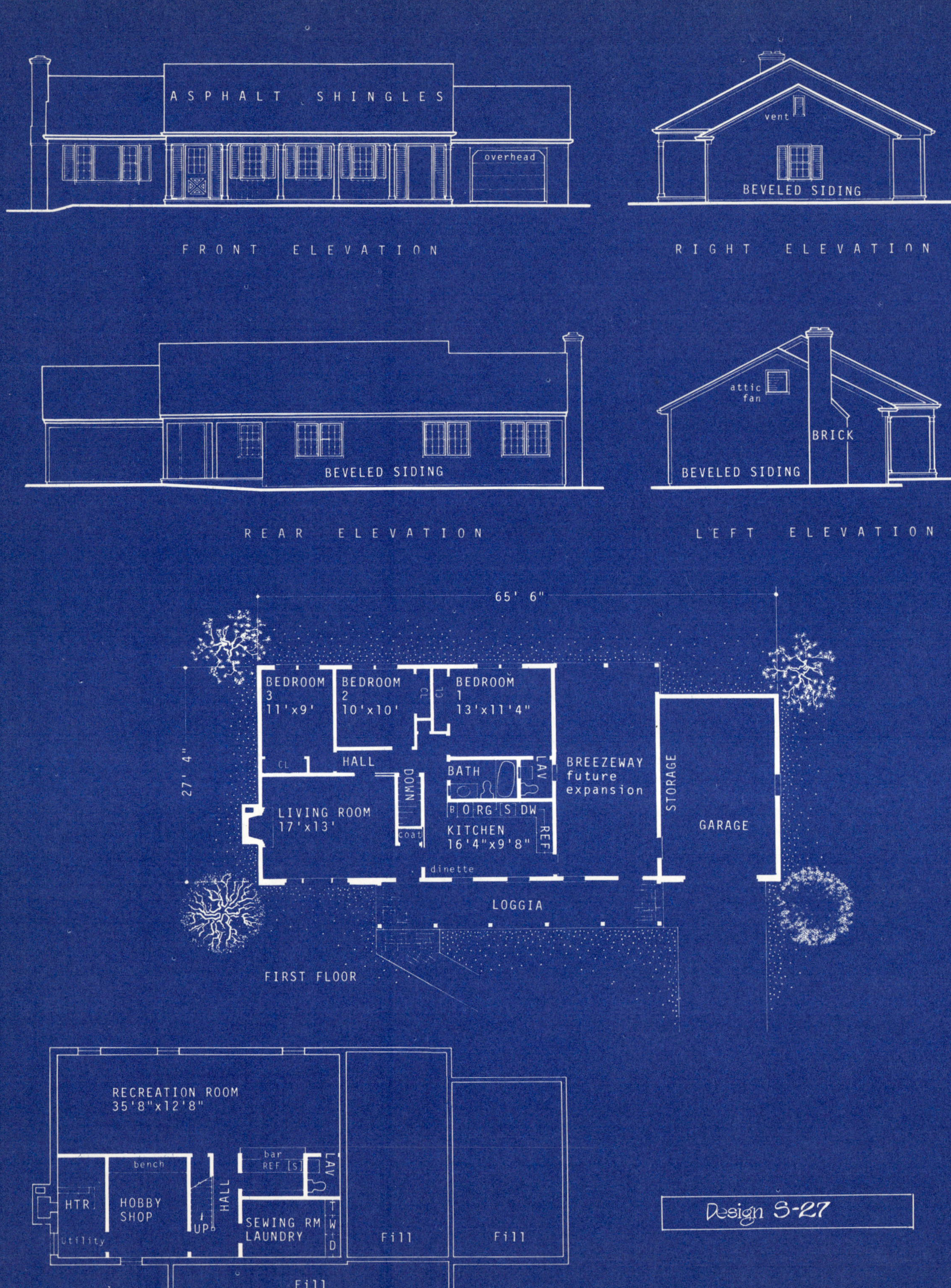
ASPHALT SHINGLES
overhead
FRONT ELEVATION
vent
BEVELED SIDING
RIGHT ELEVATION
BEVELED SIDING
REAR ELEVATION
attic fan
BRICK
BEVELED SIDING
LEFT ELEVATION
65' 6"
27' 4"
BEDROOM 3 11'x9'
BEDROOM 2 10'x10'
BEDROOM 1 13'x11'4"
CL
HALL
BATH
LAV
BREEZEWAY future expansion
STORAGE
GARAGE
LIVING ROOM 17'x13'
DOWN
coat
KITCHEN 16'4"x9'8"
REF
DW
dinette
LOGGIA
FIRST FLOOR
RECREATION ROOM 35'8"x12'8"
bench
bar
REF
LAV
HTR
HOBBY SHOP
utility
UP
HALL
SEWING RM LAUNDRY
Fill
Fill
Fill
BASEMENT
Design S-27

# Economy House With Attractive Look

A small house must very often be a compromise between what you like and what you can afford. But it need not be entirely lacking in features nor of poor design.

This ranch, with only 1030 square feet of living area has a compact plan that is economical, practical and attractive. The key rooms — the living room and the kitchen — are at the front of the house with the three bedrooms at the rear.

On the other side of the entrance foyer is the fireplaced living room, which need not be entered in order to get to any other rooms in the house. There is a door opening to the bedroom hall, but aside from the convenience of such access, the room is preserved "company-fresh" at all times.

The three bedrooms are across the back of the house. The master bedroom has a private lavatory, back-to-back with the family bathroom which is accessible from the kitchen or bedroom hall.

Utilizing the basement more than doubles the available space. A considerable economy could be effected by constructing Design S-27 on a concrete slab. In that event, the breezeway could be used to take care of the laundry and utilities.

If the house is built as shown in the plans — that is, with a full basement—then the breezeway would serve as just that, or be converted into a family room.

## Material List

|  |  |
|---|---|
| **CONCRETE** | |
| Footings | 332 cu. ft. |
| Floor slabs | 588 cu. ft. |
| **MASONRY** | |
| 12" Concrete block | 1093 sq. ft. |
| 8" Concrete block | 321 sq. ft. |
| Brick | 304 sq. ft. |
| Lally columns | 4 pcs. |
| Slate or flagstone | 216 sq. ft. |
| Fireplace brick | 19 sq. ft. |
| Flue Tile, 8 x 13 | 26 lin. ft. |
| Flue Tile, 13 x 13 | 13 lin. ft. |
| Clean out door | 2 pcs. |
| Steel lintel | 1 pc. |
| Thimble | 1 pc. |
| **CARPENTRY** | |
| Framing lumber | 6572 sq. ft. |
| Studs and plates | 3333 sq. ft. |
| Sub floor, 5/8" Plywood | 1232 sq. ft. |
| Underlayment, 5/8" Plywood | 1232 sq. ft. |
| Sheathing, 5/8" Plywood | 1884 sq. ft. |
| Roof sheathing, 5/8" Plywood | 3230 sq. ft. |
| Asphalt shingles 210# | 3230 sq. ft. |
| Siding | 1450 sq. ft. |
| Finish floors | 991 sq. ft. |
| Fascia | 208 lin. ft. |
| Insulation, 2" walls | 800 sq. ft. |
| Insulation, 3" ceilings | 1026 sq. ft. |
| **DRY WALL** | |
| Walls, 1/2" sheetrock | 3910 sq. ft. |
| Ceilings, 3/8" sheetrock | 1232 sq. ft. |
| Garage walls & ceiling, 3/8" sheetrock | 683 sq. ft. |
| or | |
| **PLASTER** | |
| House walls & ceiling, 3/8" rock lath | 5142 sq. ft. |
| Garage walls & ceiling, 3/8" rock lath | 683 sq. ft. |

FRONT ELEVATION

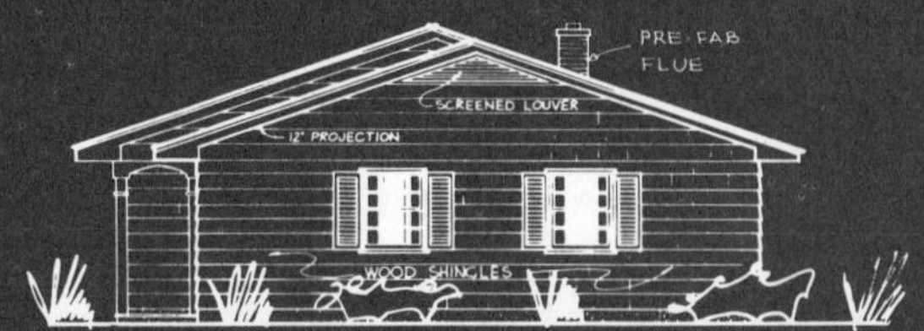

RIGHT SIDE ELEVATION

REAR ELEVATION

LEFT SIDE ELEVATION

## AREA STATISTICS

| | |
|---|---|
| HABITABLE AREA | 1064 SQ. FT. |
| LAUNDRY AREA | 116 SQ. FT. |
| GARAGE | 264 SQ. FT. |

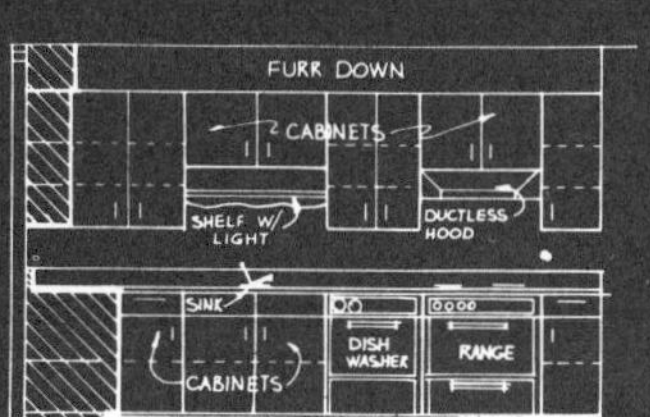


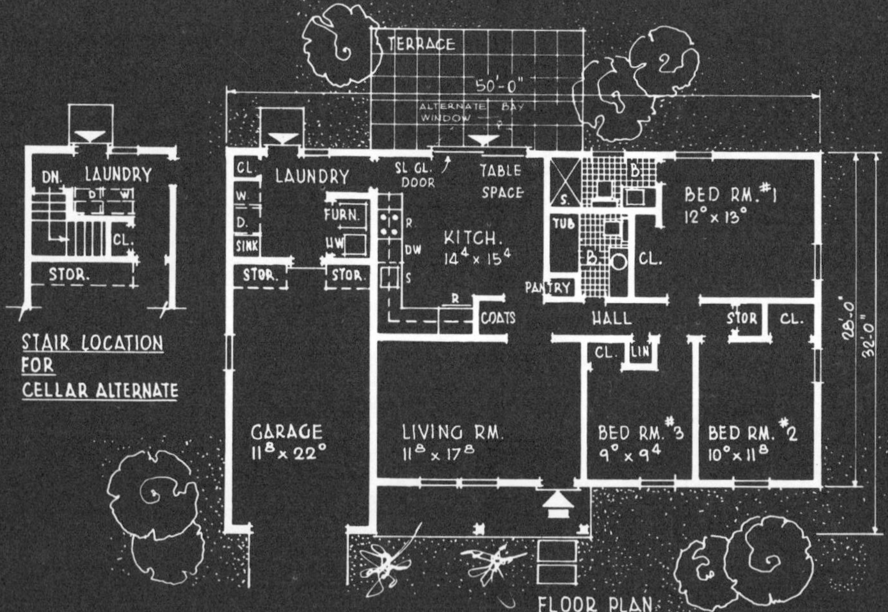


STAIR LOCATION FOR CELLAR ALTERNATE

FLOOR PLAN

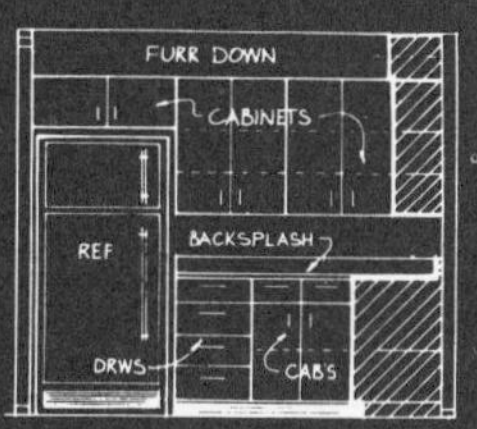


KITCHEN CABINET ELEVATIONS

DESIGN S-63

# Small House Designed for Young Family

This one-story house is for the young family with one or two children and is designed with economy of construction in mind.

By most standards it is a small house with only 1064 square feet of living area, yet it contains two full bathrooms, a family-size kitchen and a laundry-sewing area.

The front entrance door is approached beneath a covered portico, adding to the farmhouse character of the exterior. The living room, with its one end isolated, can be furnished in any one of a number of arrangements, accommodating either sectionalized pieces or more conventional larger sofas.

Among the items sure to capture the approval of the homemaker are generous kitchen counter space, a pantry closet, dinette table space on an outside wall overlooking the terrace, easy access from kitchen to the terrace, space for dishwasher, a laundry large enough to do ironing and sewing, a mud closet for outdoor clothing, raincoats and overcoats, a laundry sink and convenient access to the rear yard for outdoor drying and from the garage following a shopping trip.

Design S-63 can be built on a slab, over a crawl space or with a full cellar. There is an alternate layout for a cellar, with a stair located in the laundry. Although the plan shows the laundry on the main level, the laundry appliances can be located downstairs if the cellar version is built.

## Material List

**CONCRETE WORK**

| | |
|---|---|
| Walls | 326 cu. ft. |
| Slabs | 584 cu. ft. |
| Misc. Concrete | 45 cu. ft. |

**CARPENTRY**

| | |
|---|---|
| Framing Lumber | 1943 B.F. |
| Studs | 880 B.F. |
| Plates | 760 B.F. |
| Roof Sheathing | 1915 sq. ft. |
| Sidewall Sheathing | 1230 sq. ft. |
| Insulation Walls | 1026 sq. ft. |
| Insulation Ceilings | 1178 sq. ft. |

**MILLWORK**

| | |
|---|---|
| Exterior Doors & Frames Compl. | 2 |
| Garage Door Complete Set | 1 set |
| Sliding Glass Doors | 1 unit |
| Interior Doors & Frames Compl. | 10 |
| Bi Fold Doors | 5 |
| Window Units Complete | 9 |
| Fascia | 200 lin. ft. |
| Louvers | 2 pieces |
| Shutters | 10 pieces |
| Base | 448 lin. ft. |
| Clothes Pole | 16 lin. ft. |
| Pole Sockets | 3 pair |
| Cleats | 126 lin ft. |
| Hook Strip | 28 lin. ft. |

**KITCHEN CABINETS**

| | |
|---|---|
| Base Cabinets | 7'-9' Long |
| Wall Cabinets | 11'-0" Long |

**ROOFING**

| | | |
|---|---|---|
| Shingles | 235# asphalt | 1915 sq. ft. |
| Roofing Paper | 15# felt | 1915 sq. ft. |

**SHEET METAL**

| | |
|---|---|
| Saddle & Counter Flashing | 16 oz. |

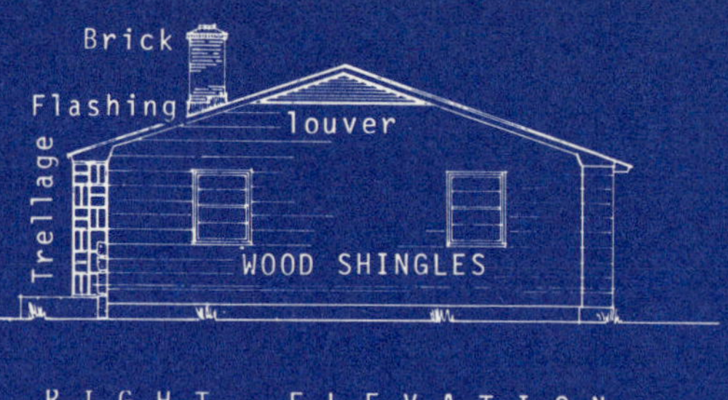


RIGHT ELEVATION

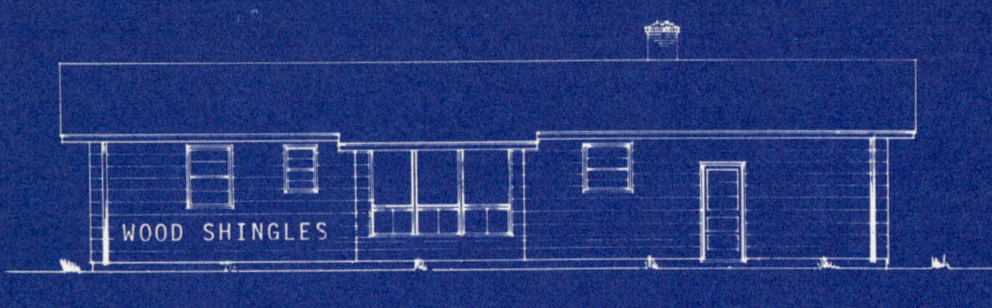

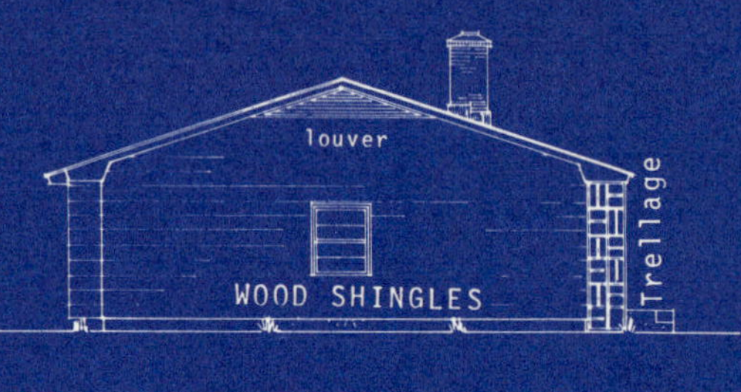


LEFT ELEVATION

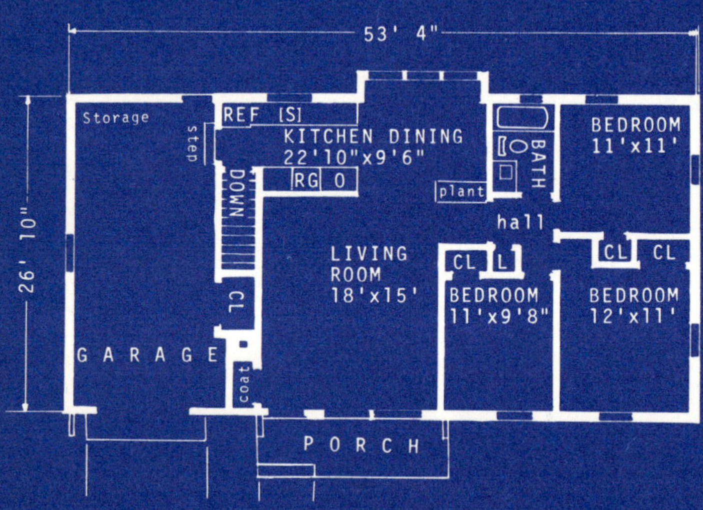


FLOOR PLAN

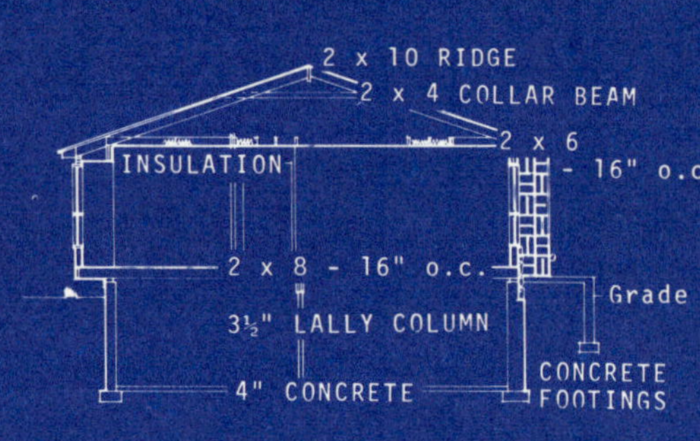

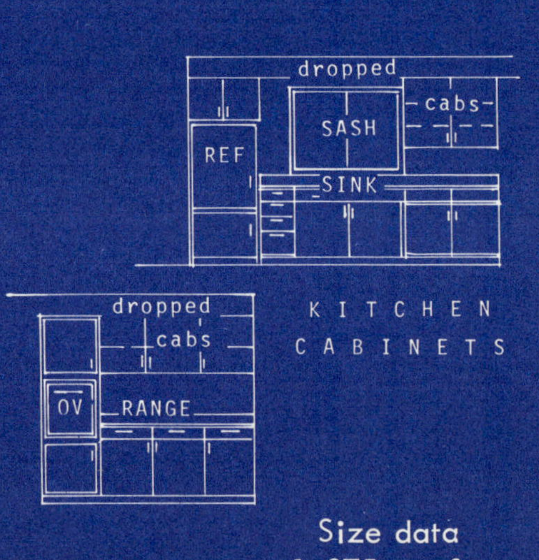


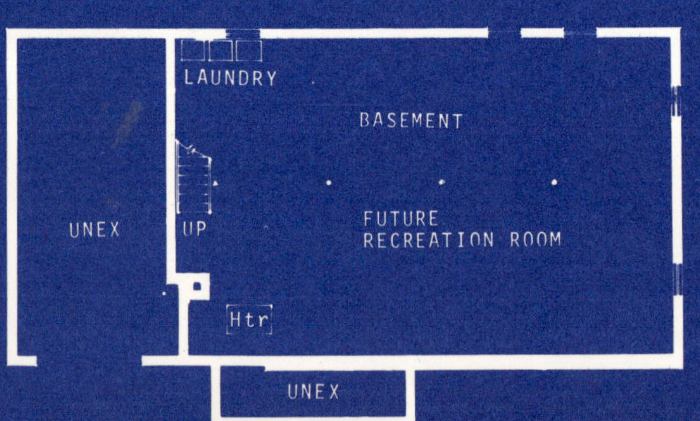


**Size data**
**1,075 sq. ft.**

***DESIGN S-51***

BASEMENT PLAN

# Small Bundle Has Attractive Packaging

This small three-bedroom ranch has an appearance of smartness seldom seen in small houses. The overall impression is one of a long, low, comfortable house that seems larger than 53′4″ by 26′10″, figures which include a one-car garage.

Adding to the pleasant appearance of the front of the house is a living-room picture window.

Visually, the living room is enlarged by the dining area, but a decorative planter sets off the dining space without blocking a view of the rear bay of triple windows.

The kitchen is a pullman type, with refrigerator, sink and counters on the back wall, and the work area, range and wall oven opposite. No space is wasted here, with a window over the sink assuring light and a cheery atmosphere.

An attached garage, a step down, at the side, has a storage area at the back. Groceries can be stored easily in the garage or basement without tracking up the kitchen.

Three bedrooms and a bathroom are at the right side of the house. There is a linen closet opposite the bathroom. In a small family, the third bedroom could be used as a combination study and den.

When finances permit and if there is a need for it, a recreation room could be created in the basement. The laundry and heater are set off and could be screened away.

## Material List

**CONCRETE WORK**

|  |  |
|---|---|
| Footings | 6.5 cu. yds. |
| Floors, basement and platforms | 16.5 cu. yds. |

**MASONRY**

|  |  |
|---|---|
| Concrete walls 8″ | 26.5 sq. ft. |
| Concrete walls 10″ | sq. ft. |
| Chimney fill | cu. ft. |
| Concrete walls 12″ | 9.5 cu. yds. |

**CARPENTRY**

|  |  |
|---|---|
| 3½″ Lally columns | 4 pieces |
| Framing lumber | 3982 B.F. |
| Studs and plates | 2456 B.F. |
| Roof sheathing | 1823 sq. ft. |
| Fascia | 184 lin. ft. |
| Soffits ⅜″ x 24″ | 120 lin. ft. |
| Exterior wall sheathing | 1324 sq. ft. |
| Roofing (230# asphalt shingles) | 19 squares |
| Insulation walls | 1150 sq. ft. |
| Insulation, ceilings | 1288 sq. ft. |
| Finished wood flooring | 997 sq. ft. |
| Resilient flooring | 44 sq. ft. |

**DRY WALL**

|  |  |
|---|---|
| Ceilings | 1200 sq. ft. |
| Walls | 3800 sq. ft. |

**MILLWORK**

|  |  |
|---|---|
| Base | 430 lin. ft. |
| Closet Pole | 18 lin. ft. |
| Shelving 12″ pine | 80 lin. ft. |
| Hook strip | 80 lin. ft. |
| Windows | 12 pieces |
| Exterior doors | 4 pieces |
| Interior doors | 11 pieces |
| Louvres | 2 pieces |

**KITCHEN CABINETS**

|  |  |
|---|---|
| Hangers | 15 lin. ft. |
| Counters | 17 lin. ft. |
| Medicine cabinets | 1 piece |

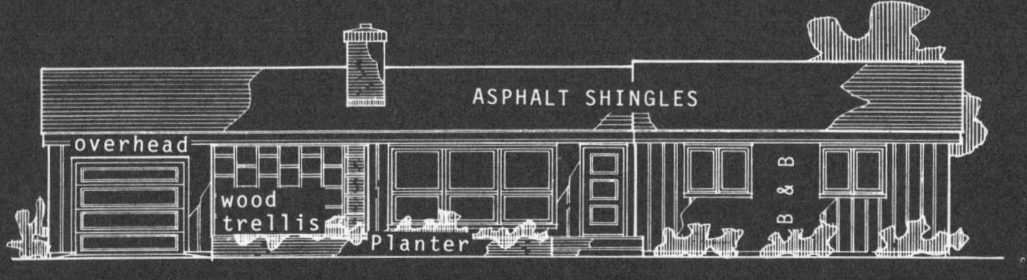


FRONT ELEVATION

LEFT ELEVATION

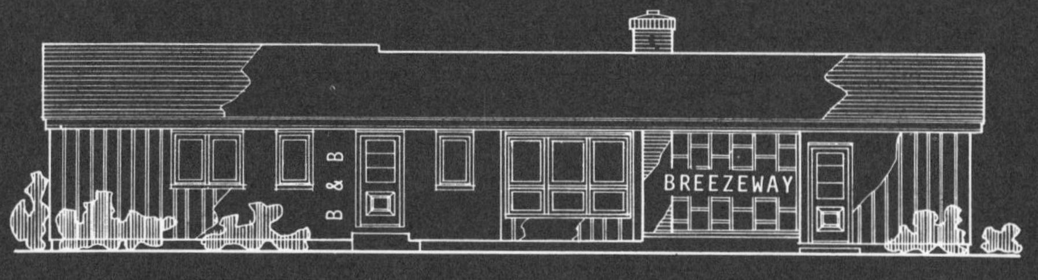


REAR ELEVATION

RIGHT ELEVATION

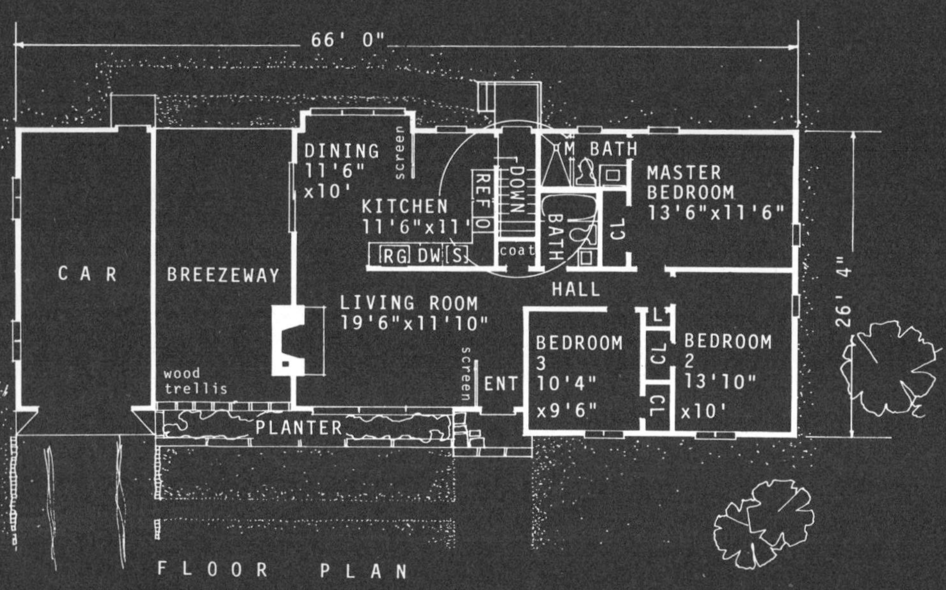


FLOOR PLAN

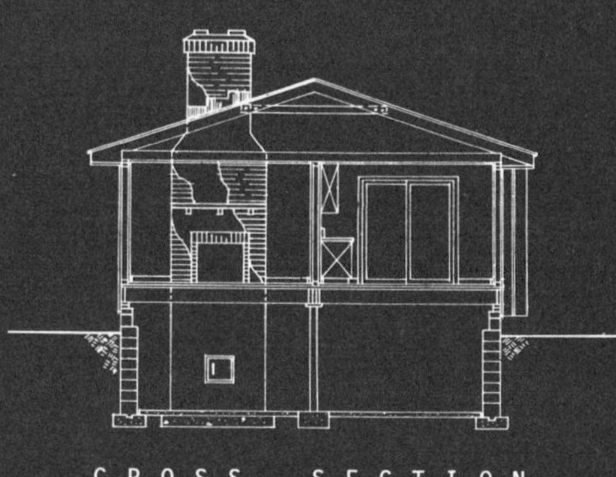

CROSS SECTION

ALTERNATE FOR BASEMENTLESS PLAN

Size data
1,078 sq. ft.

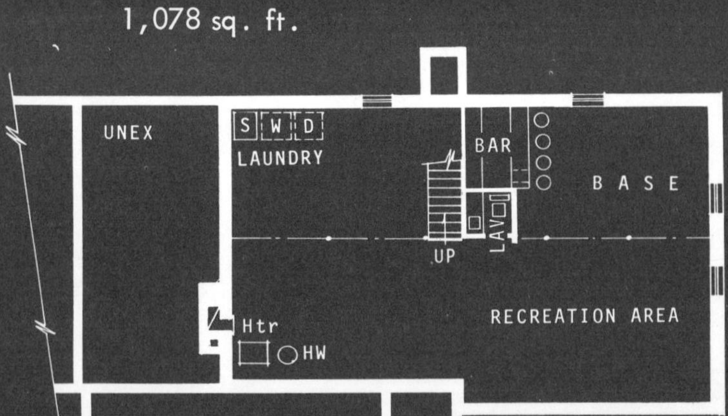


BASEMENT PLAN

KITCHEN CABINETS

DESIGN L-56

# Breezeway Among Features in Small House

When a lot is small and building costs are high, the design of a home is more important than ever. A good basic design, setting up the most livable plan possible in a limited area, can be greatly enhanced by adding flexibility and choice.

This ranch plan gains width through a reasonably inexpensive method—via a breezeway and garage. The home owner can choose from this arrangement or a two-car garage alone or eliminate both, putting the garage back on the lot. The basement offers extra utility and recreation area, but if basements are not common in the region or if the family prefers all-on-one floor living, the basement stair section can handle the heater and hot water system, conveniently located inside the back vestibule.

Screens instead of full walls are used to create a vestibule and a dining area. This expands both areas and affords privacy where needed but doesn't cut room sizes.

At the front, the living room has symmetrical window treatment for a wide view, a fireplace on the end wall, and a screen at the entry. Since the dining area is not closed off the spaces flow together naturally, making each area seem larger.

The master bedroom has a corner location, well sheltered from the living areas and with two exposures. Both front bedrooms are compact and separated by a closet wall.

## Material List

|  |  |
|---|---|
| **CONCRETE WORK** | |
| Footings | 12.75 cu. yds. |
| Floors, basement, platforms | 21.3 cu. yds. |
| **MASONRY** | |
| Cinder Block 4 x 8 x 16 solids | 230 lin. ft. |
| 4 x 8 x 16 | 100 sq. ft. |
| 8 x 8 x 16 | 900 sq. ft. |
| 12 x 8 x 16 | 600 sq. ft. |
| Chimney Brick | 30 sq. ft. |
| Masonry veneer | 400 sq. ft. |
| **CARPENTRY** | |
| Framing Lumber | 5962 bd. ft. |
| Studs, soles & plates | 2930 bd. ft. |
| Roof Sheathing | 1200 sq. ft. |
| Exterior wall sheathing | 2000 sq. ft. |
| Exterior wall covering | 2000 sq. ft. |
| Roofing | 12 squares |
| Insulation — walls | 1200 sq. ft. |
| Insulation — ceilings | 1200 sq. ft. |
| Hardwood flooring | 780 sq. ft. |
| Resilient flooring | 240 sq. ft. |
| Rough Flooring | 1200 sq. ft. |
| Sub-flooring | 325 sq. ft. |
| **DRY WALLS (includes garage)** | |
| Ceilings | 1500 sq. ft. |
| Walls | 3100 sq. ft. |
| **MILLWORK** | |
| Base | 220 lin. ft. |
| Shelving | 35 lin. ft. |
| Windows | 10 pcs. |
| Exterior doors | 5 pcs. |
| Interior doors | 10 pcs. |
| Closet Poles | 16 lin. ft. |
| **KITCHEN CABINETS** | |
| Hangers | 13 lin. ft. |
| Counters | 13 lin. ft. |

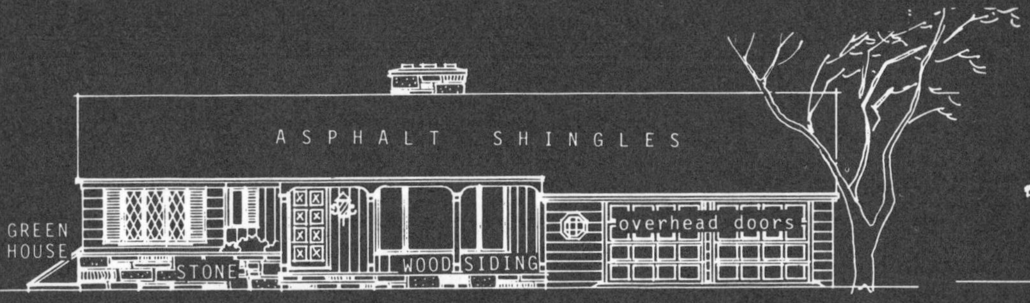

ASPHALT SHINGLES
GREEN HOUSE
STONE
WOOD SIDING
overhead doors

FRONT ELEVATION

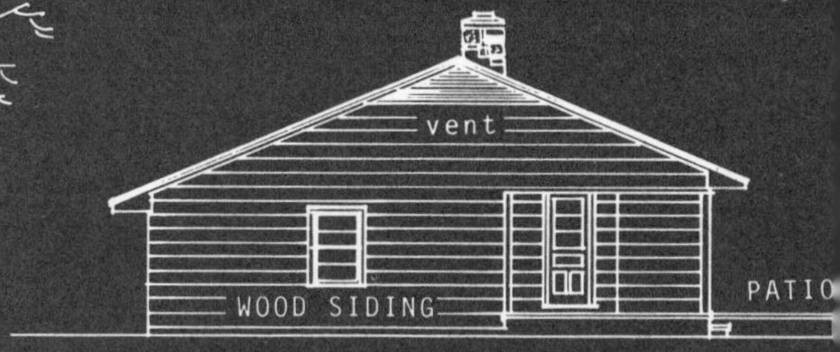

vent
WOOD SIDING
PATIO

RIGHT ELEVATION

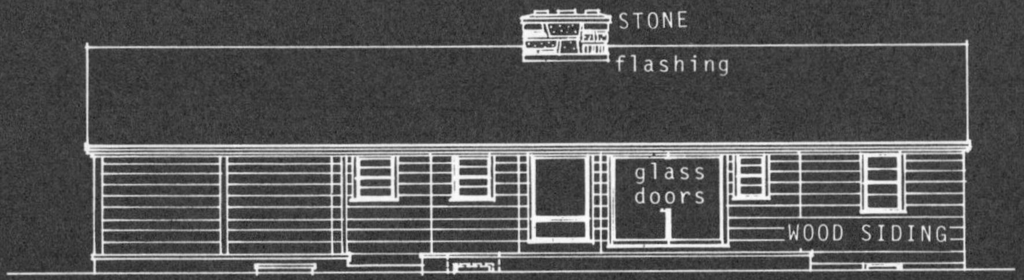

STONE
flashing
glass doors
WOOD SIDING

REAR ELEVATION

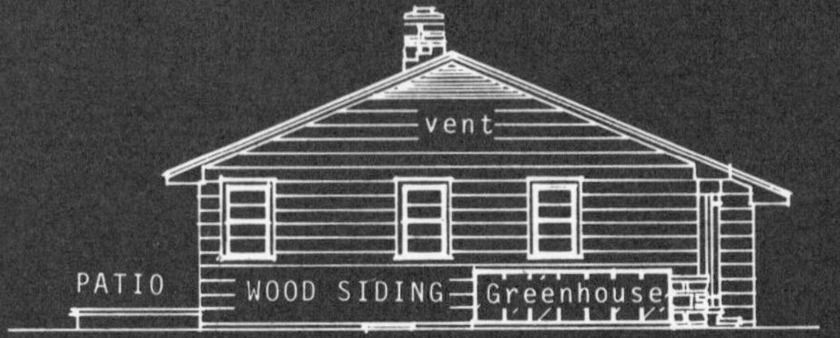

vent
PATIO
WOOD SIDING
Greenhouse

LEFT ELEVATION

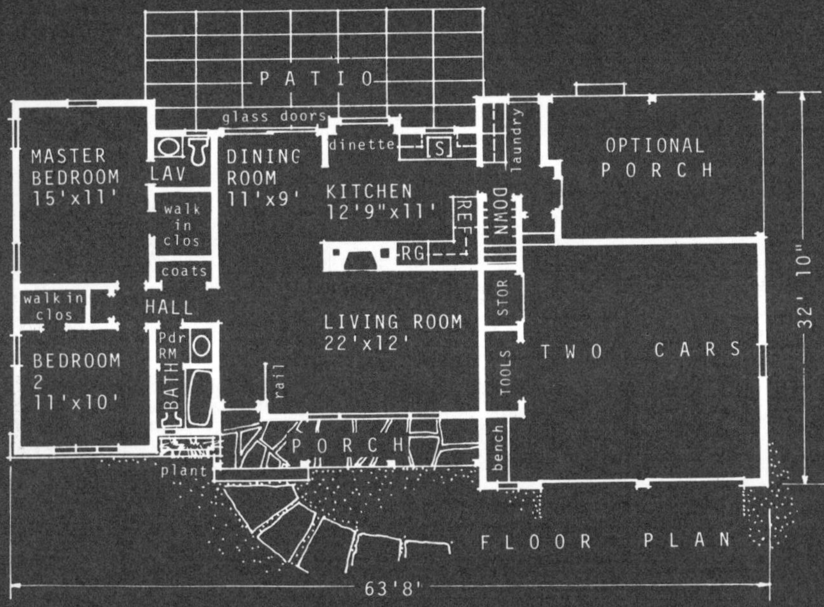

BATH
STUDY OR OPTIONAL BEDROOM 3 11'8" x10'6"
HALL
CL CL

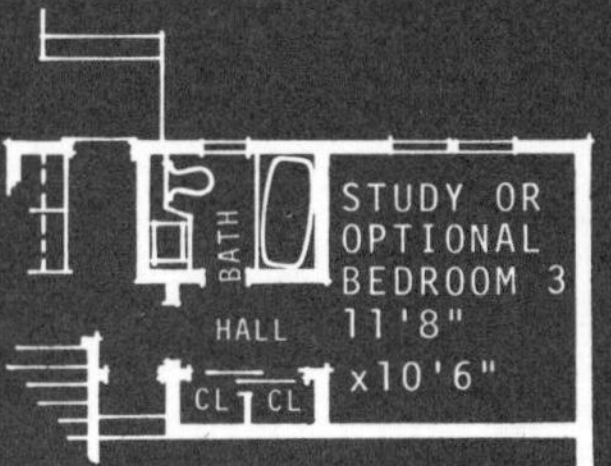

OPTIONAL THIRD BEDROOM

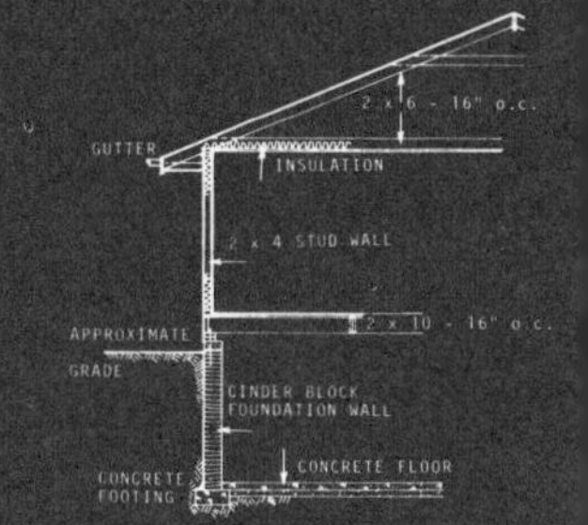

GUTTER
INSULATION
2 x 6 - 16" o.c.
2 x 4 STUD WALL
2 x 10 - 16" o.c.
APPROXIMATE GRADE
CINDER BLOCK FOUNDATION WALL
CONCRETE FLOOR
CONCRETE FOOTING

CROSS SECTIO

PATIO
glass doors
MASTER BEDROOM 15'x11'
LAV
DINING ROOM 11'x9'
dinette
S
laundry
OPTIONAL PORCH
KITCHEN 12'9"x11'
REF
DOWN
walk in clos
RG
coats
walk in clos
HALL
LIVING ROOM 22'x12'
STOR
Pdr RM
BEDROOM 2 11'x10'
BATH
rail
TWO CARS
TOOLS
PORCH
bench
plant
32' 10"

FLOOR PLAN

63'8'

dropped
cabs
REF

dropped
cabs
OV
RANGE

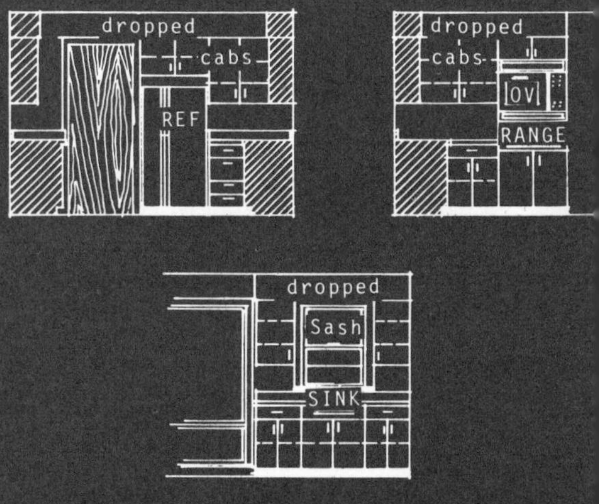

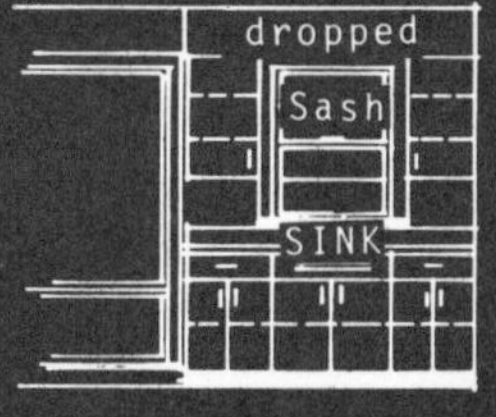


KITCHEN CABINETS

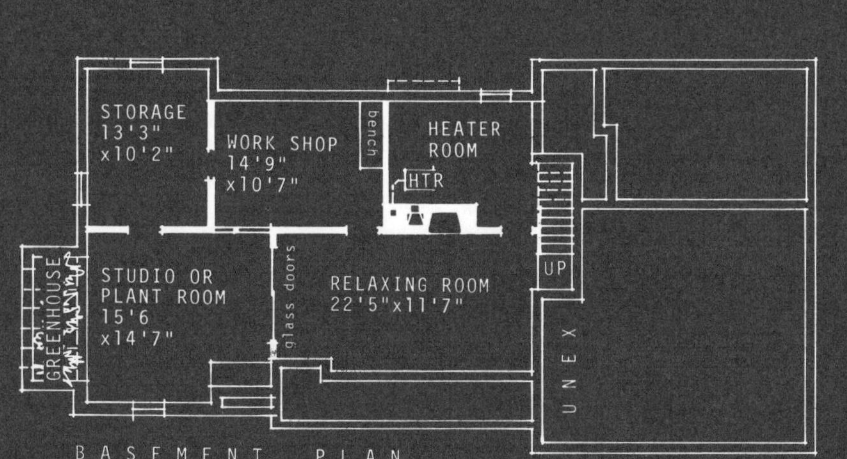


BASEMENT PLAN

Size data
1,115 sq. ft.

DESIGN S-99

# Two-Bedroom House with Optional Room

As thoughtfully designed as any larger house, this two-bedroom home makes sense for a retirement couple or small family.

It has a tasteful exterior, a compact floor plan and an expandable feature that can be utilized in any one of several ways. In addition, its full basement can be finished to provide a sizable extra living area.

The living room and dining room are full scale. A central fireplace is a handsome adjunct. A wrought iron divider rail defines the entry and routes traffic past the living room without cutting down on space.

At the back, the dining room blends in, visually gaining space for both rooms while allowing formal dining furniture. Sliding glass doors open out to the rear patio.

The kitchen is next to the dining room with a wide opening to combine the dinette and dining areas if desired. By using a folding door, the rooms can be completely separated. A full-length dinette bay window brightens the area.

That side porch, behind the two-car garage, can be made into a study or a third bedroom with closets and its own bath. Another possibility is use of the room as a rental unit for extra income.

Each of the two bedrooms in the left wing has cross ventilation. A walk-in closet and private lavatory are provided for the master bedroom. The other bathroom is "split."

## Material List

**CONCRETE WORK**

|  |  |
|---|---|
| Footings | 16 cu. yds. |
| Floors (basement & platforms) | 27 cu. yds. |

**MASONRY**

|  |  |
|---|---|
| 4 x 8 x 18 CB | 240 lin. ft. |
| 8 x 8 x 18 CB | 315 sq. ft. |
| 10 x 8 x 18 CB | 1000 sq. ft. |
| 12 x 8 x 18 CB | 175 sq. ft. |
| Chimney Fill | 410 cu. ft. |
| Flagstone - stone veneer | 20 sq. ft. |
| | 115 sq. ft. |

**CARPENTRY**

|  |  |
|---|---|
| 4" Lally columns | 4 pieces |
| Framing Lumber | 7170 B.F. |
| Stud & Plates | 3020 B.F. |
| Roof Sheathing 1 x 6 | 4590 sq. ft. |
| Plywood | 3790 sq. ft. |
| Fascia #1 pine 1 x 6 | 130 lin. ft. |
| Soffit ⅜" x 2' plywood | 130 lin. ft. |
| Shingle mould | 130 lin. ft. |
| Sheathing 1 x 6 | 2885 sq. ft. |
| Plywood | 2380 sq. ft |
| Siding | 2540 sq. ft. |
| Roofing 210# asphalt | 38 squares |
| Insulation - walls | 2380 sq. ft. |
| ceiling | 1230 sq. ft. |
| Finish wood flooring | 1220 sq. ft. |

**DRY WALL**

|  |  |
|---|---|
| Ceilings ⅜" | 1915 sq. ft. |
| Walls ½" | 5725 sq. ft. |

**MILLWORK**

|  |  |
|---|---|
| Base | 665 lin. ft. |
| Windows | 17 pieces |
| Exterior doors | 5 pieces |
| Interior doors | 18 pieces |

**KITCHEN CABINETS**

|  |  |
|---|---|
| Hangers | 16 lin. ft. |
| Counter | 12 lin. ft. |
| Vanitory | 2 pieces |

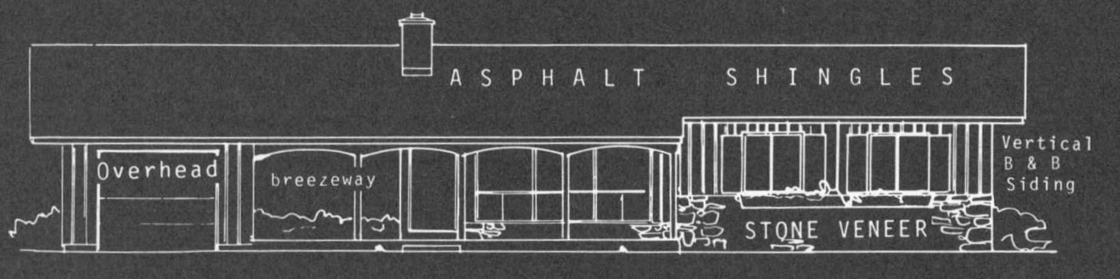


FRONT ELEVATION

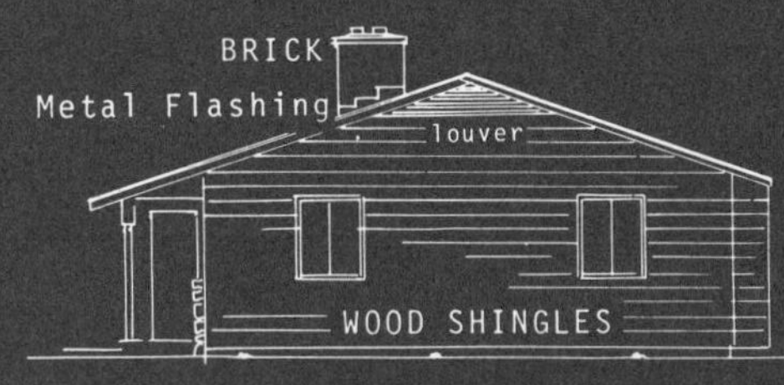


RIGHT ELEVATION

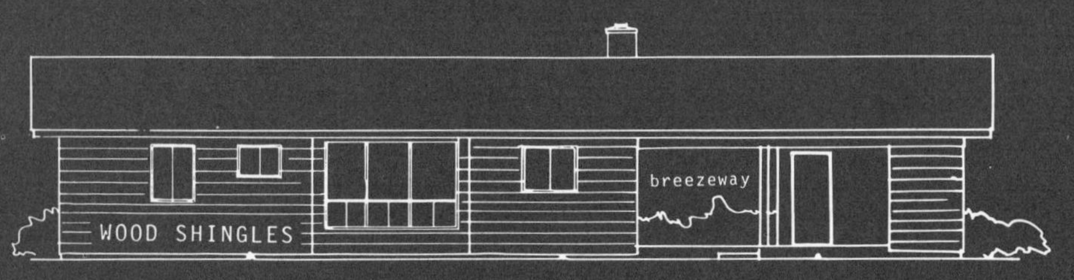


REAR ELEVATION

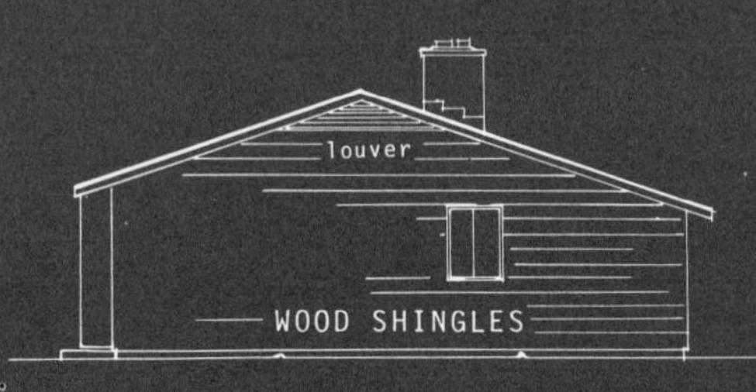


LEFT ELEVATION

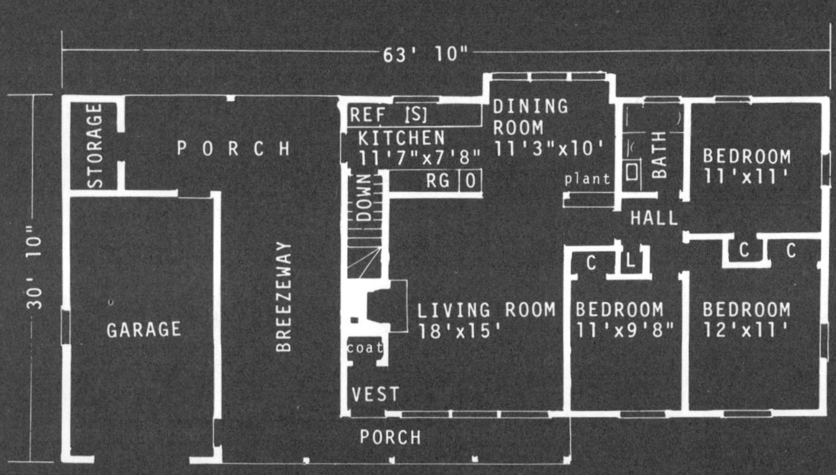


FLOOR PLAN

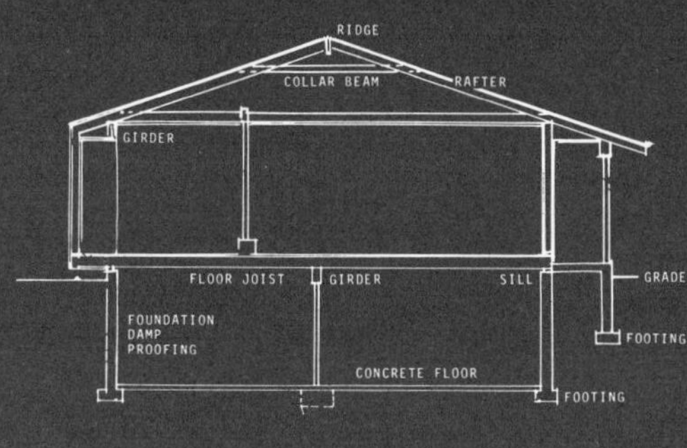


CROSS SECTION

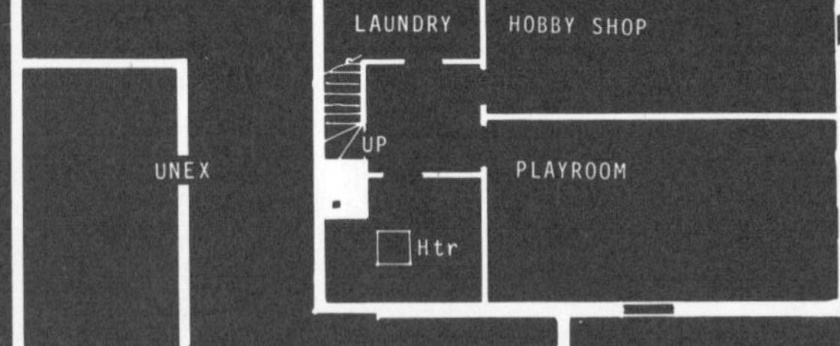


BASEMENT PLAN

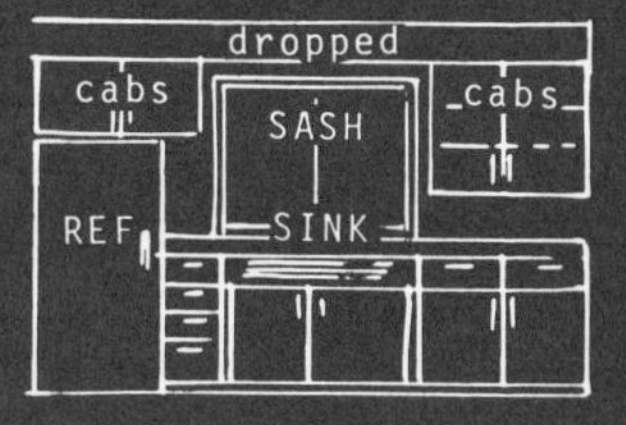


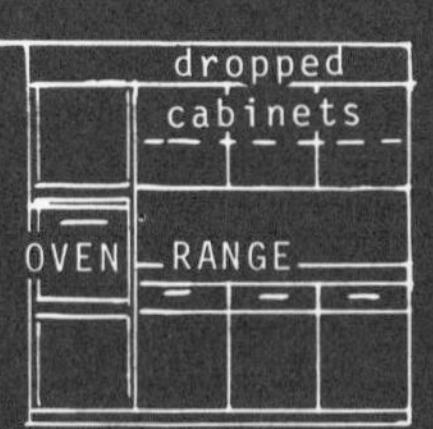


KITCHEN CABINETS

Z-92

# Breezeway Adds Length to Small House

In this ranch house with a living area of only 1117 square feet, the architect has managed to include three bedrooms within a compact floor plan yet produce an exterior with a lengthy appearance. He has accomplished this by placing a breezeway between the basic house and the garage, then extending the roof line of a long, covered front porch across the front of the garage. The house not only looks bigger, it is more interesting.

The breezeway offers an additional sheltered outdoor area and leads to a covered porch at the rear, giving a modest house extra assets at modest cost. Note from the floor plans how easy it is to get from the garage to the front door, from the kitchen to the outdoor storage closet or to the garage, or from the garage or backyard to the basement stairs.

The entrance to the living room is through a vestibule which has a coat closet. Big triple windows across the front are in the most popular styling, adding to the attractiveness of the area, as does the stove fireplace. The living room and the dining room can be completely separate or brought together. A planter delineates the dining room from the approach to the bedroom hall. Extra space is added to the dining room by a rear bay extension, all windows and a room wide.

The exterior combines stone and vertical and horizontal siding. It achieves a pleasant, traditional look with a modern-day effect.

## Material List

**CONCRETE WORK**

| | |
|---|---|
| Footings | 9.4 cu. yds. |
| Floors, basement & platforms | 11.2 cu. yds. |

**MASONRY**

| | |
|---|---|
| Conc. walls 8" | 248 sq. ft. |
| 12" | 1745 sq. ft. |
| Chimney fill | 358 cu. ft. |
| Masonry veneer | 103 sq. ft. |

**CARPENTRY**

| | |
|---|---|
| 4" Lally columns | 4 pieces |
| Framing lumber | 6083 bd. ft. |
| Studs & Plates | 3340 bd. ft. |
| Roof sheathing | 2286 sq. ft. |
| Fascia | 80 lin. ft. |
| Soffits ⅜ x 24" | 128 lin. ft. |
| Exterior wall sheathing | 1840 sq. ft. |
| Roofing (230# asph. shingles) | 23 squares |
| Insulation walls | 1840 sq. ft. |
| Insulation ceilings | 1072 sq. ft. |
| Finished wood flooring | 943 sq. ft. |
| Resilient flooring | 44 sq. ft. |

**DRY WALL**

| | |
|---|---|
| Ceilings | 1072 sq. ft. |
| Walls | 4160 sq. ft. |

**MILLWORK**

| | |
|---|---|
| Base | 520 lin. ft. |
| Shelving 12" pine | 25 lin. ft. |
| Windows | 14 pieces |
| Exterior doors | 5 pieces |
| Interior doors | 11 pieces |
| Louvres | 2 pieces |

**KITCHEN CABINETS**

| | |
|---|---|
| Hangers | 15 lin. ft. |
| Counters | 23 lin. ft. |
| Vanitory | 1 piece |
| Medicine cabinets | 1 piece |

FRONT ELEVATION

RIGHT SIDE ELEVATION

REAR ELEVATION

LEFT SIDE ELEVATION

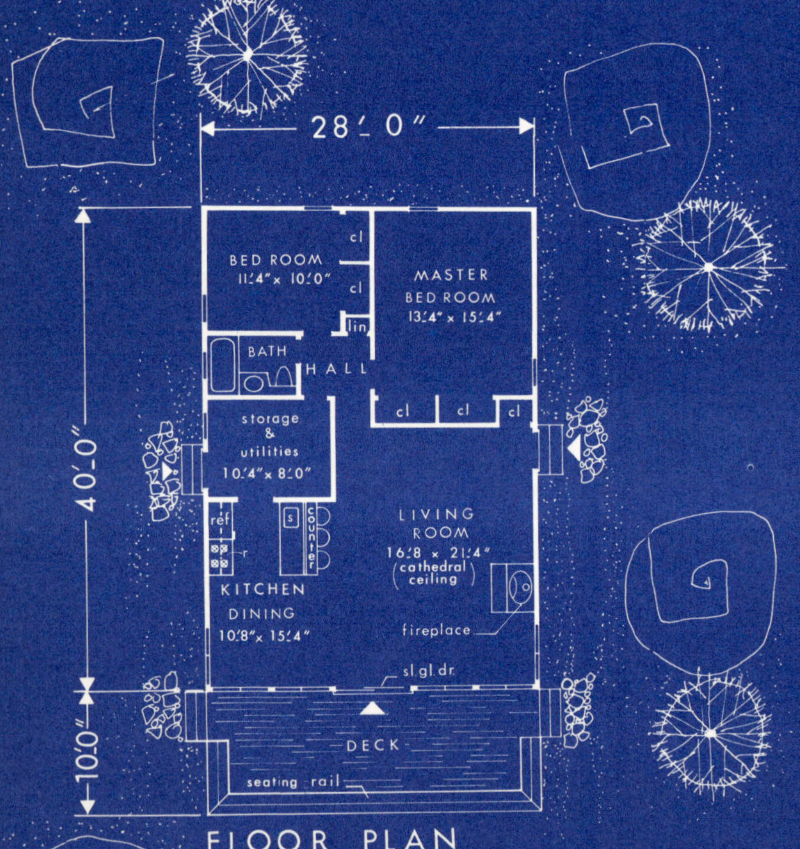


FLOOR PLAN

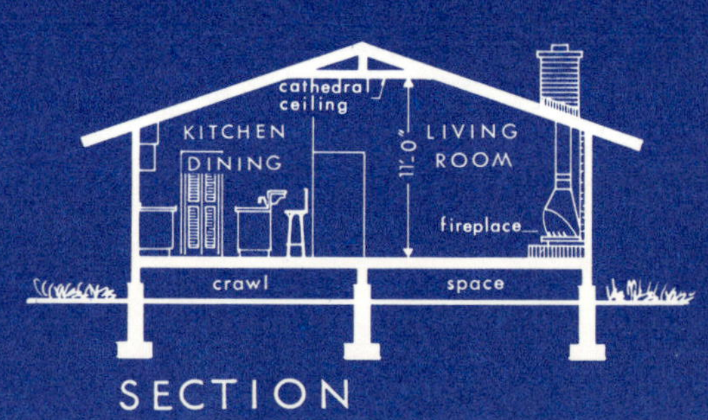


SECTION

AREA STATISTICS :

basic house ... 1,120 sq. ft.

deck .............. 280 " "

DESIGN L-46

# Vacation or Second Home in Rustic Dress

This design could make your dream of a vacation hideaway come true. A simple rectangle 40′ by 28′ — its vertical wood siding, large overhangs, protruding structural wood beams, abundance of glass and spacious wood deck all generate a rustic and warm appearance. It is easy and economical to build.

Immediately on entry you are in the main living area with a closet tucked away at the right. The living, dining and kitchen areas merge into one large space, but each is well defined.

The living room has a cathedral ceiling and a charming triangular metal fireplace. It is joined to a metal flue which projects up through the roof. The stone wall behind the fireplace and the raised stone hearth beneath it — along with the wide flooring, wood walls and the exposed rafters — give a rustic character to the inside.

The windows are so placed that you may enjoy the view. Sliding glass doors open on to the deck which can serve as an outdoor dining-living room.

Walking past the living room area, one enters the bedroom hall leading to the two bedrooms, each with cross-ventilation. There is also a bathroom with easy access to both bedrooms and to other parts of the house. There are two closets in each bedroom.

The floor is raised off the ground, creating an open crawl space underneath.

## Material List

**CONCRETE WORK**

| | |
|---|---|
| Foundations, footings, slabs, etc. | 7 cu. yds. |

**FRAMING LUMBER**

| | |
|---|---|
| Total Sills, Joists, Rafters, Studs, Plates, etc. | 6009 B.F.M. |

**SHEATHING, INSULATION**

| | |
|---|---|
| Sub Flooring | 1120 sq. ft. |
| Roof Sheathing | 1450 sq. ft. |
| Wall Insulation | 950 sq. ft. |
| Ceiling Insulation | 500 sq. ft. |

**FINISHES, INTERIOR**

| | |
|---|---|
| Finish Flooring | 1100 sq. ft. |
| Ceramic Tile Floors | 25 sq. ft. |
| Ceramic Tile Walls | 128 sq. ft. |
| Gypsum Board — house walls | 1664 sq. ft. |
| Gypsum Board — house ceilings | 1120 sq. ft. |

**FINISHES, EXTERIOR (other than masonry)**

| | |
|---|---|
| Vertical Siding | 962 sq. ft. |
| Asphalt Shingle Roofing | 1450 sq. ft. |
| Plywood Eave & Porch Soffits | 300 sq. ft. |

**WINDOW SCHEDULE**

| | |
|---|---|
| Wood Casement | 6 units |
| Wood gliding | 6 units |

**DOOR SCHEDULE**

| | |
|---|---|
| Ext. Hardwood | 2 units |
| Aluminum Sliding | 1 unit |
| Int. Hardwood, flush, staingrade | 5 units |
| Int. Hardwood, louvered, bi-folding | 6 units |

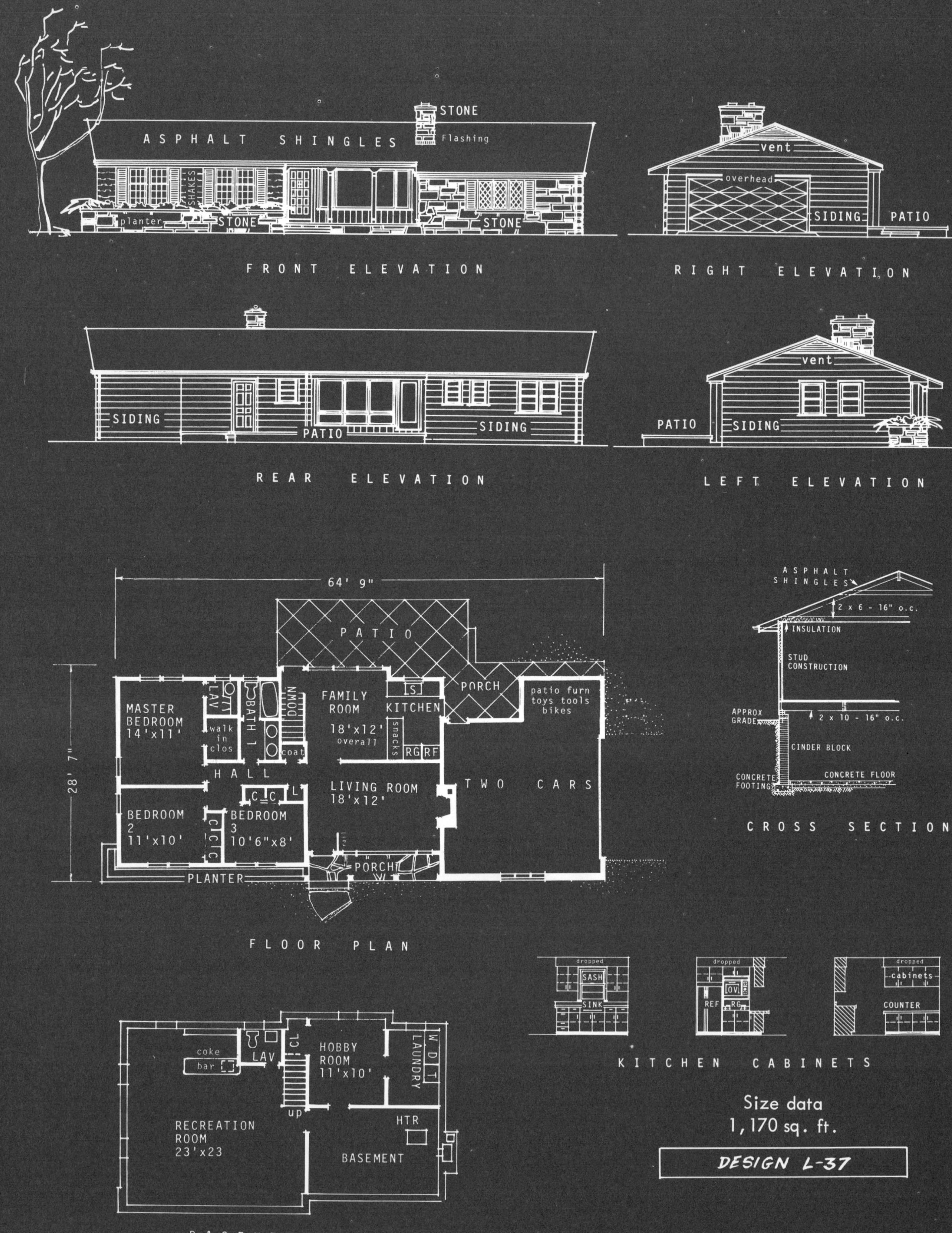


Size data
1,170 sq. ft.

DESIGN L-37

# Good Design Features in Small Ranch

Here is an example of a house that is small and economical, yet inviting in appearance and practical in its arrangement.

The compact kitchen is easy on the homemaker. The open plan of the adjacent family room removes any closed-in feeling; the snack bar adds kitchen counter-top space; the side porch can be used as a play area, or, if enclosed, it can be the utility laundry room. The storage alcove could then be made into a side porch.

The living room uses a rail separator at the front door. Access to the family room or bedroom hall is provided without cutting down on room size. The fireplace on the far wall and the big view windows at the front are good focal points for decorating. The family room can be a dining room, or it can be a breakfast or informal dining area, play space or an informal TV area. The back door opens it out to the patio area in good weather or always in a warm climate.

The master bedroom is at the back, well set off from noise by its own lavatory and walk-in closet. The two front bedrooms have enough closet area to please a family of any size. While bedroom 3 is not large, it is big enough for twin sleeping couches or double bunk beds or, in a small household, a den or study. In a retirement home, the use of extra rooms for visitors makes the three-bedroom plan just as practical.

## Material List

| CONCRETE WORK | |
|---|---|
| Footings | 14½ cu. yds. |
| Floors (basement & platforms) | 24 cu. yds. |

| MASONRY | |
|---|---|
| 4 x 8 x 18 CB | 210 lin. ft. |
| 8 x 8 x 18 CB | 345 sq. ft. |
| 10 x 8 x 18 CB | 1000 sq. ft. |
| 12 x 8 x 18 CB | 240 sq. ft. |
| Chimney fill | 300 cu. ft. |
| Flagstone | 70 sq. ft. |
| Stone Veneer | 365 sq. ft. |

| CARPENTRY | |
|---|---|
| 4" Lally Columns | 6 pieces |
| Framing Lumber | 7100 B.F. |
| Stud & Plates | 3210 B.F. |
| Roof Sheathing 1 x 6 | 2910 sq. ft. |
| Plywood | 2400 sq. ft. |
| Fascia #1 pine 1 x 6 | 130 lin. ft. |
| Soffit ⅜" x 2' Plywood | 130 lin. ft. |
| Shingle Mould | 130 lin. ft. |
| Wall Sheathing 1 x 6 | 2770 sq. ft. |
| Plywood | 2285 sq. ft. |
| Siding | 2080 sq. ft. |
| Roofing 210# Asphalt | 2400 sq. ft. |

| | |
|---|---|
| Insulation | 3050 sq. ft. |
| Finish Wood Flooring | 1350 sq. ft. |

| DRY WALL | |
|---|---|
| Ceilings ⅜" | 2575 sq. ft. |
| Walls ½" | 6320 sq. ft. |

| MILLWORK | |
|---|---|
| Base | 435 lin. ft. |
| Shelving 12" Pine | 42½ lin. ft. |
| Windows | 17 pieces |
| Exterior Doors | 5 pieces |
| Interior Doors | 16 pieces |

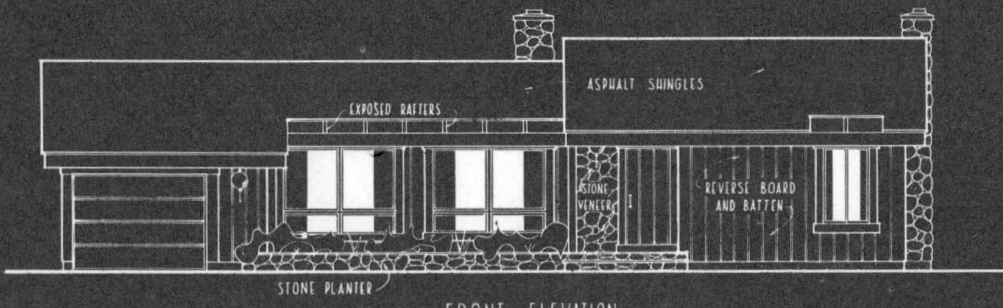

FRONT ELEVATION

RIGHT SIDE ELEVATION

REAR ELEVATION

LEFT SIDE ELEVATION

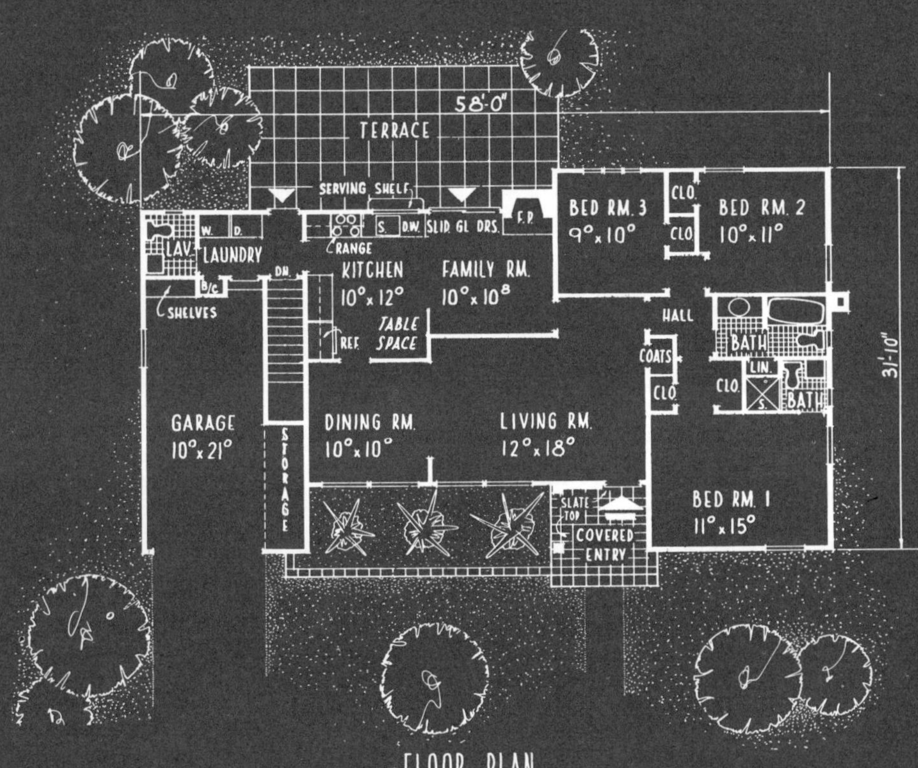

FLOOR PLAN

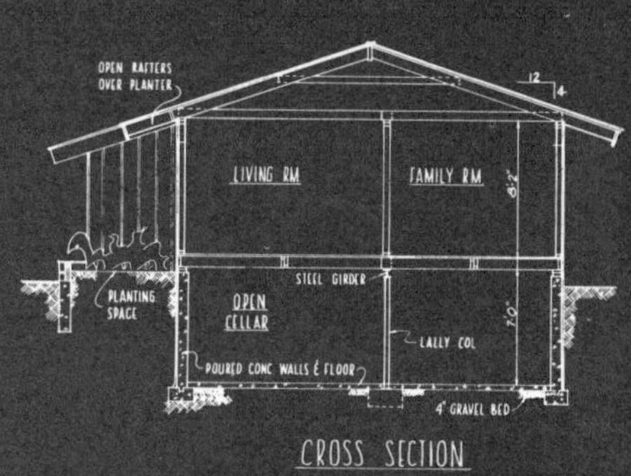

CROSS SECTION

## HOUSE DATA

| | |
|---|---|
| HABITABLE AREA | 1176 SQ. FT. |
| LAUNDRY & LAV. | 69 SQ. FT. |
| GARAGE & STOR. | 250 SQ. FT. |

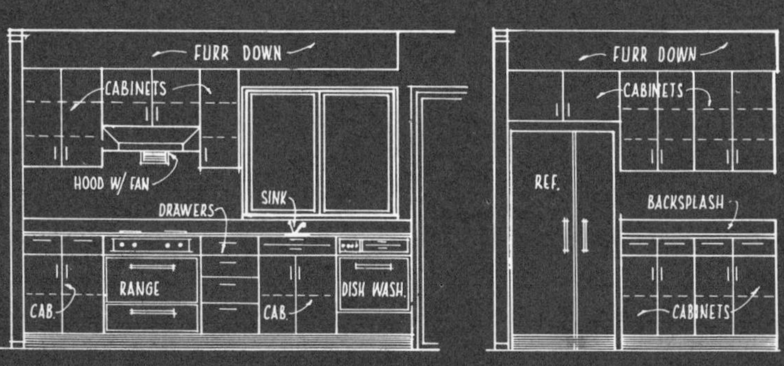

KITCHEN CABINET ELEVATIONS

DESIGN L-54

# Small Ranch Gets Contemporary Treatment

Contemporary in character — with wide overhangs, low-pitched roofs and sliding windows — this small, three-bedroom ranch nevertheless uses conventional materials.

The vertical pattern created by the reverse board and batten treatment of the exterior walls contributes to the appearance of contemporary, but the warmth of the natural wood and stone serves as the link between this style and the traditional architectural character of earlier times.

The entry is covered for protection in inclement weather. The immediate interior view is restricted to the two formal areas, the dining and living rooms. A family room to the rear is hidden, which many home owners consider an asset, since it often has a somewhat untidy "lived-in" appearance. The room has direct access through sliding glass doors to the rear terrace, is convenient to the kitchen and has a fireplace tucked into one corner.

Although designed for one car, the garage has enough storage area to handle the many pieces of power equipment, garden tools, etc., found in most structures of this kind. It can be extended 10 ft. to take two cars.

At the right side of the house are three bedrooms and two bathrooms. Hall space has been minimized. The bathrooms are located to give privacy, out of sight from any of the living areas. A full basement extends beneath the entire house.

## Material List

| CONCRETE WORK | |
|---|---|
| Concrete Walls | 1050 cu. ft. |
| Foundation Footings | 211 cu. ft. |
| Slabs | 450 cu. ft. |
| **STRUCTURAL STEEL** | |
| Lally Columns .. 3½" diam. | 8 pcs. |
| Girder .... S-6 12.5# | 46 lin. ft. |
| **BRICK WORK** | |
| Chimney & Fireplace | 121 cu. ft. |
| Flue Lining | 36 lin. ft. |
| Veneer | 35 sq. ft. |

| CARPENTRY | |
|---|---|
| Framing Lumber | 6403 B.F. |
| Studs | 2667 B.F. |
| Plates | 800 B.F. |
| Roof Sheathing | 2220 sq. ft. |
| Sub Floor | 1240 sq. ft. |
| Side Wall Sheathing | 1460 sq. ft. |
| Wall Insulation | 1050 sq. ft. |
| Ceiling Insulation | 1280 sq. ft. |
| Wood Flooring | 863 sq. ft. |
| Resilient Flooring | 300 sq. ft. |
| Underlayment Plywood | 300 sq. ft. |

| MILLWORK | |
|---|---|
| Ext. Doors & Frames Complete | 2 pcs. |
| Garage Door Complete Set | 1 set |
| Sliding Glass Door Unit | 1 unit |
| Int. Doors & Frames Complete | 12 pcs. |
| Sliding Doors | 2 pcs. |
| Bi Fold Unit | 1 unit |
| Window Units | 17 pcs. |
| Fascia | 270 lin. ft. |
| Base | 430 lin. ft. |
| **SHEET METAL** | |
| Saddle & Counter Flashing 16 oz. Copper | |

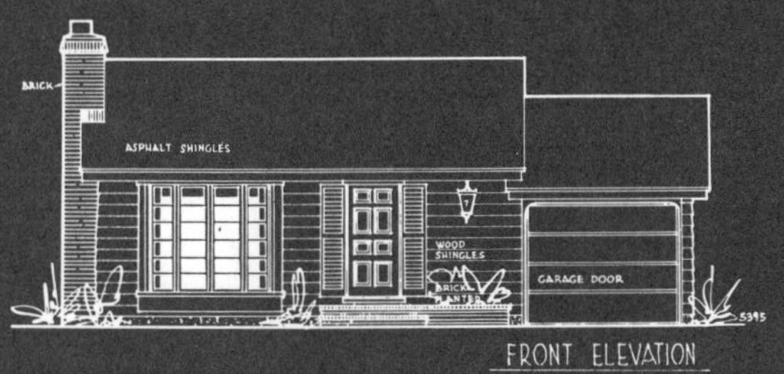


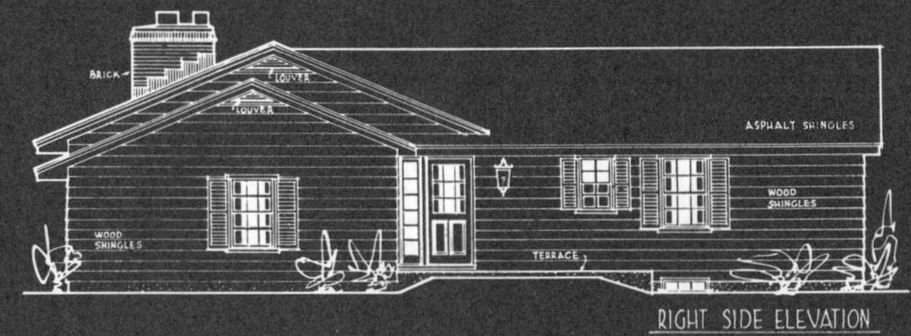


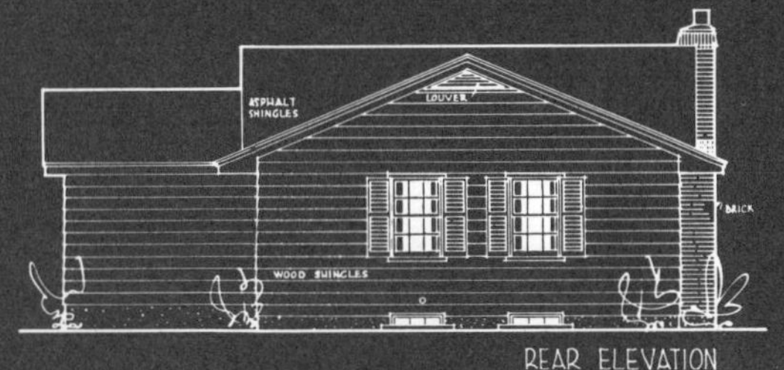

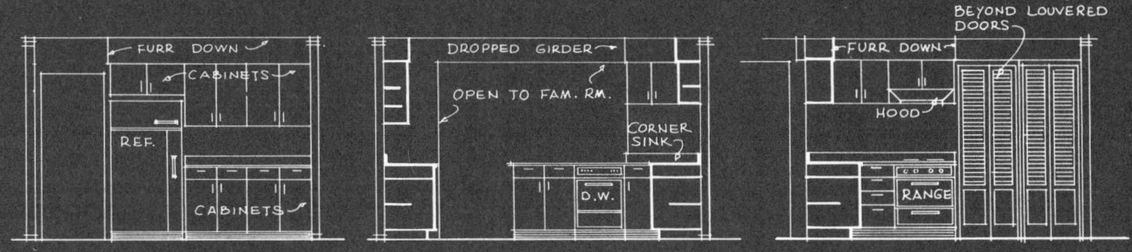


KITCHEN CABINET ELEVATIONS

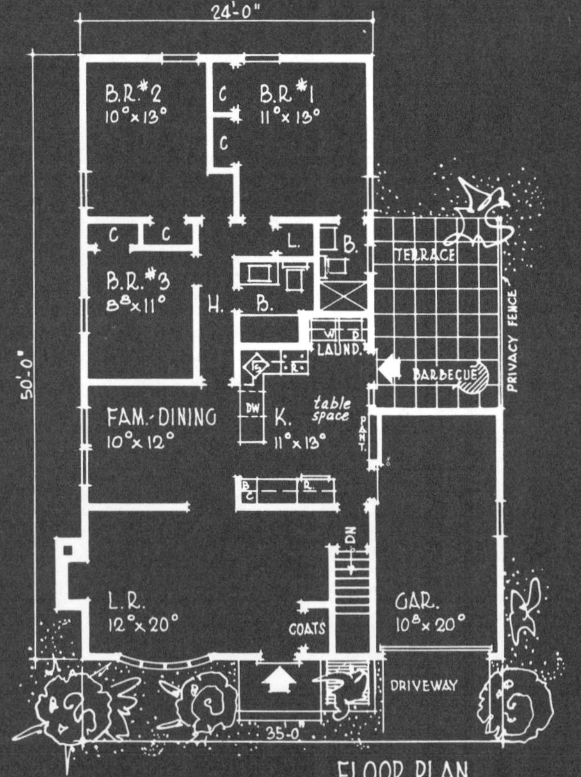


FLOOR PLAN

## HOUSE DATA

FLOOR PLAN ..... 1200 SQ. FT.
GARAGE .......... 228 SQ. FT.

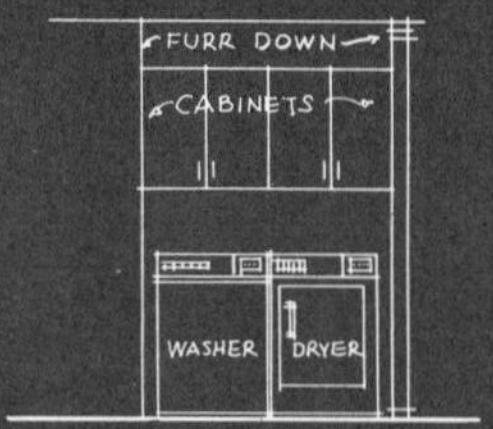


LAUNDRY EQUIPMENT ELEVATIONS

DESIGN L-30

# Designed Especially for Narrow Lot

This three-bedroom, one-story traditional can be built, complete as shown, on a plot only 50′ wide. Should the owner decide to build without the garage, the house may be constructed on a plot as narrow as 40′.

There are several planning advantages in the layout of Design L-30 which warrant examination. It can be seen how the location of the kitchen provides convenient access to several important areas. From the kitchen one can step directly to the outside terrace with its barbecue or into the garage and also to the cellar stair. The kitchen is open-planned to the dining space. The kitchen therefore becomes the control center of the house for it not only is adjacent to all of these areas but also to the laundry, which can be closed off with bi-fold doors.

Many houses which are deeper than wide need much hallway to connect the rooms, but here the architect uses the family area as part of the hall system, thus reducing wasted space.

The three bedrooms in this home are planned for maximum wall space. The main bedroom has a private bathroom with a stall shower.

Outside, the bow window, with its many panes, fits into the wood-shingled background and adds to the character of the traditional exterior. The front-entrance doorway has been kept simple, flanked by two full-length louvered shutters.

## Material List

| Item | Description | Quantity |
|---|---|---|
| **CONCRETE WORK** | | |
| Concrete Walls | | 913 cu. ft. |
| Foundation Footings | | 160 cu. ft. |
| Slabs | | 498 cu. ft. |
| Misc. Concrete | | 118 cu. ft. |
| **STRUCTURAL STEEL** | | |
| Lally Columns | 3½″ diam. | 7 pieces |
| Girder | S6 12.5# | 62 lin. ft. |
| **BRICK WORK** | | |
| Chimney | Brick | 125 cu. ft. |
| Flue Lining | T. C. | 39 lin. ft. |
| **CARPENTRY** | | |
| Framing Lumber | | 6179 B.F. |
| Studs | | 2400 B.F. |
| Plates | | 720 B.F. |
| Roof Sheathing | | 2340 sq. ft. |
| Sub Flooring | | 1200 sq. ft. |
| Side Wall Sheathing | | 1554 sq. ft. |
| Insulation Walls | | 934 sq. ft. |
| Insulation Ceilings | | 1200 sq. ft. |
| Oak Floor | | 982 sq. ft. |
| Kitchen Underlayment | | 143 sq. ft. |
| **MILLWORK** | | |
| Ext. Doors & Frames Complete | | 3 pieces |
| Garage Door Complete Set | | 1 unit |
| Int. Doors & Frames Complete | | 7 pieces |
| Bi Fold Doors | | 9 pieces |
| Bi Fold Doors | | 1 unit |
| Windows | | 15 units |
| Fascia | | 240 lin. ft. |
| Base | | 440 lin. ft. |
| **KITCHEN CABINETS** | | |
| Base Cabinets | | 9′-0″ long |
| Wall Cabinets | | 15′-9″ long |
| **ROOFING** | | |
| Shingles | 235# asphalt | 2340 sq. ft. |
| Roofing Paper | 15# felt | 2340 sq. ft. |
| **SHEET METAL** | | |
| Saddle & Counter Flashing | 16 oz. | Copper |

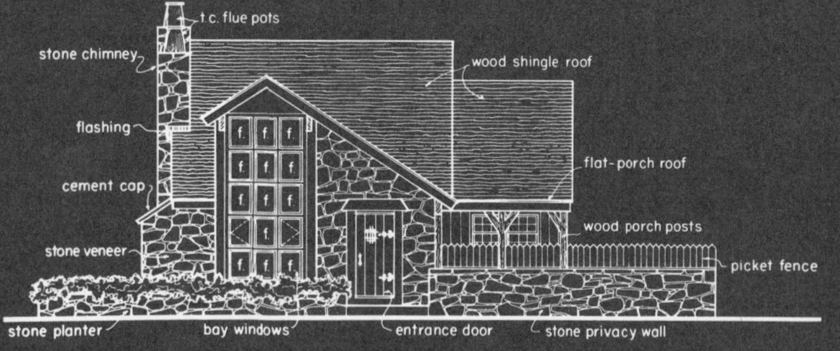


front elevation

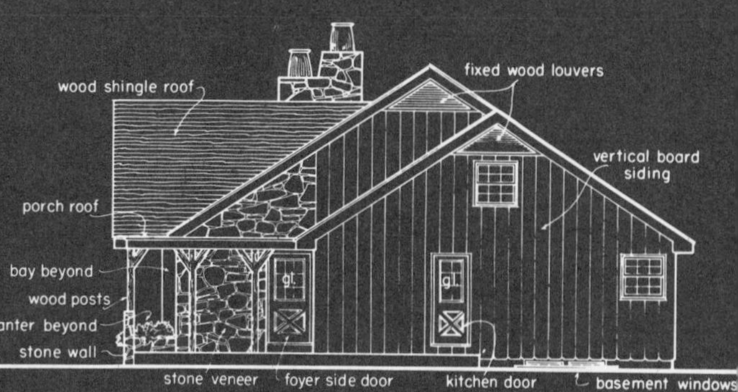


right side elevation

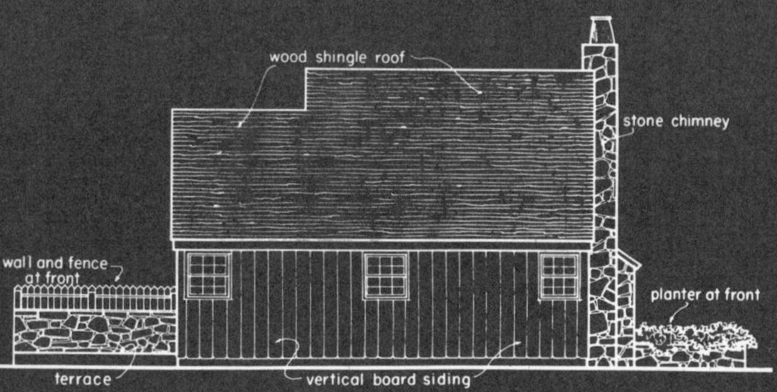


rear elevation

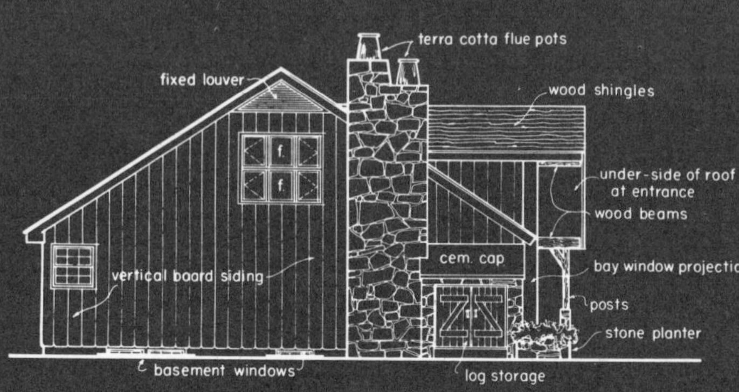


left side elevation

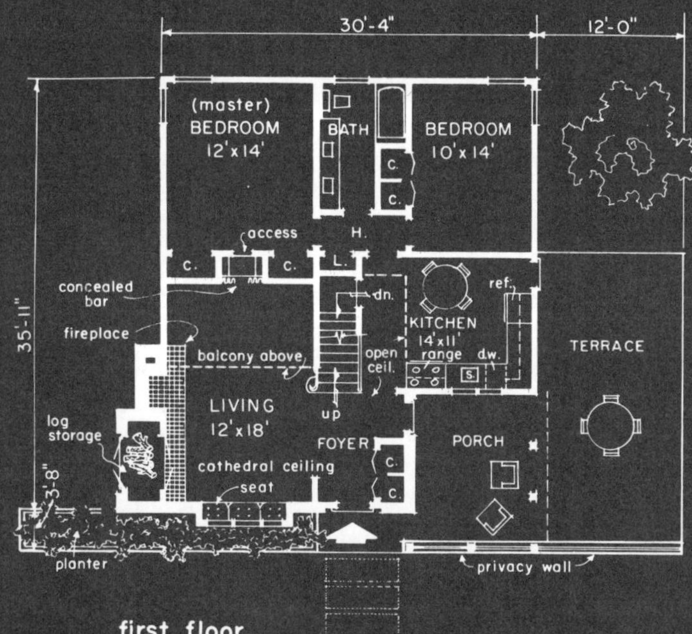


first floor

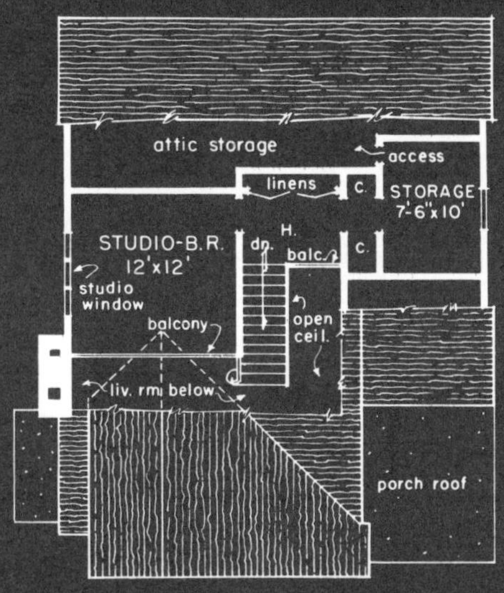


second floor

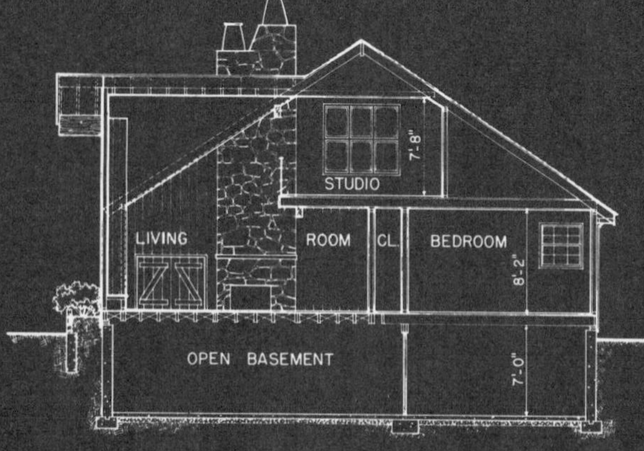


cross section

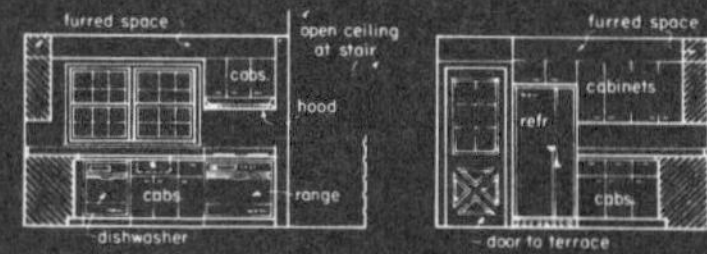


kitchen elevations

**size data**

990 square feet excluding porches and terrace ... first floor.

226 square feet excluding storage room and attic.... second floor.

DESIGN L-31

# Unusual Facade Highlights Small House

Here's a house with a unique approach to the exterior design that is minimum in size.

Note how the main front window dwarfs the front entrance door. This is because the window is $1\frac{1}{2}$ stories in height. It is the main exterior design feature of the home.

A $1\frac{1}{2}$ story design, by its very nature, has some unusable space at the second floor level where the roof slides down below headroom. The space was used without borrowing any otherwise usable room for the high window, creating a dramatic conversation piece both indoors and out.

Inside the living room, the tall window dominates the area as well as providing an excellent view of the outdoors and allowing maximum natural lighting to come indoors. Exposed ceiling beams in the room add their 17th century charm to this interesting space. In addition, a concealed bar can be used from the living room or master bedroom.

The front entry is weather-protected by the large gable roof overhang; direct access to the covered porch is provided there.

Inside, a small foyer keeps unnecessary traffic out of the living area since the kitchen and two bedrooms can be reached directly. A large closet is just inside the front door.

The kitchen is large for a home of this size. Service for outdoor eating on either the porch or terrace is efficiently provided. A portion of the kitchen ceiling is open to the area above.

## Material List

**CONCRETE**

|  |  |
|---|---|
| Walls — Piers — Chimney | 1530 cu. ft. |
| Basement Floor | 920 sq. ft. |
| Porch & Terrace Floor 6 x 6 Mesh | 460 sq. ft. |
| Water Proofing | 1000 sq. ft. |
| 5 Metal Areaways | |

**MASONRY**

|  |  |
|---|---|
| Stone Veneer | 300 sq. ft. |
| Stone Planter | 112 sq. ft. |
| Stone Wall | 100 sq. ft. |
| Stone Veneer Chimney | 390 sq. ft. |
| 46 Lin. Ft. 13 x 13 Flue Lining | |
| Common Brick Backup | 2100 |

**CARPENTRY**

|  |  |
|---|---|
| Plates — Floor Joists — Both Fls. | 3000 BM |
| 5/8 Plyscore Sub Floor & Deck | 1920 sq. ft. |
| All Studs — Headers | 3000 BM |
| Backing — Bridging | 300 BM |
| Sel. Fir Post Etc. | 400 BM |
| Ceiling Joists & Catwalk | 340 BM |
| Roof Framing | 2150 BM |
| 1/2" Plyscore Roof Boards | 1600 sq. ft. |
| O.S. Finishing Lumber Redwood | 200 BM |
| 3/8 Ext. Plywood & 3/8 Hardboard | 400 sq. ft. |
| 3/8 Plyscore Sheathing | 1600 BM |
| 7 Rolls 15 lb. Saturated Felt | |
| 1 x 12 D&M Vd Siding — Clear Redwood | 1000 BM |
| 13/16 x 2¼ Clear Oak Floor | 1380 BM |
| Basement Stairs | 210 BM |

**ROOFING**

2 Square 4 Ply Tar & Gravel
18 Square Hand-Split Cedar Shakes 24" ½ to 1¼ Thick 10" Exp.
16 Rolls 18" 30 lb. Saturated Felt

**PLASTERBOARD**

|  |  |
|---|---|
| ½ For Walls Taper Joint | 3600 sq. ft. |
| 3/8 For Ceiling Taper Joint | 1400 sq. ft. |

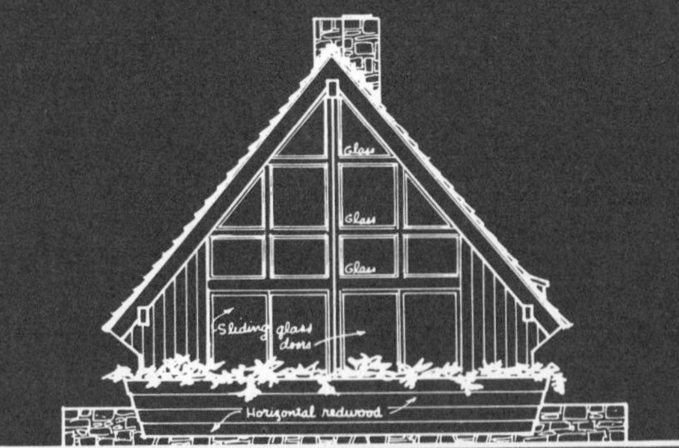


Front elevation

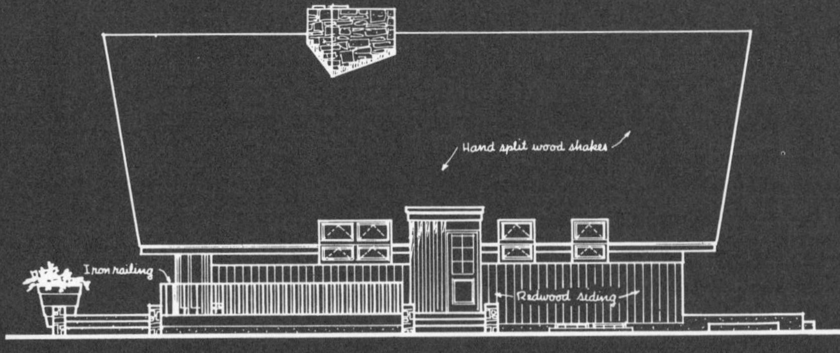


Right side elevation

Rear elevation

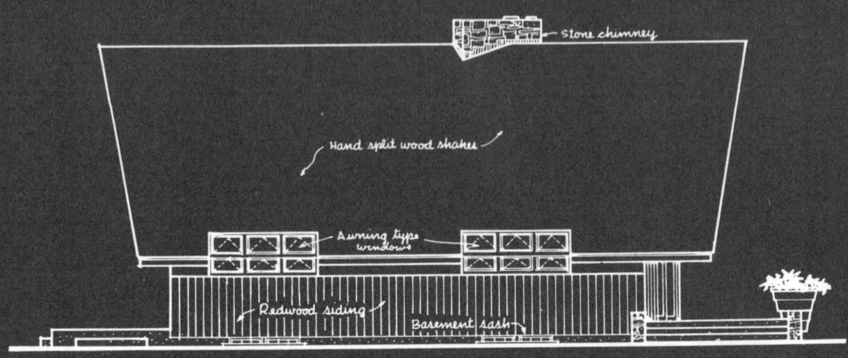


Left side elevation

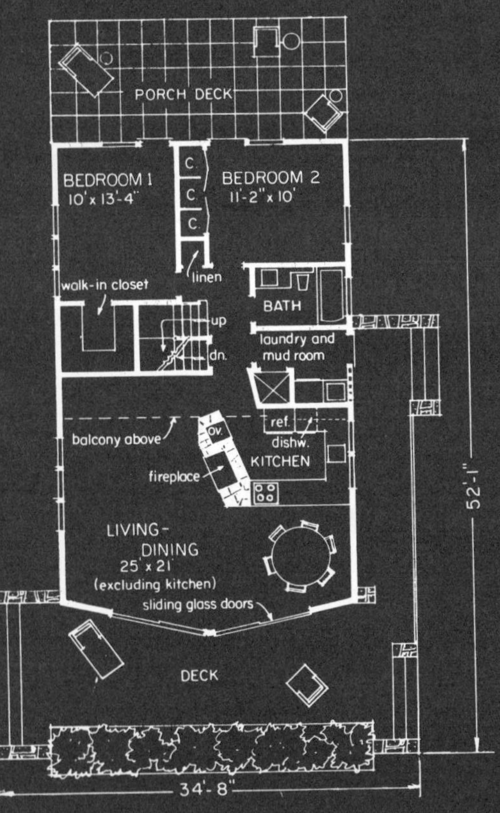


first floor plan

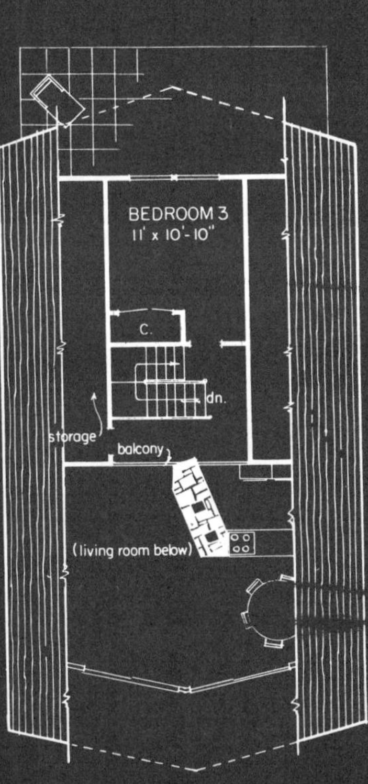


second floor plan

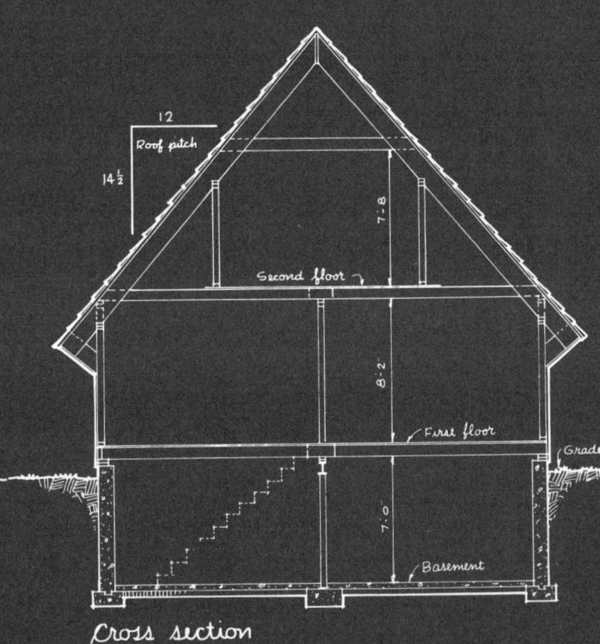


Cross section

Size data

993 square feet on first floor, excluding porches - deck.

253 square feet on second floor.

Kitchen elevations

DESIGN Z-99

# More Space in Mini-Skirted A-Frame

You might call this dramatic vacation home a mini-skirted A-frame. That's because its "skirts" or eave lines are much higher than in the usual house of this type. This provides maximum interior space on both floors with minimum roof surface.

A unique design arrangement ties the eaves into the main structure. Butterfly-type window units lap over and under. the extended eaves without interrupting the continuity. They not only catch the high sky sun, but afford both an upward and downward view.

The huge glass front area flaunts conservative design ideas and exposes the outdoors to the inside, and the indoors to the outside, without obstruction.

The overall dimensions of 52′1″ by 34′8″ include a stone-supported wooden deck that stretches across the width of the house, with a long planter at the front of it. One enters a basically open room containing a lounge, eating area and kitchen. A centrally located stone chimney shaft 20′ high creates a space separation without appearing to do so. A beam-supported balcony dramatically crosses the rear of the room.

Behind the slightly angled fireplace within the chimney shaft is a U-shaped kitchen. Two bedrooms and a bath are at the rear.

The second-floor stair leads to a front balcony that overlooks the large living room. At the rear is a bedroom with a large glass area.

## Material List

**CONCRETE**
10″ walls piers — footings ...... 1450 cu. ft.
8″ walls piers — footings ...... 750 cu. ft.
4″ concrete floor ............. 930 cu. ft.
4″ conc. floor 6 x 6 mesh-terrace 250 cu. ft.

**STEEL**
3 Lally column — 3½ - 6′8
36 Lin. 7″I 15.3 lbs. I beam
3 Double — metal areaways
20 Lin. ft. 2′8 Ornamental iron rail

**MASONRY**
100 Sq. ft. — 12″ stonewall
450 cu. ft. — stone chimney

**CARPENTRY**
Floor Framing, plates, etc. ...... 2430 BM
⅝″ Plyscore sub floor ......... 1650 sq. ft.
Studs — Plates headers ........ 2900 BM
Ceiling joists 2 x 8 36″oc Sel. fir 200 BM
Front porch framing ........... 700 BM
1 x 4 edge grain Cl. fir floor — joints full white lead ........ 630 BM
1 pc. 8 x 12 18′0) Sel. fir beam
1 pc. 8 x 12 20′0)
38 pc. 4 x 12 24′0 Sel. fir rafter & beams .................. 3900 BM
54 Lin. 2 x 16 Sel. fir ridge.... 144 BM

2560 sq. ft. — 1⅞ fiber board decking finish 1 side
2 x 8 Look outs .............. 160 BM
1 x 12 exterior trim Cl. redwood. 280 BM
⅜ exterior plywood good 1 side. 640 sq. ft.
1280 sq. ft. — ⅜ plyscore sheathing
1 x 8 D & M Vd joint Ver. Siding Cl. redwood ................ 950 BM
192 sq. ft. — ⅝ plyscore — plugged
1400 BM 13/16 x 2¼ Cl. red oak floor

**ROOFING**
26½ sq. Wood shakes 24 x ½ x 10 exposure incl. starter, etc.

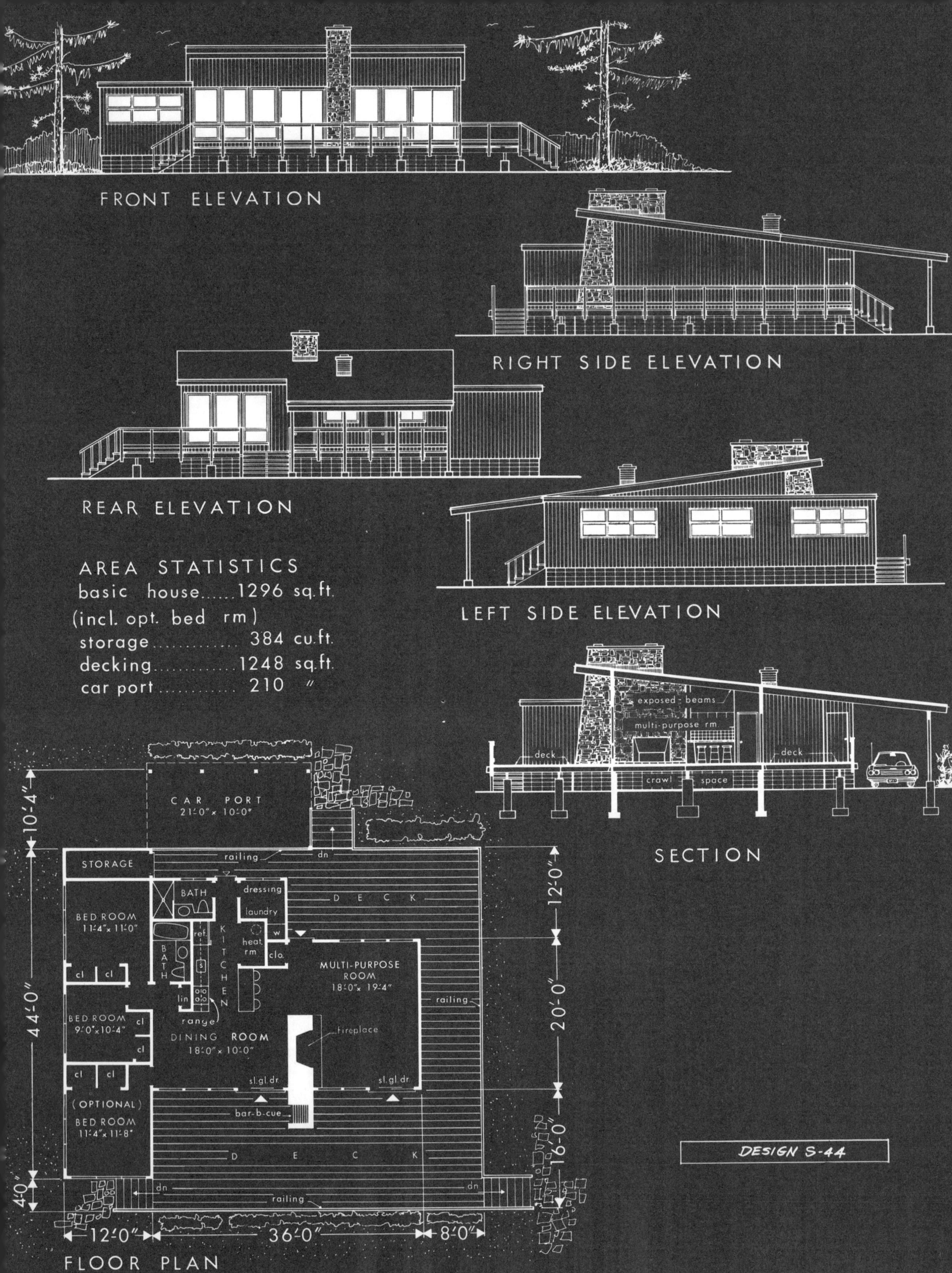
FRONT ELEVATION
RIGHT SIDE ELEVATION
REAR ELEVATION
LEFT SIDE ELEVATION
AREA STATISTICS
basic house.....1296 sq.ft.
(incl. opt. bed rm)
storage............ 384 cu.ft.
decking...........1248 sq.ft.
car port........... 210 "
exposed - beams
multi-purpose rm
deck
crawl space
SECTION
CAR PORT
21'-0" x 10'-0"
STORAGE
railing
dn
BATH
dressing
laundry
BED ROOM
11'-4" x 11'-0"
KITCHEN
ref.
heat. rm
clo.
w
D E C K
MULTI-PURPOSE ROOM
18'-0" x 19'-4"
railing
cl
lin
range
BED ROOM
9'-0" x 10'-4"
DINING ROOM
18'-0" x 10'-0"
fireplace
sl.gl.dr
(OPTIONAL)
BED ROOM
11'-4" x 11'-8"
bar-b-cue
10'-4"
44'-0"
4'-0"
12'-0"
20'-0"
16'-0"
12'-0"
36'-0"
8'-0"
FLOOR PLAN
DESIGN S-44

# Second Home Uses Simple Construction

With increasing urban congestion, many families are building vacation homes or retreats.

These range from one-room cabins to structures more pretentious than the houses in which the families spend most of their time.

This attractive house can be casual or formal. Surrounding the house on three sides is a huge deck for outdoor living. The inside flows onto this deck by means of sliding glass doors strategically placed.

A massive stone chimney contains a fireplace and acts as a divider between a large, multipurpose room and a dining room. The outside portion has a barbecue. Best of all, the chimney counteracts the strong horizontal exterior line of wood and glass.

Entering the house from either the front or rear deck, one is immediately confronted with the spaciousness of the interior and with its natural materials of wood, stone and glass.

If the budget doesn't allow the bedroom wing can be eliminated or added as money is available. With the dining room and multipurpose room stretching to more than 36′, it would be an easy matter to set up sofa beds, high risers or bunk beds for sleeping.

A desirable feature of this house is a bath and dressing room accessible from the outside. There is another bathroom for regular use convenient to the bedroom wing. A carport is attached at the rear of the house.

## Material List

**CONCRETE WORK**
Foundations, footings, slabs, etc. 37 cu. yds.

**STEEL**
Steel Angles ... 3/8" x 4" x 4" .. 10 lin. ft.

**MASONRY**
Stone Fireplace & Chimney ..... 385 cu. ft.
Concrete Block ................ 50 cu. ft.

**FRAMING LUMBER**
Total Sills, Joists, Rafters, Studs, Plates, etc. ..........4653 B.F.M.

**SHEATHING, INSULATION**
Sub Flooring ..................1392 sq. ft.
Roof Sheathing & Insulation (rigid type) .................1464 sq. ft.
Wall Insulation ...............1180 sq. ft.

**FINISHES, INTERIOR**
Vinyl Asbestos Tile (wood flooring optional) ............1294 sq. ft.
Ceramic Tile Floors ............ 50 sq. ft.
Ceramic Tile Walls ............. 210 sq. ft.
Gypsum Board — house walls' ... 415 sq. ft.

**FINISHES, EXTERIOR (other than masonry)**
Vertical Siding Texture 1-11 ....1426 sq. ft.
Built up Roofing ..............1815 sq. ft.
Plywood Eave & Soffits ........ 210 sq. ft.

**WINDOW SCHEDULE**
Gliding ...................... 2 units
Flexivent awning ............. 30 units
Fixed Sash ................... 8 units

**DOOR SCHEDULE**
Ext. Hardwood ................ 3 units
Aluminum Sliding ............. 2 units
Int. Hardwood, flush, staingrade . 6 units
Int. Hardwood, louvered, bi-folding 6 units

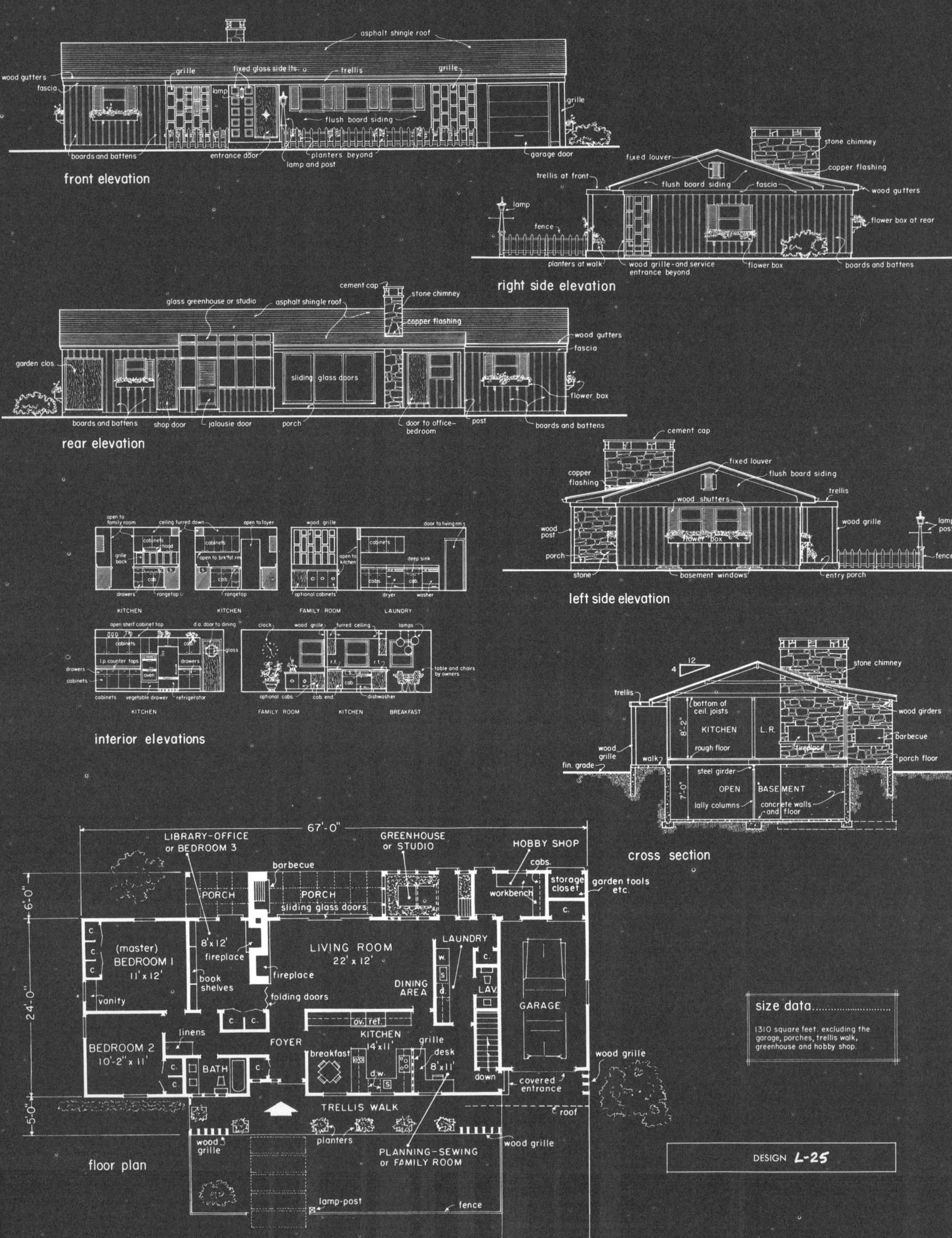

asphalt shingle roof
wood gutters
fascia
grille
fixed glass side lts.
trellis
grille
lamp
flush board siding
grille
boards and battens
entrance door
planters beyond
lamp and post
garage door
front elevation
stone chimney
fixed louver
copper flashing
trellis at front
flush board siding
fascia
wood gutters
lamp
fence
flower box at rear
planters at walk
wood grille-and service entrance beyond
flower box
boards and battens
right side elevation
cement cap
stone chimney
glass greenhouse or studio
asphalt shingle roof
copper flashing
wood gutters
fascia
garden clos.
sliding glass doors
flower box
boards and battens
shop door
jalousie door
porch
door to office-bedroom
post
boards and battens
rear elevation
cement cap
fixed louver
copper flashing
flush board siding
trellis
wood shutters
wood post
wood grille
lamp post
porch
flower box
fence
stone
basement windows
entry porch
left side elevation
KITCHEN
KITCHEN
FAMILY ROOM
LAUNDRY
KITCHEN
FAMILY ROOM
KITCHEN
BREAKFAST
interior elevations
stone chimney
trellis
bottom of ceil. joists
wood girders
KITCHEN
L.R.
barbecue
fireplace
rough floor
wood grille
walk
porch floor
fin. grade
steel girder
OPEN
BASEMENT
lally columns
concrete walls and floor
cross section
67'-0"
LIBRARY-OFFICE or BEDROOM 3
GREENHOUSE or STUDIO
HOBBY SHOP
barbecue
cabs.
PORCH
PORCH
workbench
storage closet
garden tools etc.
sliding glass doors
6'-0"
(master) BEDROOM 1 11' x 12'
8'x12'
fireplace
LIVING ROOM 22' x 12'
LAUNDRY
fireplace
book shelves
DINING AREA
LAV.
GARAGE
vanity
folding doors
24'-0"
linens
ov. ref.
BEDROOM 2 10'-2" x 11'
FOYER
KITCHEN 14'x11'
grille
breakfast
desk 8'x11'
BATH
d.w.
down
covered entrance
wood grille
5'-0"
TRELLIS WALK
roof
wood grille
planters
wood grille
PLANNING-SEWING or FAMILY ROOM
lamp-post
fence
floor plan
size data
1310 square feet, excluding the garage, porches, trellis walk, greenhouse and hobby shop.
DESIGN L-25

# House for Active Family, Young or Old

Hidden behind this homey exterior of trellises, vines and picket fences is a modest-sized shelter for a new family or retired couple with emphasis on activity.

For the young, the compact plan allows room for expansion. For the older, it offers enough space to have overnight guests or married childrens' visit. The house is designed to handle several activities.

There are many features usually found in larger homes. Among them are front entrance foyer, kitchen with breakfast room, full-sized laundry, service area lavatory, two fireplaces, a huge barbecue, a wall of bookshelves and fireplace in the third bedroom and a vanity in the master bedroom.

A library or office, which can easily double as a third bedroom, has a fireplace, wall of bookshelves and its own private porch.

A room adjoining the kitchen has a planning desk full of doors and drawers.

At the rear is a greenhouse with exposure to the outdoors, the porch and the living-dining area.

The hobby room or workshop has access to the laundry and outdoors as well as a door leading to the garage.

Entirely aside from its activity areas, Design L-25 has a well laid-out floor plan with all sections reachable from the central foyer.

## Material List

**INSULATION**
1200 sq. ft. Semi Thick Batts
1400 sq. ft. Full Thick Batts

**ROOFING**
27 Sq. King Size - 300 lb. Asphalt Shingles
12 Rolls 15 lb. Sat. Felt 2 Ply
1 Sq. Tar & Gravel

**PLASTER BOARD**

| | |
|---|---|
| ½″ Plaster Board | 4000 sq. ft. |
| ⅜″ Plaster Board | 1600 sq. ft. |
| ⅝″ Plaster Board | 800 sq. ft. |

**CONCRETE**

| | |
|---|---|
| 8″ & 10″ Concrete Wall - Footing, etc. | 1980 cu. ft. |
| Basement & Garage Floor | 1670 sq. ft. |
| Porches - 6″ Mesh | 470 sq. ft. |

**MASONRY**

| | |
|---|---|
| Stone Veneer Chimney | 450 sq. ft. |
| Common Brick Backup | 3200 Brick |

**CARPENTRY**

| | |
|---|---|
| Plates - Joists, etc. | 2250 BM |
| Sub Floor, laid diag. | 1660 BM |
| Backing Bridging - Catwalk | 680 BM |
| All Studs - Headers | 3660 BM |
| Ceiling Joists | 1100 BM |
| Roof Framing | 2400 BM |
| Fin. Lumber - Clear Red Wood | 320 BM |
| Roof Boards - ½ Plyscore | 2500 sq. ft. |
| ⅜ Plyscore Sheathing | 1800 sq. ft. |
| 9 Roll - 15 lb. Saturated Felt | |
| 1 x 10 Vert. Siding S4S Clear Red Wood | 1900 BM |
| 800 Lin. 1 x 3 S4S - Battens | |
| 13/16 x 2¼ Clear Red Oak Fl. | 1070 BM |
| 4′ - 0 x 8′0 ⅝ Plugged Plyscore | 560 sq. ft. |
| 5/16 Hardboard Siding - Good 1 Side | 550 sq. ft. |

FRONT ELEVATION

RIGHT SIDE ELEVATION

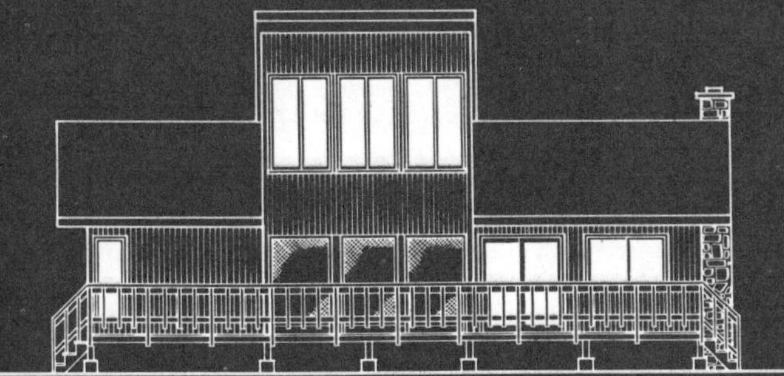

REAR ELEVATION

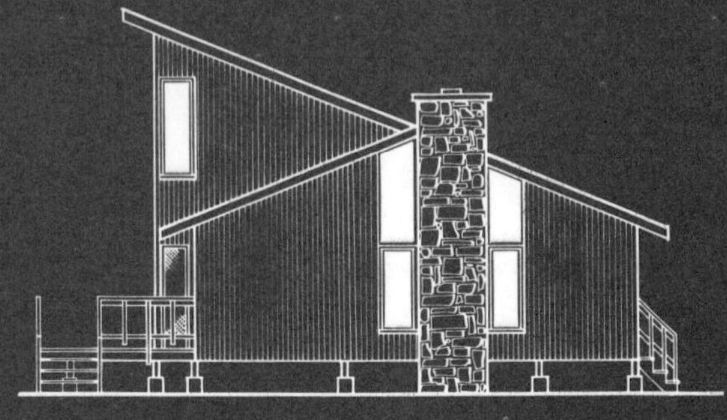

LEFT SIDE ELEVATION

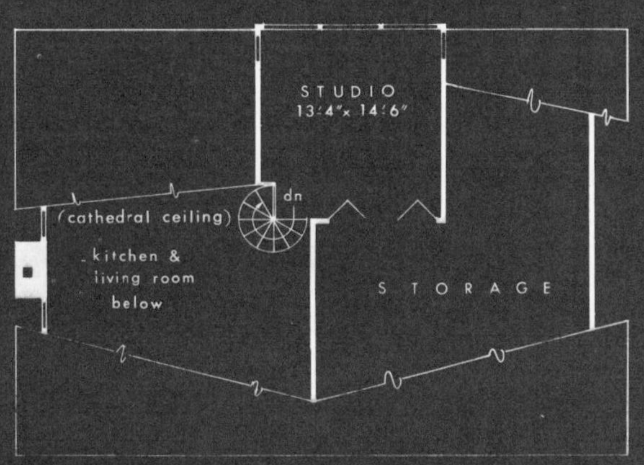


SECOND FLOOR PLAN

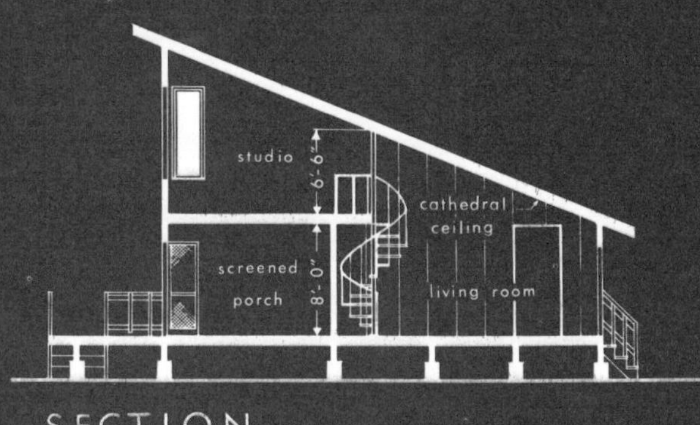


SECTION

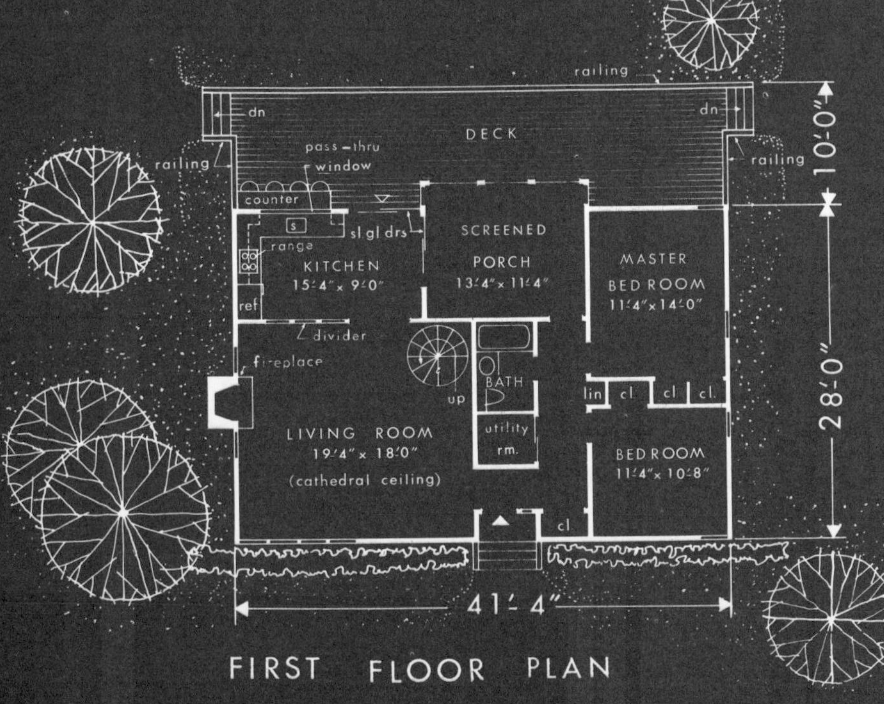


FIRST FLOOR PLAN

## AREA STATISTICS:

| | | |
|---|---|---|
| 1st floor ............... | 1019 | sq.ft. |
| 2nd floor -(studio).. | 193 | " |
| screened porch ...... | 151 | " |
| storage ............... | 476 | " |
| deck .................... | 392 | " |

DESIGN S-94

# Two-Bedroom Home for All Seasons

Finding a suitable home with two bedrooms that will fit on a small lot is not easy.

That fact was kept in mind in designing this house while recognizing the trend toward second homes. The result is a pleasant yet practical house that fulfills either need.

Trim and contemporary, the exterior is a combination of vertical siding and glass, with the use of standard lumber and window sizes helping to keep down construction costs.

Going through a recessed front entry to the central foyer, one gets an instant view of a dramatic living room with a cathedral ceiling, a stone fireplace flanked by two windows and a spiral staircase leading to a second-floor studio.

The kitchen is large and functional. A large window over the kitchen sink is strategically located to pass food through to the counter on the huge rear deck for outdoor dining. Sliding glass doors in the kitchen lead to this deck. To the right of the kitchen is a screened porch reached by sliding doors.

The master bedroom overlooks the rear deck and has two closets. The second bedroom faces the front of the house. Both rooms have cross ventilation.

The rear deck stretches the entire width of the house, is completely railed and has a short stairway at each end. Because of its dimensions, 41′4″ by 10′, it is an ideal place for entertaining either small or large gatherings.

## Material List

**CONCRETE WORK**
Foundations, footings, slabs, etc. 25 cu. yds.

**MASONRY**
Brick Fireplace & Chimney ..... 180 cu. ft.
Stone Chimney ................ 206 cu. ft.

**FRAMING LUMBER**
Total Sills, Joists, Rafters, Studs, Plates, etc. .......... 13,000 BFM

**SHEATHING, INSULATION**
Sub Flooring .................. 1368 sq. ft.
Roof Sheathing ................ 1558 sq. ft.
Wall Insulation ............... 1200 sq. ft.
Ceiling Insulation ............ 1500 sq. ft.
Floor Insulation .............. 1148 sq. ft.

**FINISHES, INTERIOR**
Vinyl Tile or Oak Flooring ...... 1368 sq. ft.
Ceramic Tile Floors ............ 25 sq. ft.
Ceramic Tile Walls ............ 60 sq. ft.
Gypsum Board (or paneling) house walls ............... 1704 sq. ft.
Gypsum Board - House Ceilings.. 1170 sq. ft.

**FINISHES, EXTERIOR (other than masonry)**
Vertical Siding (Texture 1-11) ... 1212 sq. ft.
Asphalt Shingle Roofing ........ 1158 sq. ft.
Plywood Eave & Porch Soffits ... 358 sq. ft.

**WINDOW SCHEDULE**
Wood Casement .............. 15 units
Fixed Glass ................. 2 units
Gliding ..................... 1 unit

**DOOR SCHEDULE**
Ext. Hardwood & sidelite ....... 1 unit
Ext. Hardwood ............... 1 unit
Aluminum Sliding ............ 2 units
Int. Hardwood, flush, staingrade . 5 units
Int. Hardwood, louvered, bi-folding 5 units
Screening (fixed or operable).... 5 units

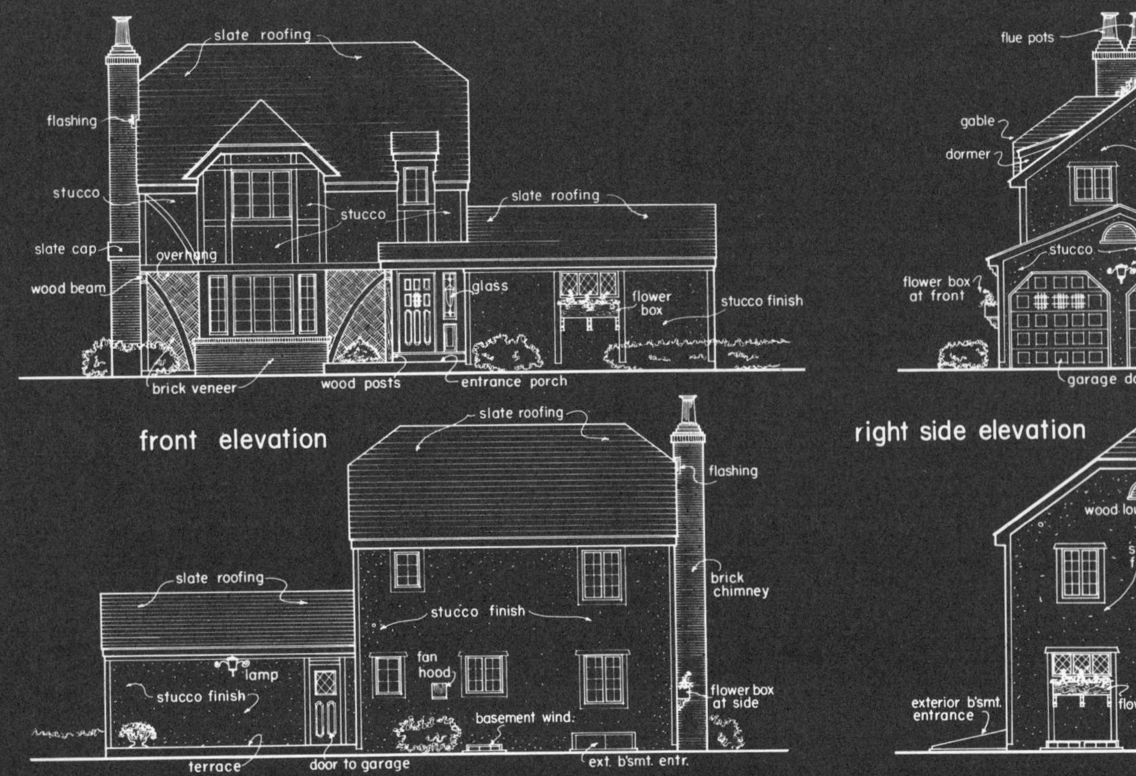


front elevation

rear elevation

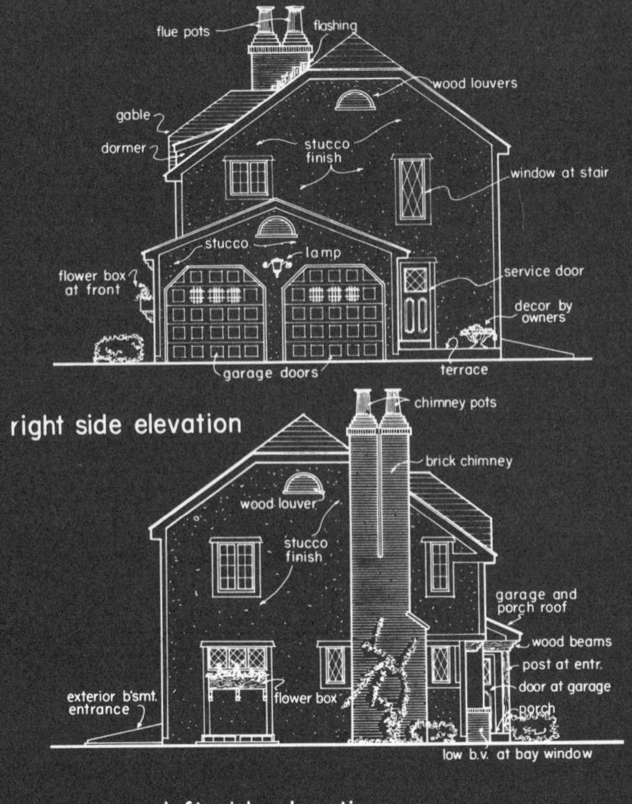


right side elevation

left side elevation

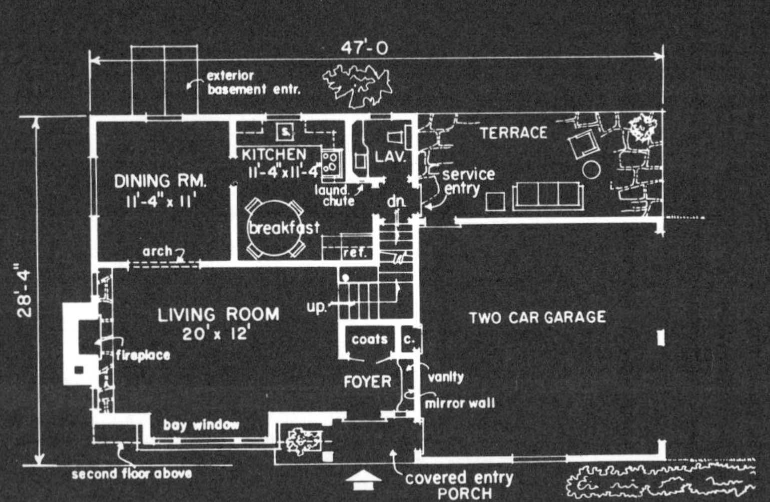


first floor plan

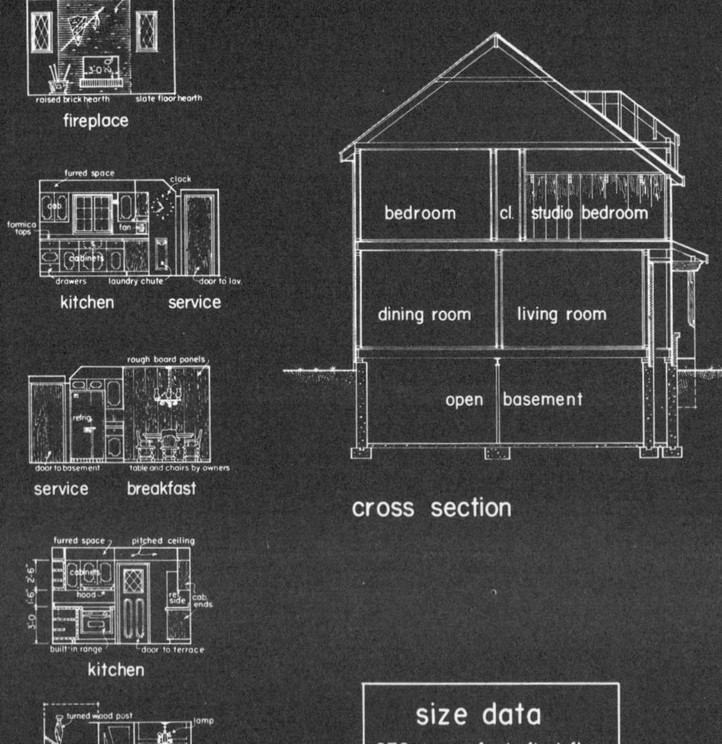


26'-8"
mirror wall
master BEDROOM 1 16'-6" x 11'
BATH
laund. chute
dn.
26'-4"
C. C. C. C.
center HALL
L.
C. C.
raised floor
studio BEDROOM 2 13'-6" x 11'-10"
BEDROOM 3 12'-2" x 9'-2"
slate roof

second floor plan

furred space
lamps
vanitory
mirror
ceramic tile
W.C.
tub
cab. under

main bath

interior elevations

**size data**

672 square feet · first floor excluding garage porch and terrace

701 square feet · second floor

DESIGN S-77

# Economy Stressed in English Tudor

Impressive in its English Tudor styling, which gives it the appearance of comfortable solidity, this two-story house has two other major plus factors — livability and construction costs.

With only 672 square feet of habitable area on the first floor, it competes most favorably in the modest-cost housing field. Attaching the two-car garage to the livable portion makes the exterior seem like anything but a moderate-priced home.

Putting no more than the basic three rooms on the first floor kept the second floor the proper size to accommodate three bedrooms, their closets and a full bathroom. The omission of a center hall gives more room to the living area.

Two exterior features are the 7-foot covered front entrance porch and the 11-foot living room bay window.

Although a separate dining room is provided, the kitchen has a roomy breakfast table space. It also has 16 lineal feet of cabinet space. The lavatory location is established for its step-saving efficiency. It cuts down on dirt tracking, serves the kitchen, basement, outdoors and garage equally well, and isn't too far to be used as a powder room for guests in the living room.

The private terrace behind the garage can easily be used for outdoor eating, since the kitchen is close by.

## Material List

**FOUNDATION (Basement) and MASONRY**
8" and 10" Walls, etc. ......... 1470 cu. ft.
4" Concrete floor .............. 1290 sq. ft.
Flagstone .................... 119 sq. ft.
Sill bolts - waterproofing & 4 areawalls
3 Lally Columns
23 Lin. ft. 7" I 15.3 lb. Steel
1 Ext. Entrance door & stair
336 Sq. ft. 4" Brick veneer
110 Cu. ft. Brick backup
Firebrick angle iron, etc.

**CARPENTRY**
All framing house and garage .. 10,710 BM
All sheathing & decking,
avg. 1/2" thick .............. 5780 sq. ft.
9 Rolls - 15 lb. Saturated Felt
13/16 x 2¼ Clear Red Oak Floor. 1380 BM
All shelving & paneling ........ 460 BM

**INSULATION**
2200 Sq. ft. of batt type

**PLASTERBOARD**
For walls, ceiling & garage,
avg. 1/2" thick .............. 6090 sq. ft.

**ROOFING**
20 Sq. Colorbestos slate shingles
5 Rolls - 15 lb. Saturated Felt

**MILLWORK**
1 Front door & frame
4 OS doors & frames 2'6 avg.
21 Case sash & frames
3 Base sash & frame
7 Ins. Hollow Core Birch
12 - 2 panel louver - trim 2 sides
1020 Lin. ft. of finish moldings
40 Lin. casing
3 Half Circle louvers
1 Flight main box stairs
14 Lin. vanities - 2'0 x 2'7
9 Lin. base cabinets
13 Lin. wall cabinets

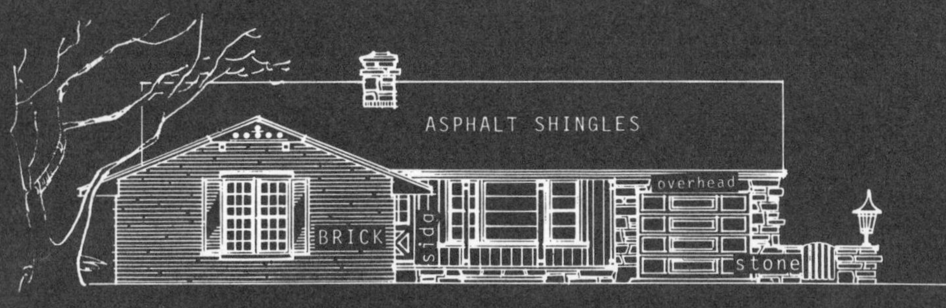


FRONT ELEVATION

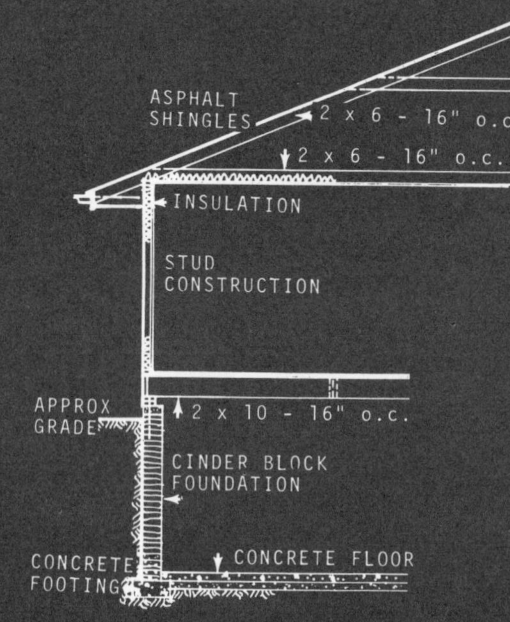


CROSS SECTION

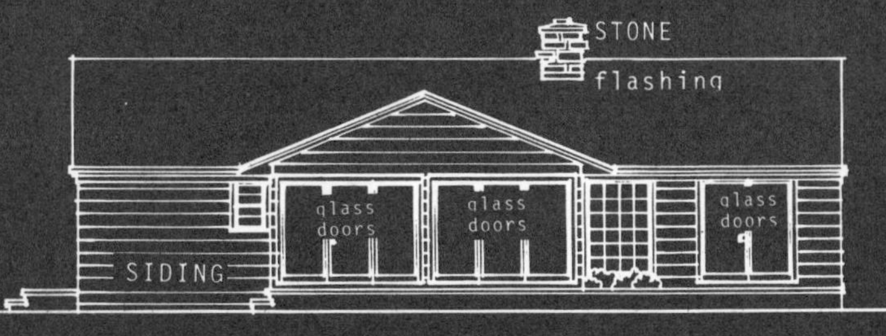


REAR ELEVATION

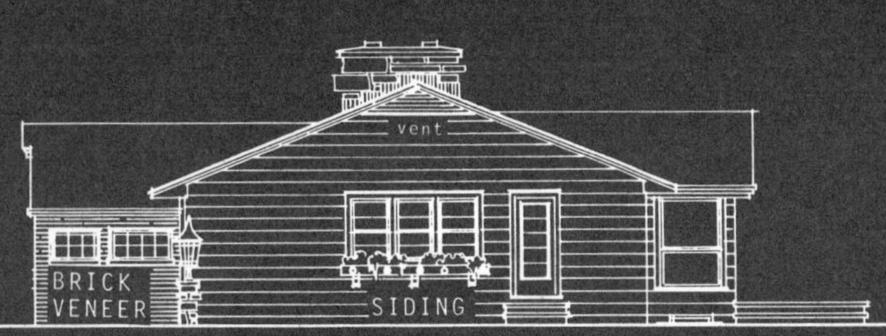


RIGHT ELEVATION

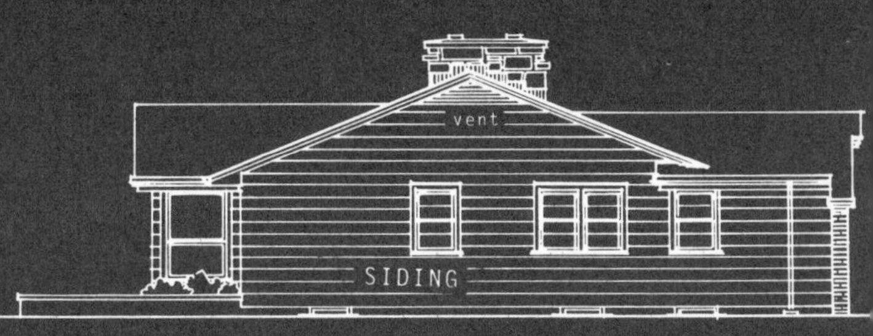


LEFT ELEVATION

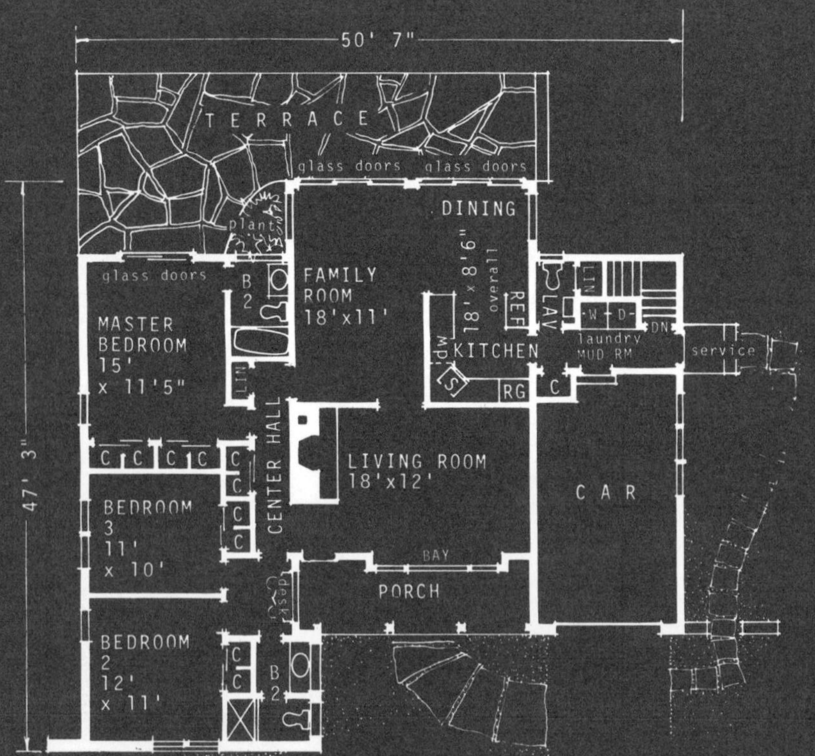


FLOOR PLAN

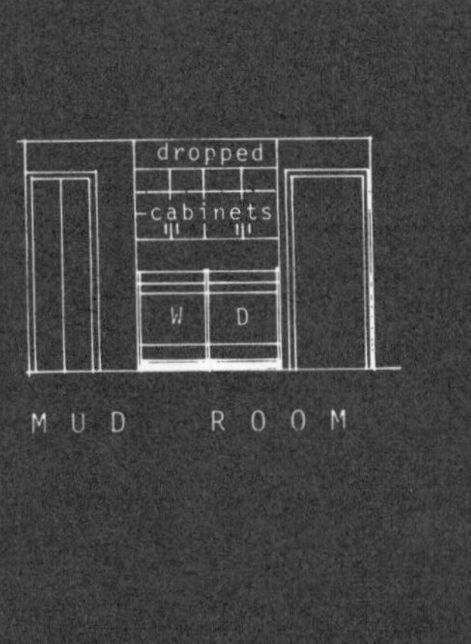


MUD ROOM

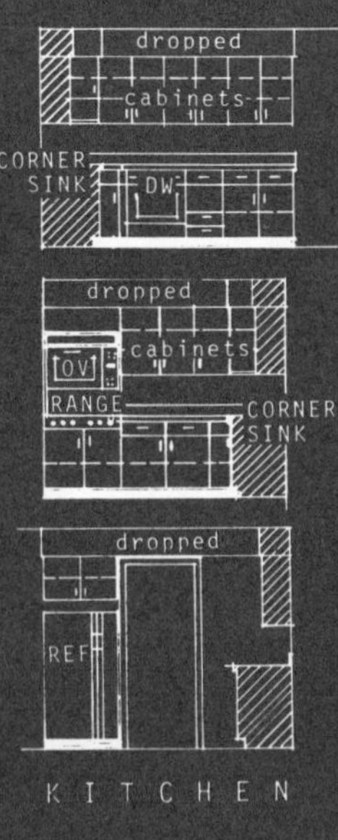


KITCHEN

DESIGN L-13

# Small Lot Can Handle This Inviting Ranch

With overall dimensions of 50′7″ by 47′3″, including the garage, this house is practical for a small lot, yet is so carefully planned that even if property size need not be considered, it would be a wise choice for its livability and economy features.

For informal occasions, the rear terrace is open from the planning area or the family room so that an outdoor get-together can be served, with guests or youngsters coming and going freely without disturbing the more formal front area.

A plus feature that has special advantages for the parents is the private alcove terrace opening directly from the master bedroom where the older folks can enjoy their privacy while the youngsters use the informal areas for their friends.

As for the budget factor, the habitable area is 1388 square feet with minimum waste space and no fancy gimmicks, making it available for maximum financing.

The living room has details such as the bay window overlooking the front porch and the fireplace corner forming an extension as well as a divider from the center hall. A door to the family room means that the two areas can be utilized conveniently if extra large entertaining space is needed.

The kitchen, dining area and family room are treated as a unit. The kitchen has the private spot with a corner sink.

## Material List

**CONCRETE WORK**

| | |
|---|---|
| Footings | 14½ cu. yds. |
| Floors (basement & platforms) | 27½ cu. yds. |

**MASONRY**

| | |
|---|---|
| 4 x 8 x 18 CB | 235 lin. ft. |
| 8 x 8 x 18 CB | 330 sq. ft. |
| 10 x 8 x 18 CB | 1020 sq. ft. |
| 12 x 8 x 18 CB | 310 sq. ft. |
| Brick veneer — chimney fill | 350 sq. ft. |
| | 465 cu. ft. |
| Stone veneer | 160 sq. ft. |

**CARPENTRY**

| | |
|---|---|
| 4″ Lally columns | 7 pieces |
| Framing lumber | 7280 B.F. |
| Stud & Plates | 3370 B.F. |
| Roof sheathing 1 x 6 | 3500 sq. ft. |
| Plywood | 2900 sq. ft. |
| Fascia #1 pine 1 x 6 | 95 lin. ft. |
| Soffit ½″ x 2′ plywood | 95 lin. ft. |
| Shingle mould | 95 lin. ft. |
| Wall sheathing 1 x 6 | 3125 sq. ft. |
| Plywood | 2580 sq. ft. |
| Siding | 2630 sq. ft. |
| Roofing 210# asphalt | 29 squares |
| Insulation — walls | 1890 sq. ft. |
| Ceiling | 1530 sq. ft. |
| Roof | 475 sq. ft. |
| Finish wood flooring | 1590 sq. ft. |

**DRY WALL**

| | |
|---|---|
| Ceilings ⅜″ | 1940 sq. ft. |
| Walls ½″ | 4805 sq. ft. |

**MILLWORK**

| | |
|---|---|
| Base | 480 lin. ft. |
| Shelving 12″ pine | 53 lin. ft. |
| Windows | 17 pieces |
| Exterior doors | 6 pieces |
| Interior doors | 19 pieces |

FRONT ELEVATION

RIGHT SIDE ELEVATION

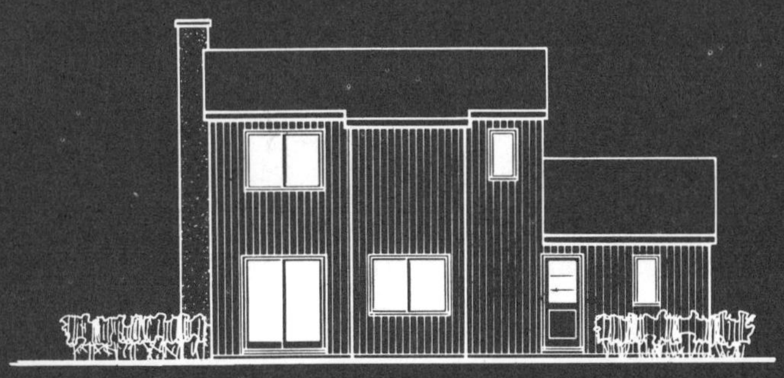

REAR ELEVATION

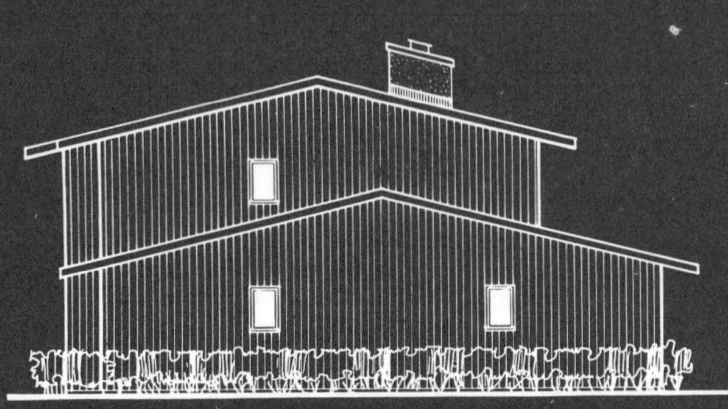

LEFT SIDE ELEVATION

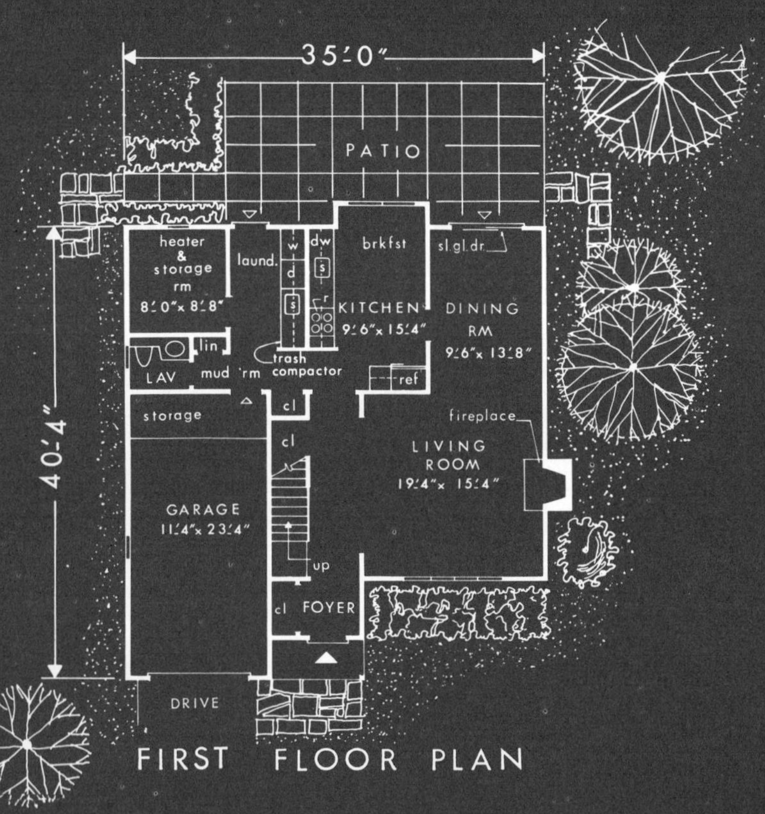


FIRST FLOOR PLAN

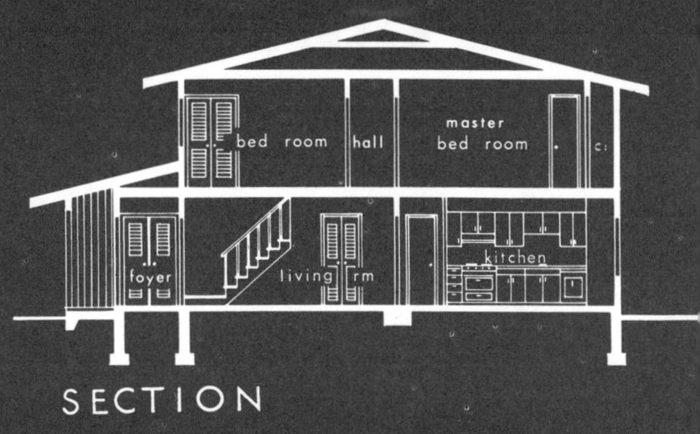


SECTION

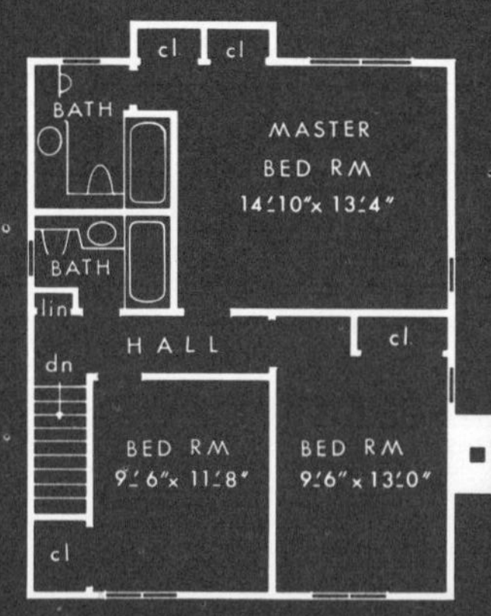


SECOND FLOOR PLAN

AREA STATISTICS

| | | |
|---|---|---|
| 1st floor | 692 | sq. ft. |
| 2nd floor | 709 | ″ |
| laundry & lav | 118 | ″ |
| heater & storage | 98 | ″ |
| garage | 226 | ″ |

DESIGN L-22

# Two-Story Fits Easily On Small Lot

Soaring construction and land costs have made it necessary for the architect to keep in sight the economical limits of most families. Designing a home which meets these needs, without any sacrifice in the quality of living, is a real challenge.

The two-story house shown here was designed to comply with these requirements. The size of the dwelling is such that it can be placed on a relatively small lot. It is only 35′ wide and could easily fit on a lot 50′ in width, bearing in mind individual zoning regulations. Total habitable area is approximately 1400 square feet, about 900 on each floor.

Entrance into the house is by means of a foyer, off which is a living room made more spacious by its merger with a dining room in an L-shaped fashion. Sliding glass doors in the dining room lead to a rear patio. The kitchen, also off the foyer, is more than 15′ long and is adjacent to the dining room. The eating area of the kitchen is further enhanced by an attractive bay.

Although the architect intended that this house be all electric, other types of heating may be used at the owner's option.

The second floor, with its three bedrooms and two baths, is reached by a decorative stair leading up from the foyer on the first floor level. The large master bedroom has double exposure and is equipped with its own private bath.

## Material List

**CONCRETE WORK**

| | |
|---|---|
| Foundations, footings, slabs, etc. | 42 cu. yds. |

**STEEL**

| | |
|---|---|
| Steel Plate ..... ½" x 11½" .. | 10 lin. ft. |
| Reinforcing Mesh .............. | 1080 sq. ft. |

**MASONRY**

| | |
|---|---|
| Brick Fireplace & Chimney ..... | 70 cu. ft. |

**FRAMING LUMBER**

| | |
|---|---|
| Total Sills, Joists, Rafters, Studs, Plates, etc. .......... | 6717 B.F.M. |

**FINISHES, INTERIOR**

| | |
|---|---|
| Vinyl Tile .................... | 340 sq. ft. |
| Oak Flooring .................. | 887 sq. ft. |
| Ceramic Tile Floors ............ | 85 sq. ft. |
| Ceramic Tile Walls ............ | 370 sq. ft. |
| Gypsum Board - house walls ..... | 3022 sq. ft. |
| Gypsum Board - house ceilings ... | 1464 sq. ft. |

**FINISHES, EXTERIOR (other than masonry)**

| | |
|---|---|
| Vertical Siding (Texture 1-11).... | 1700 sq. ft. |
| Asphalt Shingle Roofing ........ | 1631 sq. ft. |
| Plywood Eave & Porch Soffits ... | 220 sq. ft. |

**WINDOW SCHEDULE**

| | |
|---|---|
| Wood Casement ............... | 10 units |
| Wood Gliding ................. | 2 units |

**DOOR SCHEDULE**

| | |
|---|---|
| Ext. Hardwood, paneled & glazed . | 1 unit |
| Ext. Hardwood Flush ........... | 1 unit |
| Aluminum Sliding ............ | 1 unit |
| Wood Overhead garage ......... | 1 unit |
| Int. Hardwood, flush, staingrade . | 8 units |
| Int. Hardwood, louvered doors ... | 4 units |
| Int. Hardwood, louvered, bi-folding | 8 units |

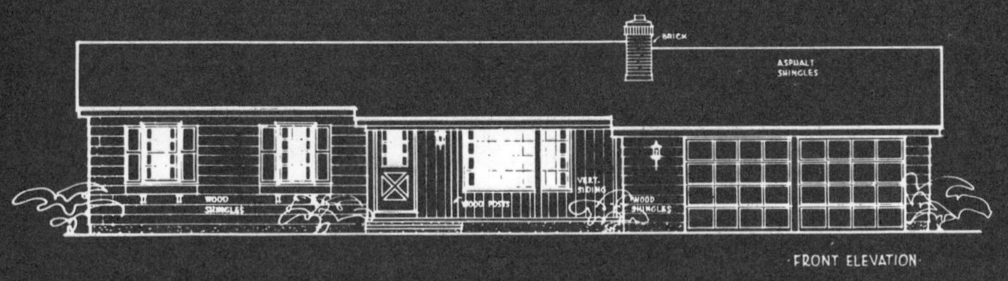

·FRONT ELEVATION·

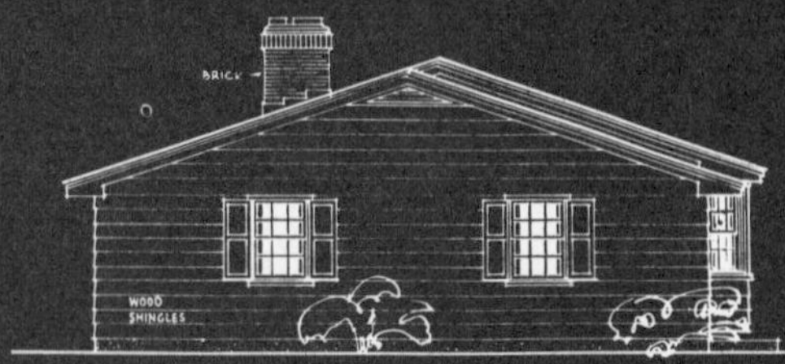

·RIGHT SIDE ELEVATION·

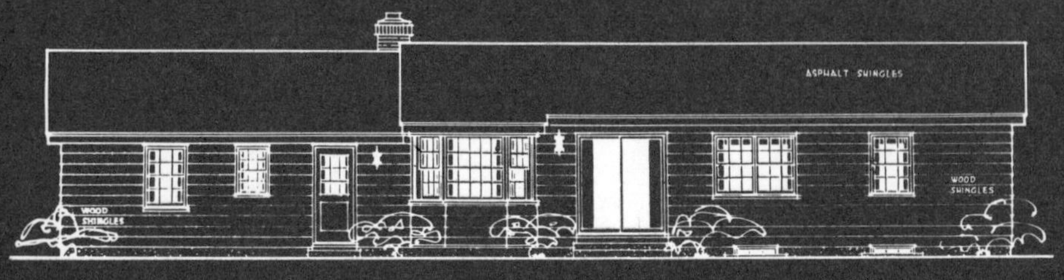

·REAR ELEVATION·

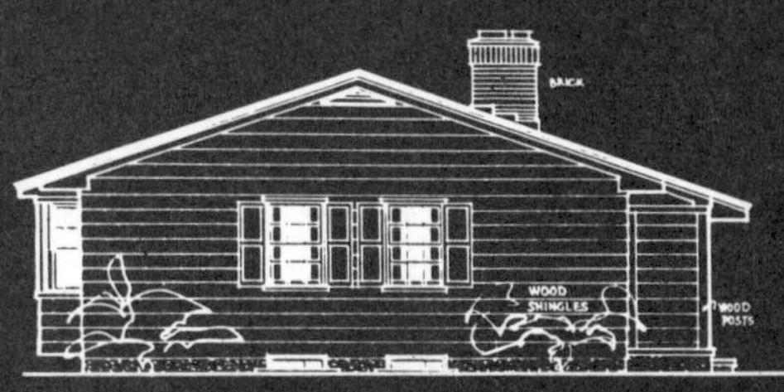

·LEFT SIDE ELEVATION·

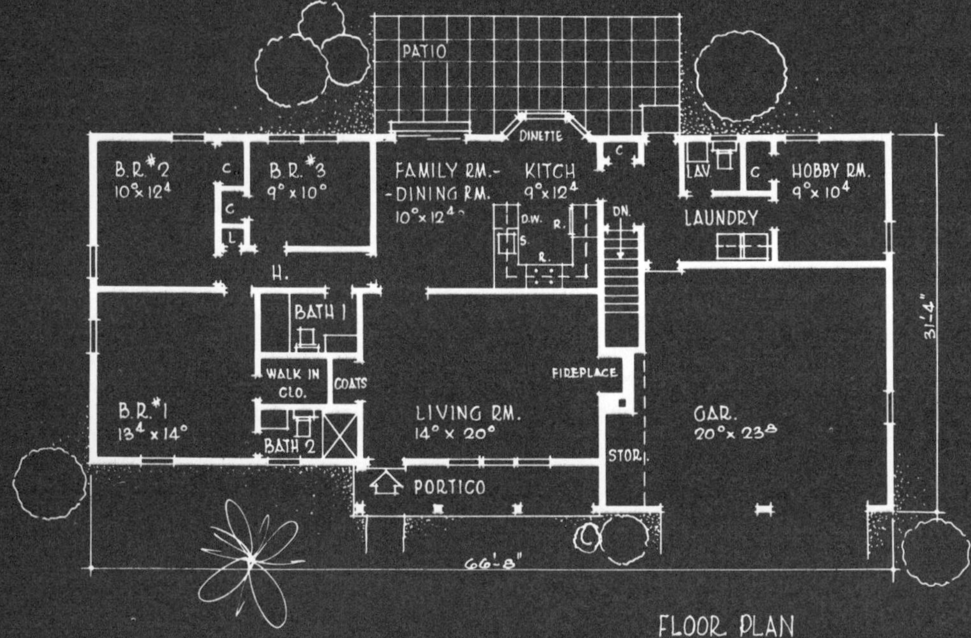

FLOOR PLAN

## HOUSE DATA

FLOOR PLAN ....... 1175 SQ. FT.
LAUND. & HOBBY RM..... 257 SQ. FT.
GARAGE ............ 487 SQ. FT.

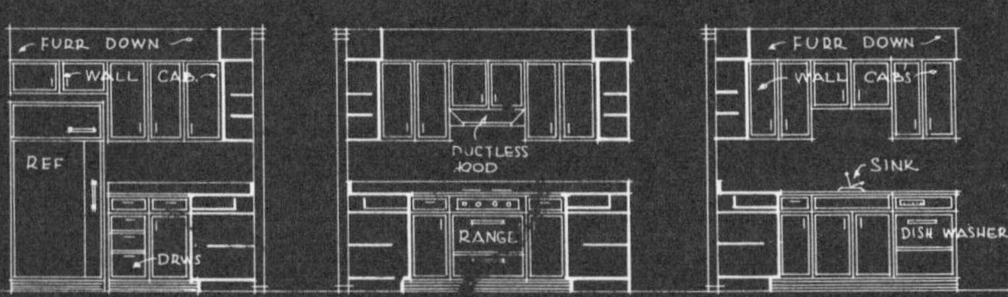

KITCHEN CABINET ELEVATIONS

Design L-18 —

# Small Ranch is Big in Livability

Small in area, but totally adequate in the number and size of rooms, this ranch provides sufficient livability to satisfy a family requiring three bedrooms as well as being a fine house for a retirement couple.

Design L-18 has only 1175 square feet of habitable area in the main section, plus 257 square feet for the laundry and hobby room, but care has been taken to insure generous wall spaces for furniture placement.

On entering the house, one sees a fireplace designed in keeping with the basic traditional character. Directly ahead is a view through sliding glass doors of the rear patio and garden. The traffic through the living room is minimized, all of it being on one side and across the shortest distance.

The kitchen layout is in the popular U-shape, all appliances being in the desirable work triangle. The garage is oversized with ample storage space.

Note that the hall space in the bedroom area has been held to a minimum. Generous windows provide sunlit bedrooms. The master bedroom has a private bath and a large walk-in closet.

Wood shingles and vertical siding make up the entire side wall treatment of the exterior, giving it an early New England flavor. The entrance portico adds a sense of luxury and provides the exterior detail needed to create a focal point where it is most needed.

## Material List

**MILLWORK**

| | |
|---|---|
| Exterior Doors & Frames Compl. | 2 pieces |
| Garage Door Complete Set | 2 units |
| Interior Doors & Frames Compl. | 11 pieces |
| Sliding Glass Door Units | 1 unit |
| Sliding Doors | 4 pieces |
| Bi Fold Doors | 1 unit |
| Windows | 15 units |
| Base | 500 lin. ft. |

**ROOFING**

| | | |
|---|---|---|
| Shingles | 235# asphalt | 2380 sq. ft. |
| Roofing Paper | 25# felt | 2380 sq. ft. |

**CONCRETE WORK**

| | |
|---|---|
| Concrete Walls | 959 cu. ft. |
| Foundation Footings | 222 cu. ft. |
| Slabs | 641 cu. ft. |
| Misc. Concrete | 141 cu. ft. |

**STRUCTURAL STEEL**

| | | |
|---|---|---|
| Lally Columns | 3½" Diam. | 5 pieces |
| Girder | 6" I 12.5# | 43 lin. ft. |

**BRICK WORK**

| | | |
|---|---|---|
| Chimney | Brick & Block | 110 cu. ft. |
| Flue Lining | T. C. | 32 lin. ft. |

**CARPENTRY**

| | |
|---|---|
| Framing Lumber | 7519 B.F. |
| Studs | 2667 B.F. |
| Plates | 800 B.F. |
| Roof Sheathing | 2380 sq. ft. |
| Sub Flooring | 1157 sq. ft. |
| Side Wall Sheathing | 1747 sq. ft. |
| Insulation Walls | 1287 sq. ft. |
| Insulation Ceilings | 1420 sq. ft. |
| Wood Flooring | 867 sq. ft. |
| Kitchen Plywood | 235 sq. ft. |
| Vinyl Flooring | 221 sq. ft. |

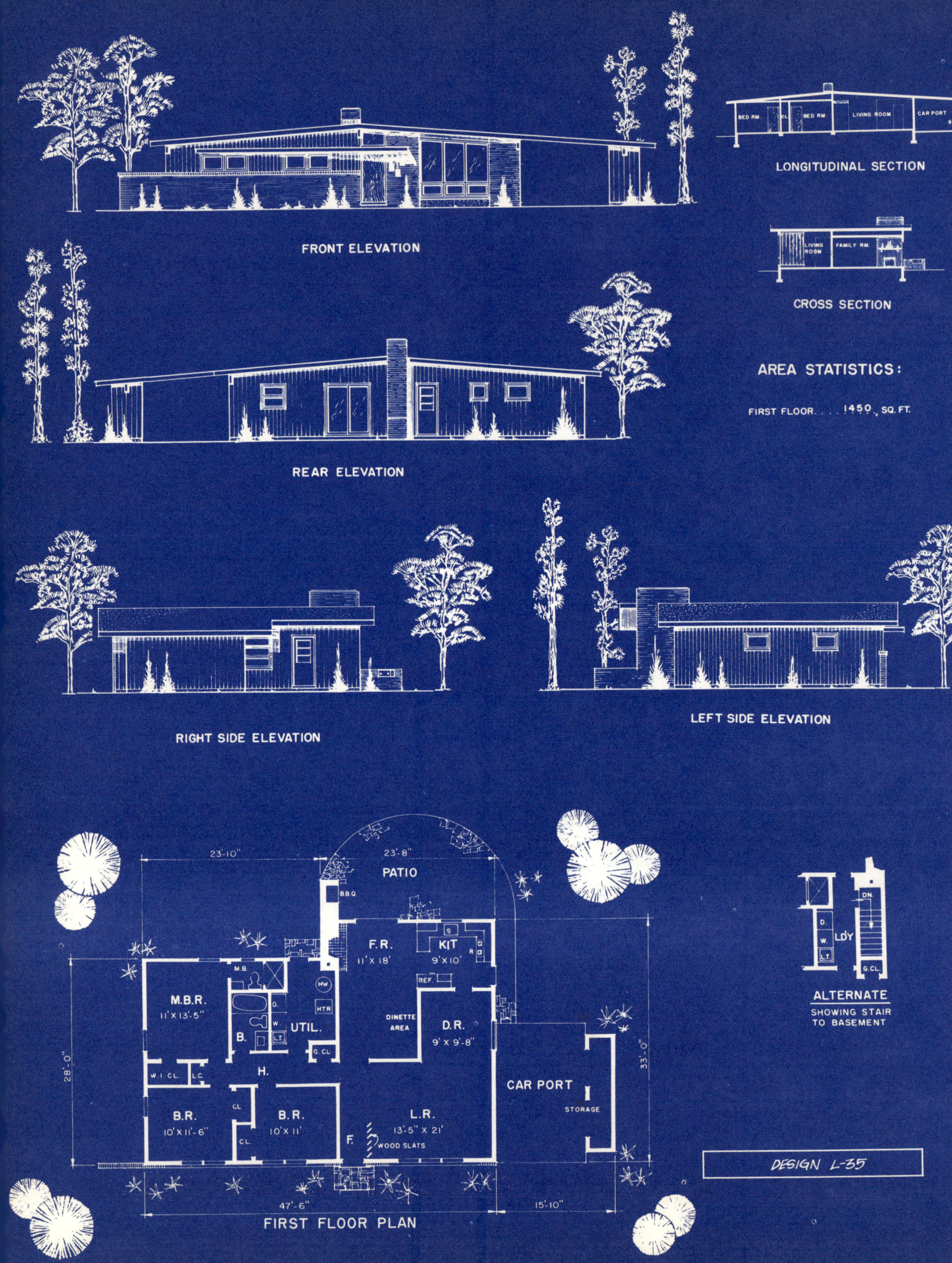
BED RM.
CL
BED RM.
LIVING ROOM
CAR PORT
LONGITUDINAL SECTION
FRONT ELEVATION
LIVING ROOM
FAMILY RM.
CROSS SECTION
AREA STATISTICS:
FIRST FLOOR. . . . 1450 SQ. FT.
REAR ELEVATION
RIGHT SIDE ELEVATION
LEFT SIDE ELEVATION
23'-10"
23'-8"
PATIO
B.B.Q.
F.R.
11' X 18'
KIT
9' X 10'
S
REF.
M.B.
M.B.R.
11' X 13'-5"
HW
HTR
D.
W.
LT.
B.
UTIL.
G.CL.
DINETTE AREA
D.R.
9' X 9'-8"
W.I. CL.
L.C.
H.
28'-0"
33'-0"
CAR PORT
STORAGE
B.R.
10' X 11'-6"
CL
CL.
B.R.
10' X 11'
F.
L.R.
13'-5" X 21'
WOOD SLATS
47'-6"
15'-10"
FIRST FLOOR PLAN
DN.
D.
W.
LDY
LT.
G.CL.
ALTERNATE
SHOWING STAIR
TO BASEMENT
DESIGN L-35

# Contemporary Ranch with Eye Appeal

Low, sweeping roof lines are combined with floor-to-ceiling glass areas, interesting windows, vertical V-joint siding, well-placed brick and other attractive touches to make this three-bedroom contemporary ranch house something special.

An overhanging roof canopy extends over the front bedroom windows and provides shelter from the weather to the front entrance. Inside, the interest continues. Tapered slats temporarily screen the entry from the L-shaped studio-ceiling living and dining areas while straight ahead is the family entertaining center.

The wide expanse of the family room-kitchen area does not stop at the walls due to the location of the sliding glass doors. Rather, it allows the rear paved patio, accentuated by the outdoor brick barbecue grille and gardens, to become part of the house.

Each of the front two bedrooms has ample closets and a linen closet is in the hall. The master bedroom enjoys a double outside exposure, a walk-in closet and a private tiled shower-stall bathroom.

A good-sized outdoor closet is adjacent to the carport for storing outdoor furniture, garden tools, garage supplies, etc. The carport could easily be enclosed in areas which require more car protection.

Although this is a basementless design, the house can be built with a full or partial cellar.

## Material List

**CONCRETE WORK**
Footings, floors, etc. .......... 30 cu. yds.

**MASONRY**
12" Concrete Block ............ 120 blocks
8" Concrete Block ............ 650 blocks
4" Brick Veneer .............. 450 sq. ft.

**FRAMING LUMBER**
Sills, joists, rafters, etc. ....... 4800

**EXTERIOR SHEATHING**
½" Exterior Plywood ........... 1600 sq. ft.
⅝" Exterior Plywood ........... 2200 sq. ft.

**WINDOW SCHEDULE**
(1) 4-0 x 4-2 Awning
(1) 3-0 x 3-0 Awning
(10) 3-0 x 2-0 Awning
(1) 2-0 x 2-0 Awning
(3) 3-0 x 6-0 Fixed

**DOOR SCHEDULE**
(1) 3-0 x 6-8 x 1¾ Front Ent. Door
(2) 2-8 x 6-8 x 1¾ Sash Door
(4) 2-6 x 6-8 x 1¾ Fl. Ext. Door
(1) 6-0 x 6-8 Glass Sliding Door Unit
(5) 2-6 x 6-8 x 1⅜ Fl. Int. Door
(2) 2-4 x 6-8 x 1⅜ Fl. Int. Door
(6) 2-0 x 6-8 x 1⅜ Fl. Int. Door

**EXTERIOR FINISHES**
Built-Up Roofing .............. 22 squares
Vertical Siding ................ 1400 sq. ft.

**INTERIOR FINISHES**
Ceramic Tile Floor ............. 45 sq. ft.
Ceramic Tile Walls ............ 220 sq. ft.
Vinyl Tile .................... 160 sq. ft.
Sheetrock .................... 5300 sq. ft.
Oak Flooring .................. 1185 sq. ft.

FRONT ELEVATION

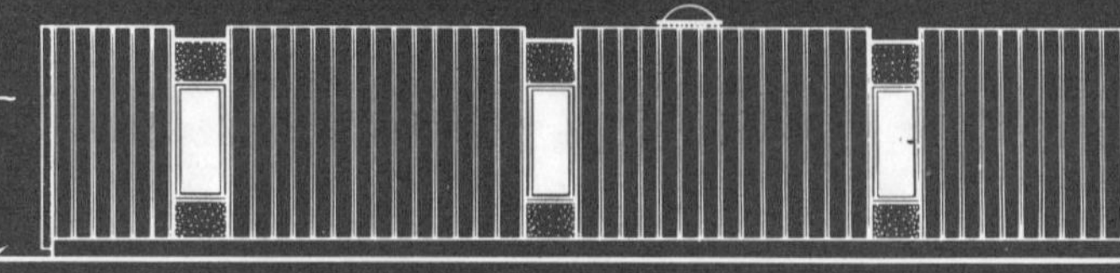

RIGHT SIDE ELEVATION

REAR ELEVATION

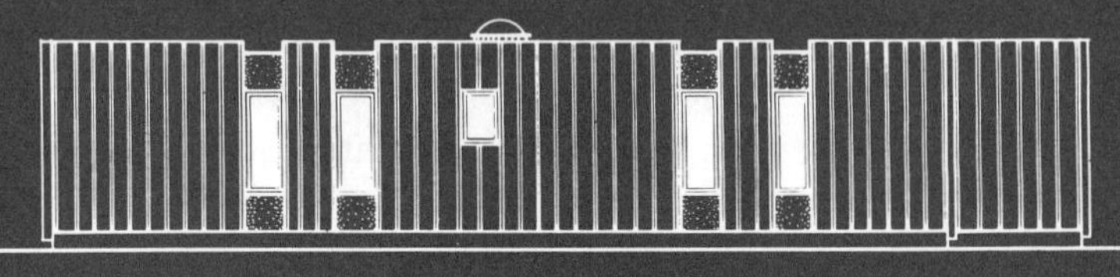

LEFT SIDE ELEVATION

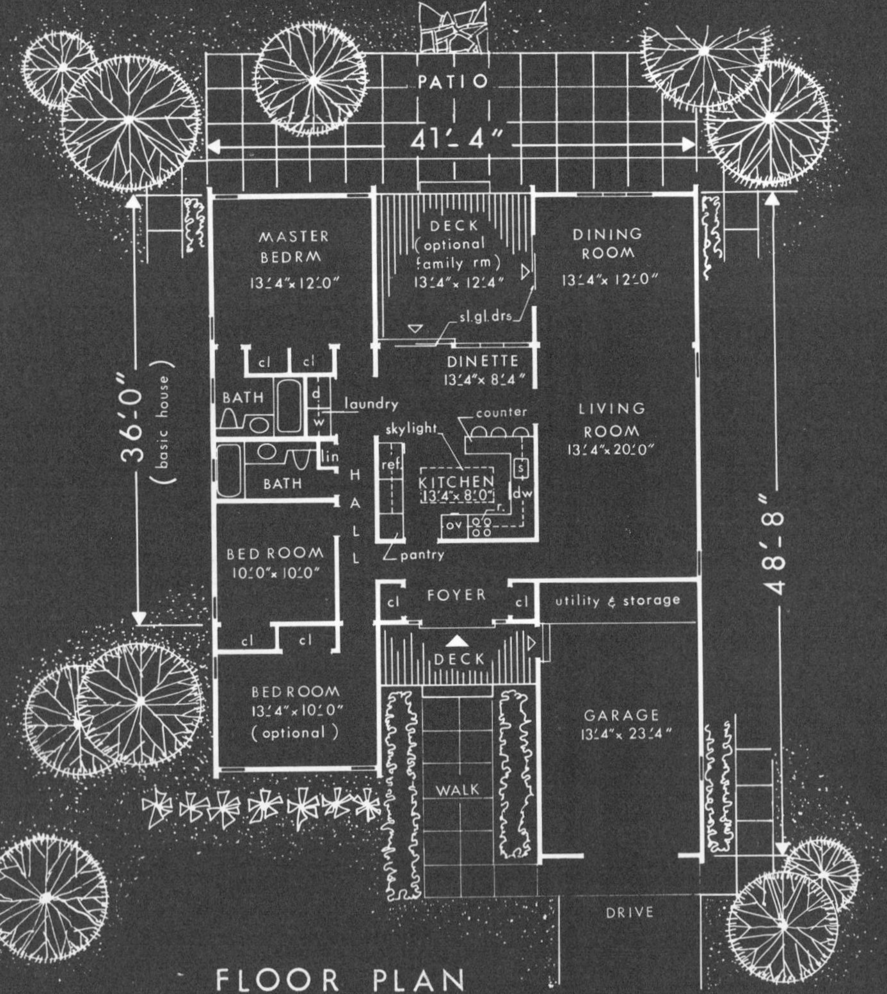


FLOOR PLAN

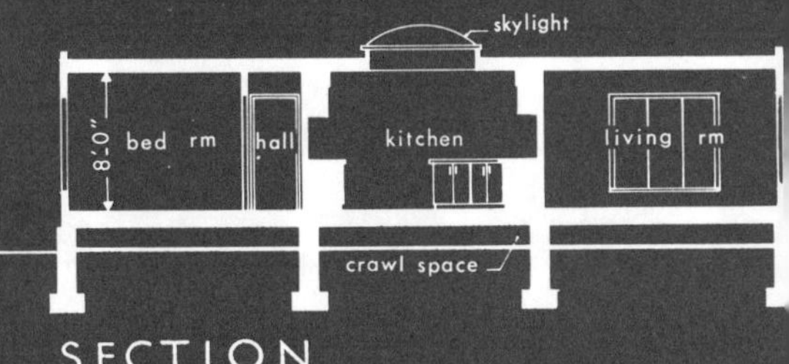


SECTION

## AREA STATISTICS :

basic house ........... 1280 sq.ft.
optional bed room .. 172 "
garage & storage ..... 330 "
decking ................. 230 "

DESIGN L-40

# Economy Construction in Contemporary

Innovative yet unpretentious in its exterior appearance, this contemporary house has a crisp design that affords an opportunity for a life style of warmth, flexibility and comfort.

Despite its many features, it is economical to build and can fit on a modest-sized lot. Built over a crawl space, it is of simple construction with all floor and roof beams of stock length so there is no waste.

Although designed as a three-bedroom residence with all rooms on one floor, the third bedroom may be added at a later date. Or it may be used as a study, studio or library.

The plan is H-shaped. The bedroom wing is separated from the living-dining garage wing by the combination of central kitchen, dinette and central foyer. The entrance to the house is through a courtyard contained by the two wings. Double entrance doors open on to a central foyer with two closets.

The efficient kitchen is the hub of the house; it has recessed lighting around its perimeter and a domed skylight. The dinette, which merges with the kitchen, flows onto a porch partially enclosed by the rear portion of the two wings.

The bedroom wing to the left of the foyer contains three bedrooms (third room optional) and two baths.

The living and dining rooms are to the right of the foyer. The latter has access to the deck through sliding glass doors.

## Material List

**CONCRETE WORK**

| | |
|---|---|
| Foundations, footings, slabs, etc. | 24 cu. yds. |

**FRAMING LUMBER**

| | |
|---|---|
| Total Sills, Joists, Rafters, Studs, Plates, etc. | 5881 B.F.M. |

**SHEATHING, INSULATION**

| | |
|---|---|
| Sub Flooring | 1680 sq. ft. |
| Wall Sheathing | 2080 sq. ft. |
| Roof Sheathing | 1680 sq. ft. |
| Wall Insulation | 1800 sq. ft. |
| Ceiling Insulation | 1830 sq. ft. |

**FINISHES, INTERIOR**

| | |
|---|---|
| Vinyl Asbestos Tile | 1680 sq. ft. |
| Ceramic Tile Floors | 45 sq. ft. |
| Ceramic Tile Walls | 148 sq. ft. |
| Gypsum Board — house walls | 2416 sq. ft. |
| Gypsum Board — house ceilings | 1608 sq. ft. |
| Gypsum Board — garage | 550 sq. ft. |

**FINISHES, EXTERIOR (other than masonry)**

| | |
|---|---|
| Textured Plywood Panels | 277 sq. ft. |
| Vertical Siding | 2080 sq. ft. |
| Built-up Roofing | 1680 sq. ft. |

**WINDOW SCHEDULE**

| | |
|---|---|
| Wood Casement | 13 units |

**DOOR SCHEDULE**

| | |
|---|---|
| Ext. Hardwood, paneled & glazed | 2 units |
| Ext. Hardwood, paneled | 1 |
| Aluminum Sliding | 2 |
| Int. Hardwood, flush, staingrade | 5 |
| Int. Hardwood, louvered, bi-folding | 11 |
| Tilt-up Garage Door | 1 |

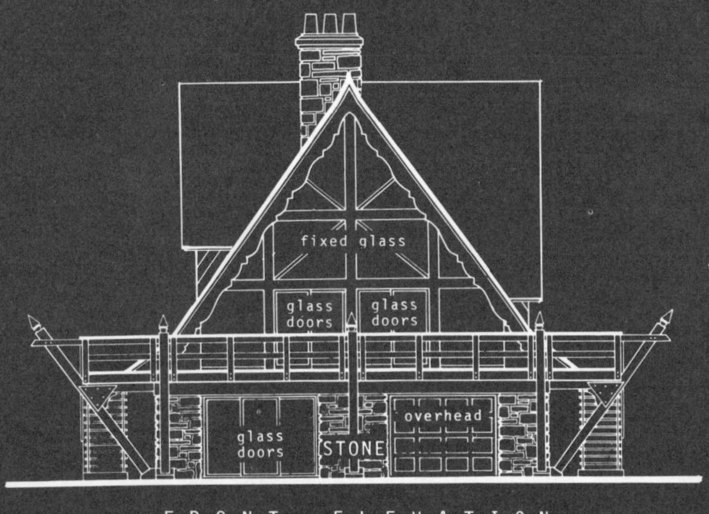


STONE

RIGHT ELEVATION

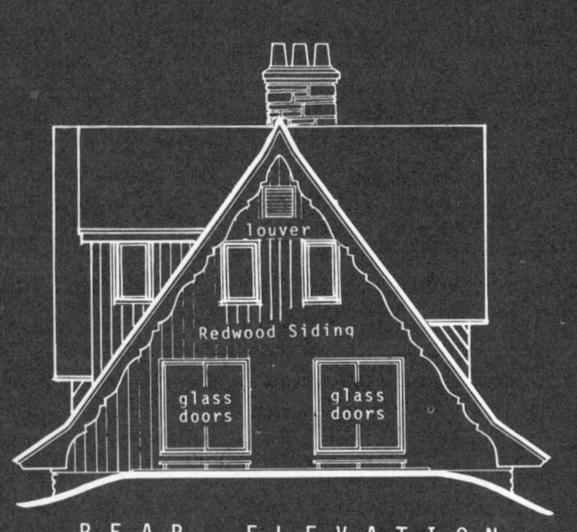


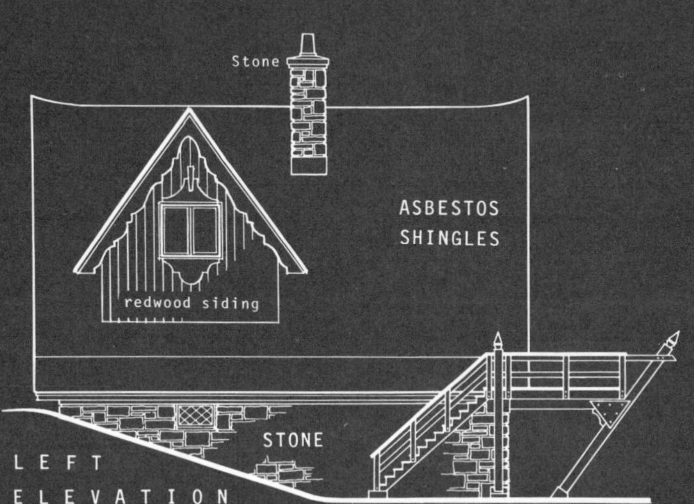


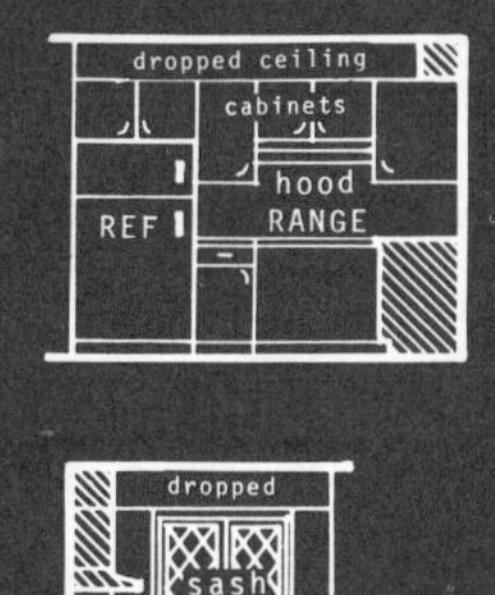

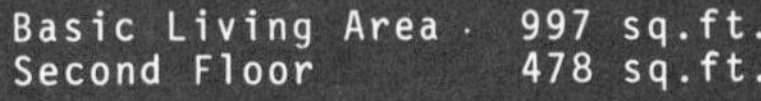
Basic Living Area 997 sq.ft.
Second Floor 478 sq.ft.

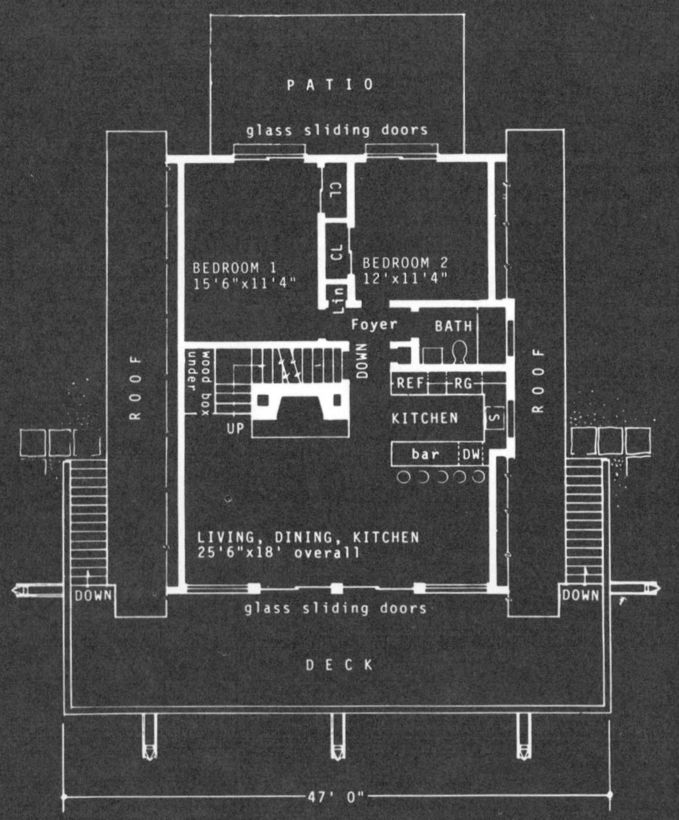


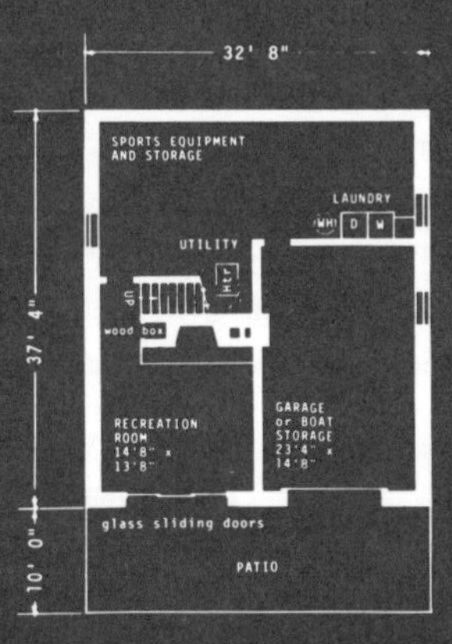


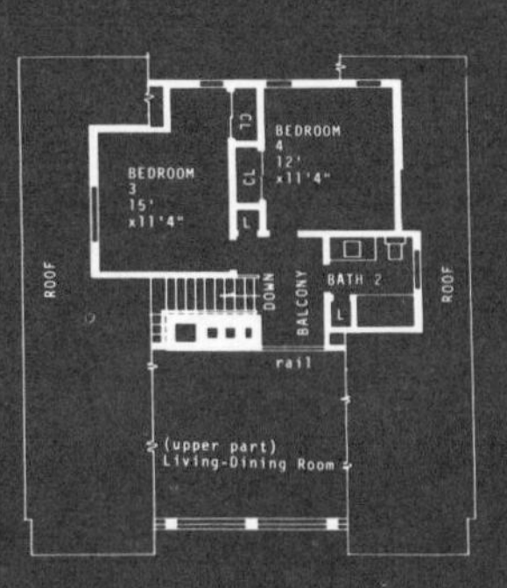


DESIGN L-51

# Expanded Version of Popular A-Frame

This A-frame, with its chalet styling, was created to meet the needs of a family requiring four bedrooms in a second home.

Since even some suburban communities are beginning to accept A-frame styling, the "one-house" family can enjoy the advantages of a vacation atmosphere all the time.

Depending on the slope of the property to be used, the ground level can also be the entry level with the sliding glass doors opening into the recreation room off the front patio. The comfortable room has a large fireplace for informal chilly evenings.

The deck on the main level acts as play, lounging and entry area with steps on either side for ready access from below. The sky-high windows are a background for the outdoor area and a striking highlight for a broad expanse inside. The area includes the open kitchen where a two-way counter serves as snack bar into the dining area or work counter for the kitchen itself. A second fireplace is just as massive as the one in the recreation room.

Two bedrooms on the main level are fairly well self-contained at the back. A nice feature is the use of sliding glass doors in both bedrooms with views all the time and an easy way out to the private patio beyond.

On the upper level, the stairway ends at the balcony landing, and past the protective, decorative railing is the living area below.

## Material List

**CONCRETE WORK**

| | |
|---|---|
| Footings | 16½ cu. yds. |
| Ground Floor | 13 cu. yds. |

**MASONRY**

| | |
|---|---|
| Cinder Blocks 8" x 8" x 16" | 299 sq. ft. |
| 16" x 16" x 16" | 379 sq. ft. |
| 12" x 8" x 16" | 805 sq. ft. |
| Chimney Fill (block & brick) | 444 cu. ft. |
| Stone Veneer (Interior) | 555 sq. ft. |
| (Exterior) | 771 sq. ft. |
| Bluestone Sills | 12 lin. ft. |

**CARPENTRY**

| | |
|---|---|
| Lally Columns 4" | 2 pcs. |
| Steel Gussets ¼" | 72 sq. ft. |
| Redwood Exterior lumber | 6580 bd. ft. |
| Redwood 1" x 10" siding | 955 sq. ft. |
| Exterior ¾" plywood | 550 sq. ft. |
| Exterior ⅜" plywood | 600 sq. ft. |
| ⅝" Plyscore | 9150 sq. ft. |
| Framing Lumber | 9500 bd. ft. |
| Insulation | 3050 sq. ft. |
| Roofing Shingles | 2770 sq. ft. |
| Resilient Tile Flooring | 2950 sq. ft. |
| ½" Gypsum Board | 6800 sq. ft. |

**EXTERIOR OPENINGS**

1 — 3 light Pella sliding doors
4 — 2 light Pella sliding doors
6 Casement windows — screens & storm sash
1 Overhead Garage Door

**MILLWORK**

13 Interior Flush Doors
Kitchen cabinets, 24 lineal feet
2 Medicine Cabinets
Base & Shoe Moldings, 360 lin. ft.
2 Flights Int. Stairs, 14 risers ea.
4 Sliding Doors

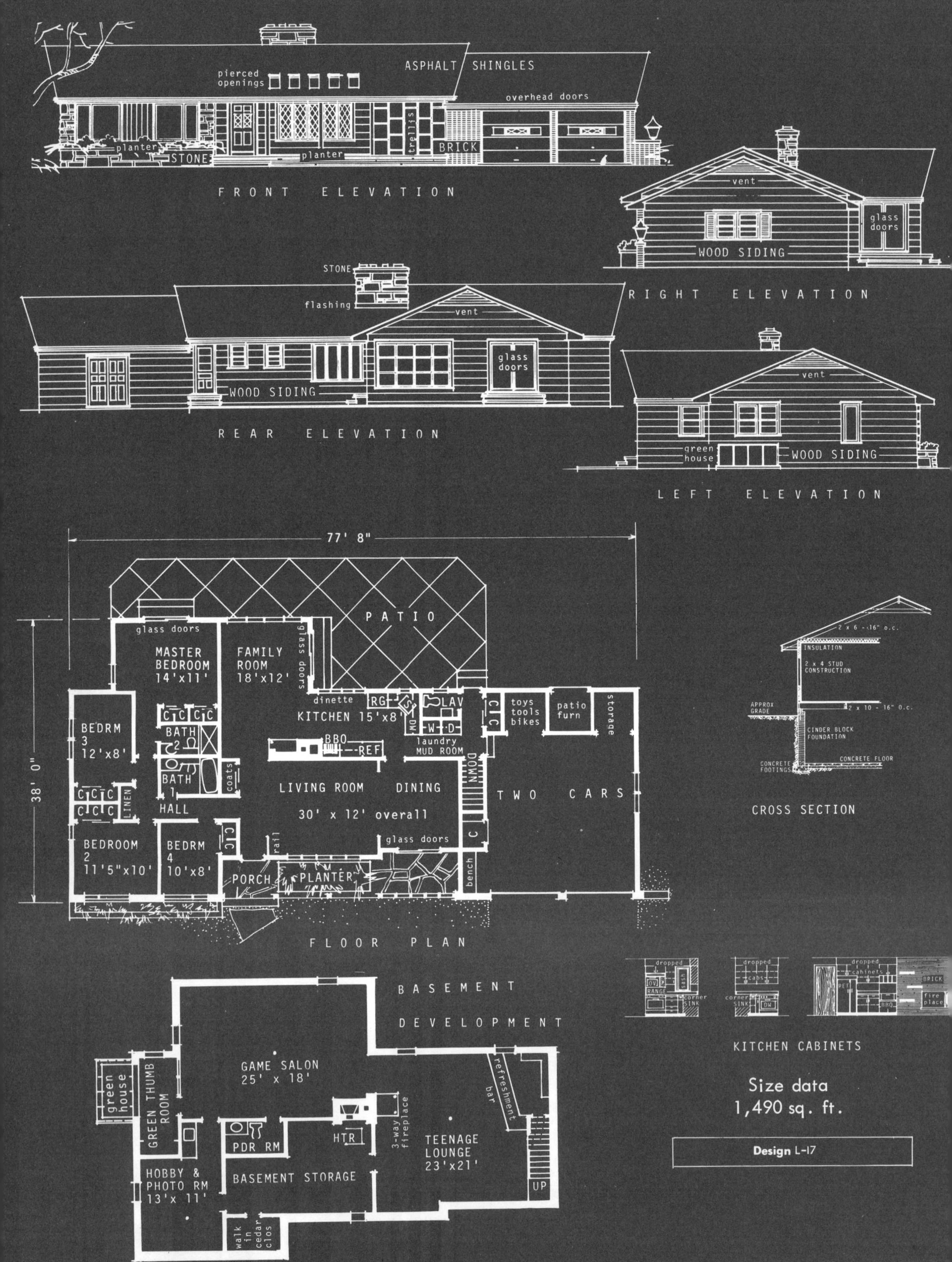


Size data
1,490 sq. ft.

**Design L-17**

# Four-Bedroom House in Practical Size

The open plan concept has been effectively utilized in this handsome ranch.

In creating a design for a four-bedroom house while keeping costs within a reasonable budget limit, there must be no waste space. By permitting certain rooms to flow together visually in typical open planning fashion, the plan provides larger living areas while keeping the overall size below 1500 square feet.

The living room and the dining room are to the right of the entrance hall and stretch 30′ along the front of the house. The living room windows are diamond-paned. With a rail on one side and the wall of the dining room on the other, a charming alcove is formed. Through the open arch, the dining room, with access from the kitchen, has a sliding glass window wall at the front, opening to a sheltered snack porch shielded by pierced lattice from the street.

At the rear of the house are the family room and kitchen, separated but without a wall between them. The family room is living size with a fireplace next to a kitchen barbecue. This is a lovely room with its windows looking out to the back patio and adjacent sliding glass doors forming a window wall plus access to the big patio on the side.

The left wing of the house fits in four bedrooms. The master bedroom at the back has two exposures, one of which is a sliding door unit out to a private end of the patio.

## Material List

| CONCRETE WORK | |
|---|---|
| Footings | 19 cu. yds. |
| Floors (basement & platforms) | 27½ cu. yds. |
| **MASONRY** | |
| 4 x 8 x 18 CB | 285 lin. ft. |
| 8 x 8 x 18 CB | 255 sq. ft. |
| 10 x 8 x 18 CB | 1355 sq. ft. |
| 12 x 8 x 18 CB | 220 sq. ft. |
| Brick Veneer - Chimney Fill | 130 sq. ft. |
| | 440 cu. ft. |
| Flagstone - Stone Veneer | 190 sq. ft. |
| | 245 sq. ft. |

| CARPENTRY | |
|---|---|
| 4″ Lally Columns | 8 pieces |
| Framing Lumber | 9143 B.F. |
| Stud & Plates | 2702 B.F. |
| Roof Sheathing 1 x 6 | 4500 sq. ft. |
| Plywood | 3720 |
| Fascia # pine 1 x 6 | 145 lin. ft. |
| Soffit ⅜″ x 2′ plywood | 145 lin. ft. |
| Shingle Mould | 145 lin. ft. |
| Wall Sheathing 1 x 6 | 3070 sq. ft. |
| Plywood | 2530 |
| Siding | 2520 sq. ft. |
| Roofing 210# asphalt | 38 squares |

| Insulation - Walls | 1720 sq. ft. |
|---|---|
| Ceilings | 1640 sq. ft. |
| Finish Wood Flooring | 1485 sq. ft. |
| **DRY WALL** | |
| Ceilings ⅜″ | 3330 sq. ft. |
| Walls ½″ | 9615 sq. ft. |
| **MILLWORK** | |
| Base | 960 lin. ft. |
| Shelving 12″ pine | 60 lin. ft. |
| Windows | 16 pieces |
| Exterior Doors | 8 pieces |
| Interior Doors | 27 pieces |

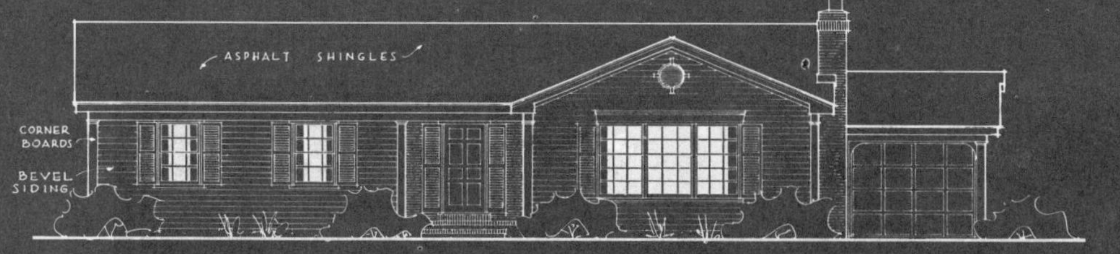


- FRONT ELEVATION -

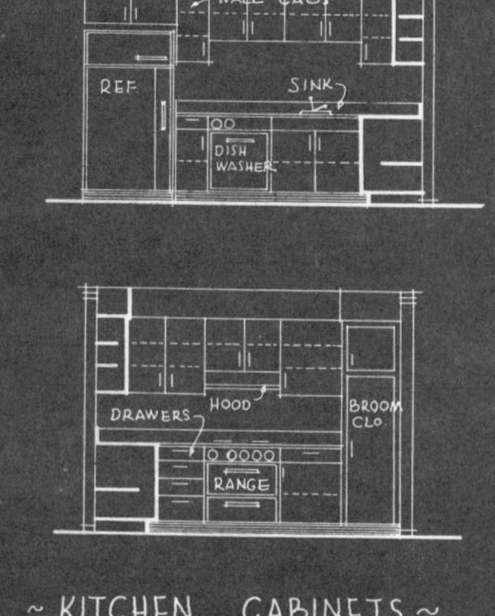


~ KITCHEN CABINETS ~

- REAR ELEVATION -

- RIGHT SIDE ELEVATION -

- LEFT SIDE ELEVATION -

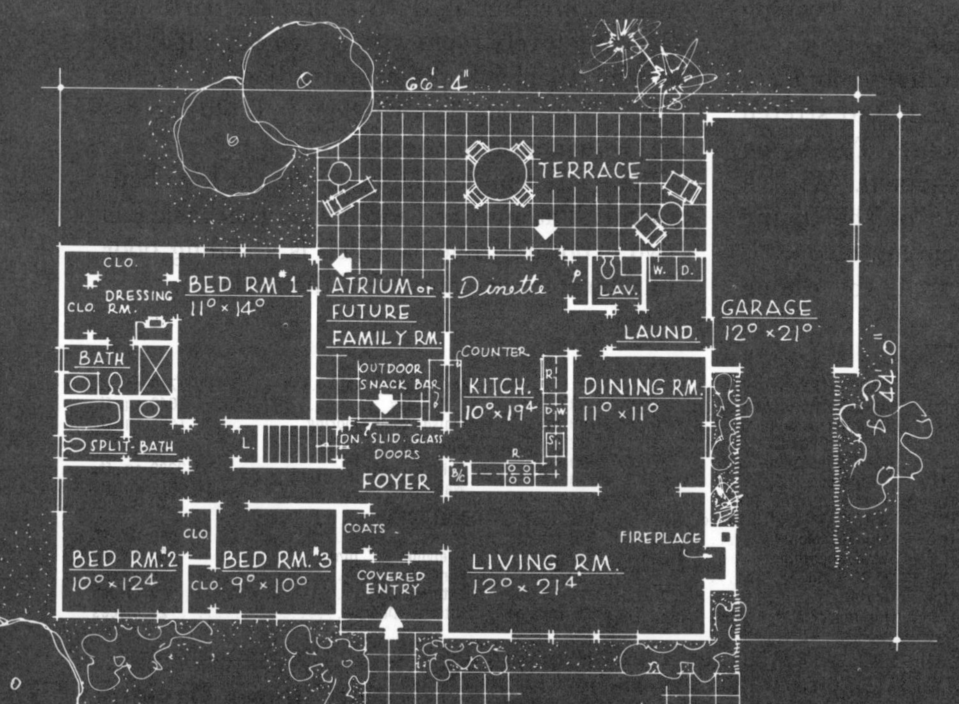


## HOUSE DATA

| | |
|---|---|
| FIRST FLOOR | 1491 SQ. FT. |
| GARAGE | 267 SQ. FT. |

**Design L-26**

# Roman Atrium in an American Ranch

In this plan the architect has designed a small three-bedroom ranch house with an atrium as a center of interest, indicating on the layout that should the owner prefer it, the atrium can be enclosed as a family room. Although the atrium has Roman origins, the exterior of this house is a one-story version of a Colonial New England residence.

The atrium makes possible several highly desirable circulation features, the most important being the direct view and immediate access to it from the entrance foyer. There are large glass areas overlooking this space from the kitchen. Featured is an outdoor snack bar pass-through.

With its three bedrooms, two bathrooms, living room, dining room and toilet-wash up room, Design L-26 is a compact, livable unit which could be attractive to a retired couple who may want a degree of luxury without the maintenance requirements of a huge house.

From the area around the front doors, both the kitchen and hall bathroom are hidden. The immediate impact of the atrium, seen through sliding glass doors, plus the huge living room with its fireplace on the far end give the first-time visitor precisely the impression an owner desires.

The dining room is an L pattern with the living area open-planned to make these two rooms flow toward each other, an advantage when giving informal dinner parties.

## Material List

**CONCRETE WORK**

| | | |
|---|---|---|
| Concrete Walls | | 1192 cu. ft. |
| Foundation Footings | | 265 cu. ft. |
| Slabs | | 647 cu. ft. |
| Misc. Concrete | | 120 cu. ft. |

**STRUCTURAL STEEL**

| | | |
|---|---|---|
| Lally columns | 3½" diam. | 9 pieces |
| Girder | 6" I Beam | 52 lin. ft. |

**BRICK WORK**

| | | |
|---|---|---|
| Chimney | Brick | 126 cu. ft. |
| Flue Lining | T. C. | 38 lin. ft. |

**CARPENTRY**

| | |
|---|---|
| Framing Lumber | 7442 B.F. |
| Studs | 3600 B.F. |
| Plates | 1080 B.F. |
| Roof Sheathing | 2688 sq. ft. |
| Sub Flooring | 1495 sq. ft. |
| Side Wall Sheathing | 2560 sq. ft. |
| Insulation Walls | 1370 sq. ft. |
| Insulation Ceilings | 1495 sq. ft. |
| Wood Flooring | 1188 sq. ft. |
| Kitchen Plywood | 200 sq. ft. |

**MILLWORK**

| | |
|---|---|
| Exterior Doors & Frames Compl. | 4 pieces |
| Slid. Glass Doors | 1 unit |
| Garage Door Complete Set | 1 set |
| Interior Doors & Frames Compl. | 12 units |
| Bi Fold Doors | 13 units |
| Window Units | 20 units |
| Fascia | 260 lin. ft. |
| Base | 540 lin. ft. |

**KITCHEN CABINETS**

| | |
|---|---|
| Base Cabinets | 8'6" lin. ft. |
| Wall Cabinets | 13'0" lin. ft. |

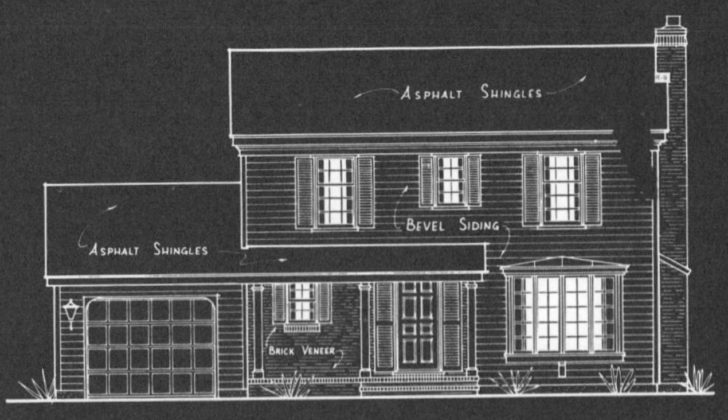


FRONT ELEVATION

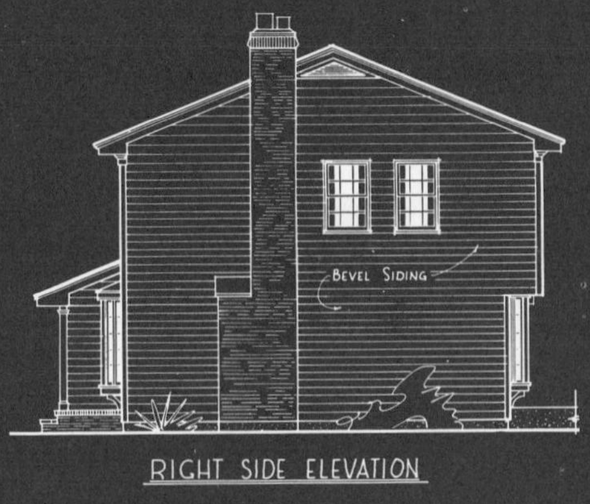


RIGHT SIDE ELEVATION

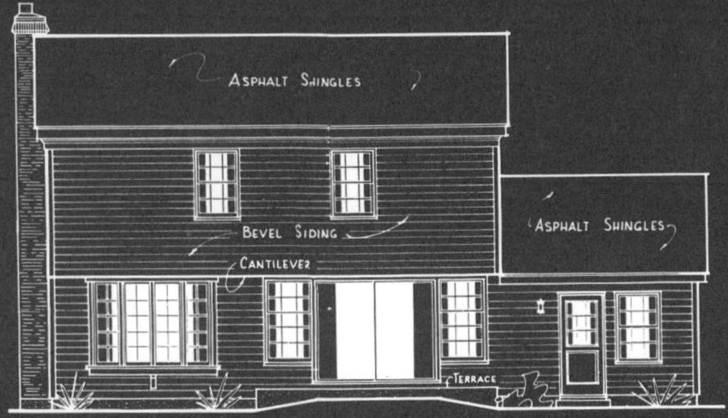


REAR ELEVATION

LEFT SIDE ELEVATION

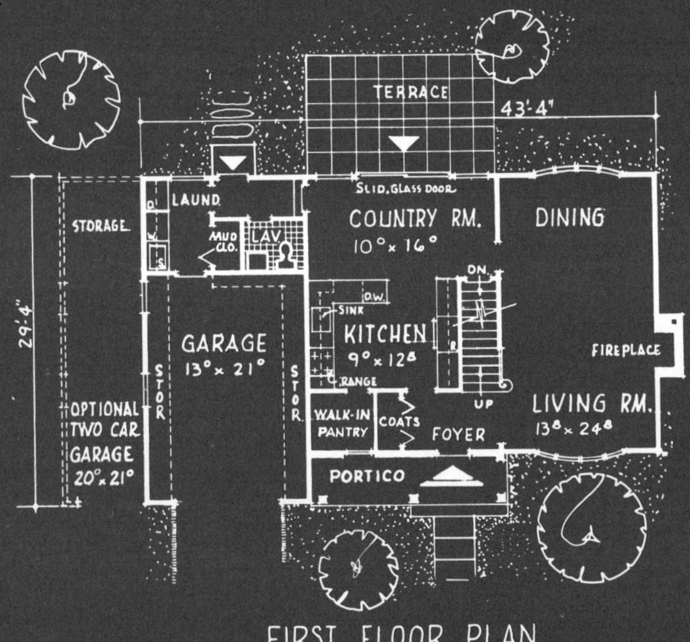


FIRST FLOOR PLAN

BED RM. 3
10° x 11⁶
CLO. CLO.
BED RM. 2
10° x 13°
STATUARY NICHE
HALL
DN.
SPLIT BATH
BED RM. 1
13° x 14°
CLO.
RAIL
LIN
BATH
S.
BED RM. 4
9° x 9⁴
CLO.
CLO.

SECOND FLOOR PLAN

HOUSE DATA

| | |
|---|---|
| FIRST FLOOR | 730 SQ. FT. |
| SECOND FLOOR | 819 SQ. FT. |
| LAUNDRY & LAV. | 110 SQ. FT. |
| GARAGE | 292 SQ. FT. |

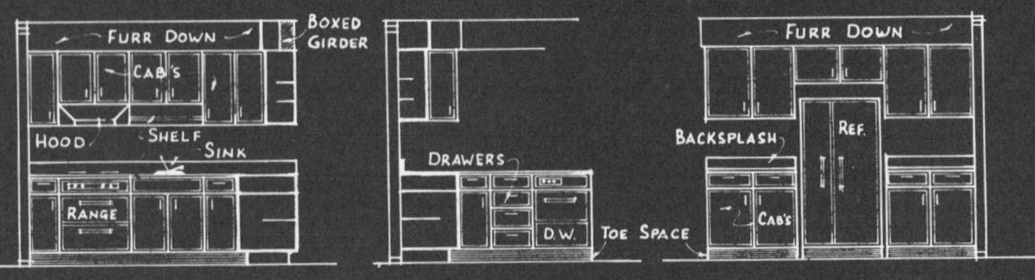


KITCHEN CABINET ELEVATIONS

DESIGN L-44

# Kitchen Layout Star of Small Two-Story

Here's a house whose overall dimensions of 43'4" by 29'4", including a one-car garage, make it suitable for a fairly small lot — yet it actually makes a comfortable home for a large family.

There are two reasons for this. First, it's a two-story with four bedrooms and two bathrooms on the scond floor. Secondly, the architect has done an excellent job of utilizing the available space to provide gratifying livability within a traditional exterior.

A most interesting first-floor arrangement is evident immediately on entering the foyer. Almost straight ahead is the kitchen, open-planned to a family room (called a country room on the floor plan) and tied in through a wall of glass to a rear terrace. Slightly to the right of the foyer is a combination living room-dining room, also stretching from the front to the rear of the house and integrated by identical bow windows at opposite ends. The first glimpse of the living room is enhanced by a fireplace on the facing wall. A stairway with open balustrade enhances the foyer.

On the second floor are four bedrooms with a split bath for three of the rooms. The owners have their own private bath with generous clothes closets nearby. All the bedrooms have generous closets and wall space.

There is a full basement under the main part of the house.

## Material List

| Item | Specification | Quantity |
|---|---|---|
| **CONCRETE WORK** | | |
| Concrete Walls | | 735 cu. ft. |
| Slabs | | 445 cu. ft. |
| Foundation Footings | | 180 cu. ft. |
| Misc. Concrete | | 295 cu. ft. |
| **STRUCTURAL STEEL** | | |
| Lally Columns | 3½" diam. | 6 pieces |
| Girder | S6 — 12.5# | 38 lin. ft. |
| Girder | S7 — 15.3# | 10 lin. ft. |
| **BRICK WORK** | | |
| Chimney | Brick | 132 cu. ft. |
| Flue Lining | T. C. | 57 lin. ft. |
| Veneer | 4" Brick | 93 sq. ft. |
| **CARPENTRY** | | |
| Framing Lumber | | 6443 B.F. |
| Studs | | 3334 B.F. |
| Plates | | 1000 B.F. |
| Roof Sheathing | | 1630 sq. ft. |
| Sub Flooring | | 1572 sq. ft. |
| Side Wall Sheathing | | 1616 sq. ft. |
| Insulation Walls | | 1254 sq. ft. |
| Insulation Ceilings | | 987 sq. ft. |
| Wood Flooring | | 1352 sq. ft. |
| Kitchen Plywood | | 114 sq. ft. |
| **MILLWORK** | | |
| Exterior Doors & Frames Compl. | | 2 pieces |
| Garage Door Complete Set | | 1 unit |
| Sliding Glass Door Unit | | 1 unit |
| Interior Doors Complete Set | | 15 pieces |
| Bi Fold Doors | | 11 units |
| Window Units | | 22 units |
| Fascia | | 222 lin. ft. |
| Base | | 570 lin. ft. |
| Stairs | 13 risers | 1 set |
| Stairs | 12 risers | 1 set |
| Louvers | | 3 pieces |
| Shutters | | 14 pieces |

Front elevation

Wood shakes

Stucco on brick ven.

Right side elevation

Rear elevation

Wood shakes

Sliding glass doors

Stucco

Garage doors

Left side elevation

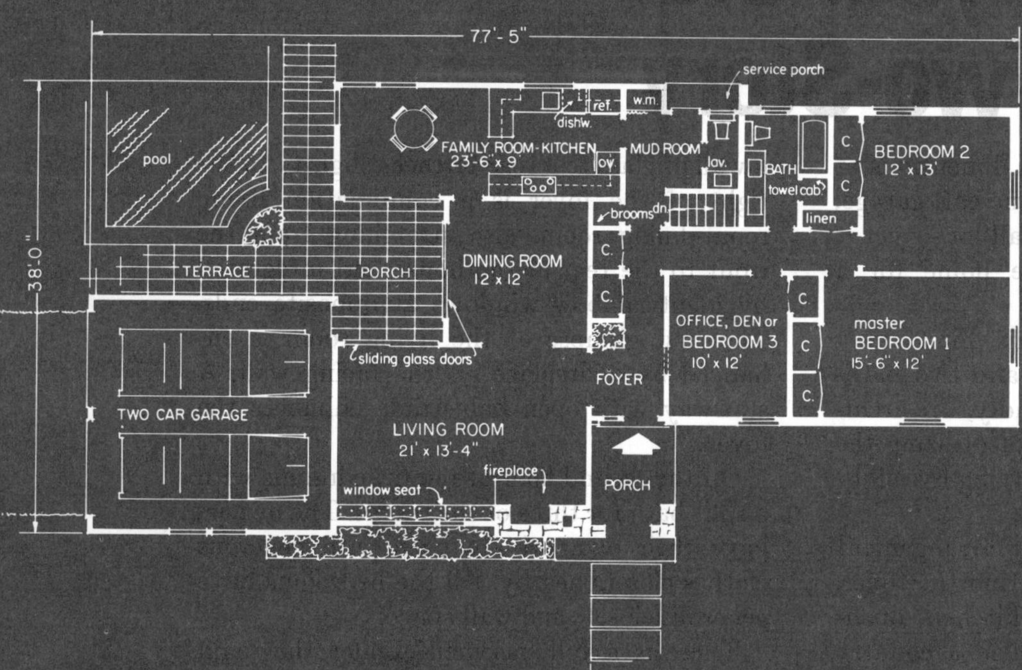


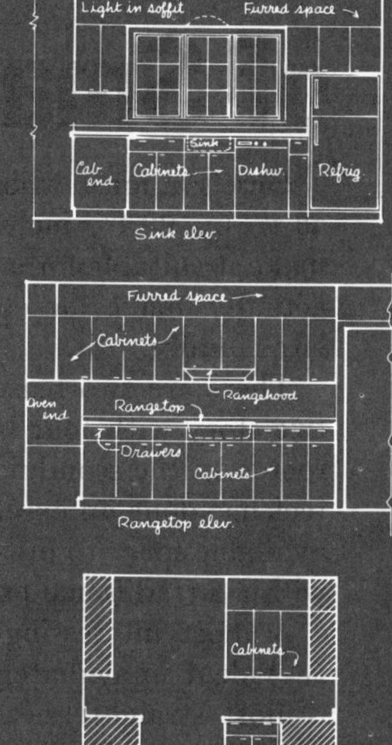


Kitchen elevation

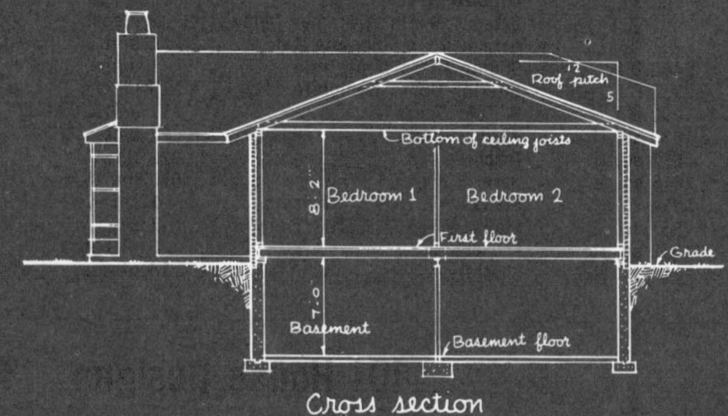


Cross section

## Size data

1,655 square feet, not incl. porches and garage.

DESIGN S-2

# Old World Charm in Cotswold Cottage

The Cotswold is associated with charming, nostalgic country cottages with unusual design twists that give them individual character. Although English in origin, some of their features were borrowed from Irish and Scottish dwellings.

Like the original Cotswolds, this one features a stone chimney at the front of the house. Adding to the character of the entranceway is a trellis dropping down from a point along the chimney front to a planter.

The living room, to the left of the foyer, has a fireplace and sliding glass doors leading to a porch and a large outdoor area.

To the rear of the living room is the dining room, also with sliding glass doors to the porch. And on the other side of the dining room is a combined family room-kitchen 23′ long with access to the porch. Thus, the three principal living areas of the house not only lead to the porch but gain visual spaciousness because the large expanses of glass seem to bring them together. That's why this house, with its habitable space of 1655 square feet, appears to be much bigger that it really is.

Three bedrooms and a bathroom are housed in a wing at the right with plenty of closet space. Bedroom number 3 can be made into an office or den if not needed for sleeping quarters. By installing a door in one of the walls, direct access to the front foyer can be created.

## Material List

**CONCRETE**
10″ walls — piers chimney, etc. . 1700 cu. ft.
8″ walls .................... 400 cu. ft.
4″ Conc. porch floor 6 x 6 mesh . 350 sq. ft.
4″ Base conc. floor ........... 1560 sq. ft.

**STEEL**
5 - 3½ - 6′8 lally columns
40 Lin. 8″ WF 20 lb. — I beam
46 Lin. 4″ x 3½ — 5/16 angle iron

**MASONRY**
4″ Brick veneer — with some scattered stones ............ 800 sq. ft. 1 ton stone
Stone wall at porch 12″ with arch 80 sq. ft.
5″ Stone for chimney veneer .... 310 sq. ft.
Backup for chimney — 154 cu. ft. 2700 com. brick
110 Fire Brick
32 Lin 13 x 13 Flue lining
200 Pressed brick for wall — at rear entry

**CARPENTRY**
Floor framing plates, etc. ...... 3800 BM
⅝ Plyscore — sub floor ........ 1632 sq. ft.
All 2 x 4 studs, etc. .......... 3300 BM
Ceil. joists, catwalk .......... 2460 BM
Rafters, lookouts, etc. ......... 4840 BM
½ Plyscore roof boards ........ 3500 sq. ft.
⅜ Plyscore sheathing .......... 2220 sq. ft.
⅜ Ext. plywood good 1 side ... 300 sq. ft.
10 roll 15 lb. sat. felt
300 sq. ft. ⅝ plyscore plugged
1540 BM — 13/16 x 2¼ Cl. red oak flooring

**ROOFING**
34 sq. hand-split shakes
180 Lin. starter
150 Lin. Hips & ridges red cedar

**PLASTER BOARD**
½ for walls .................. 4900 sq. ft.
⅜ for ceilings ................ 1800 sq. ft.

FRONT ELEVATION

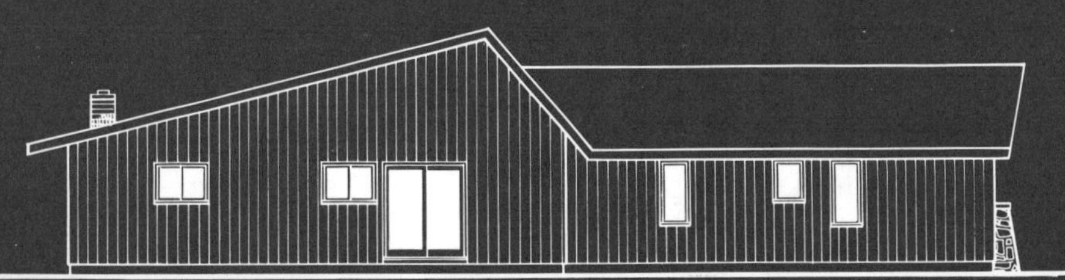

REAR ELEVATION

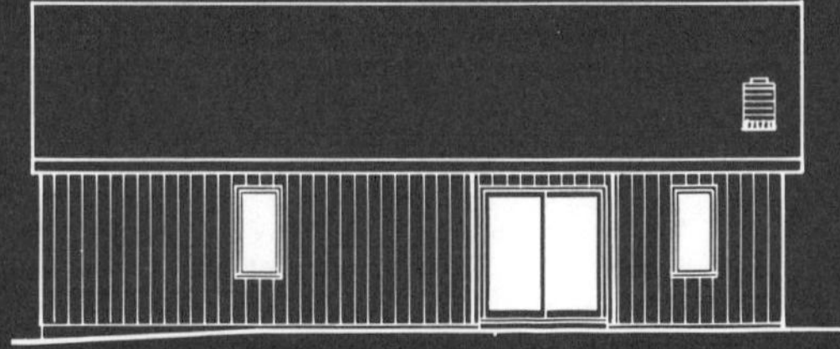

RIGHT SIDE ELEVATION

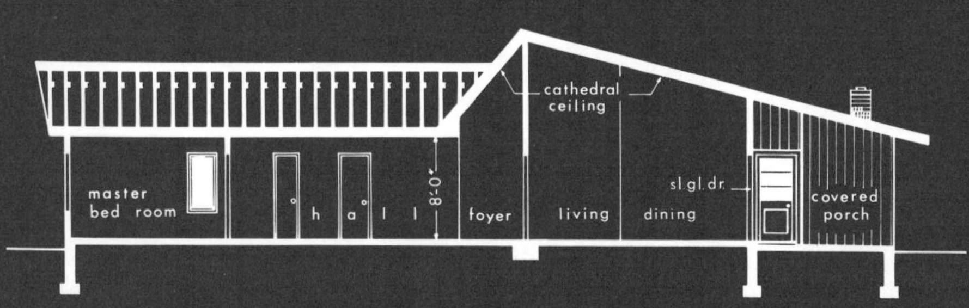


SECTION

LEFT SIDE ELEVATION

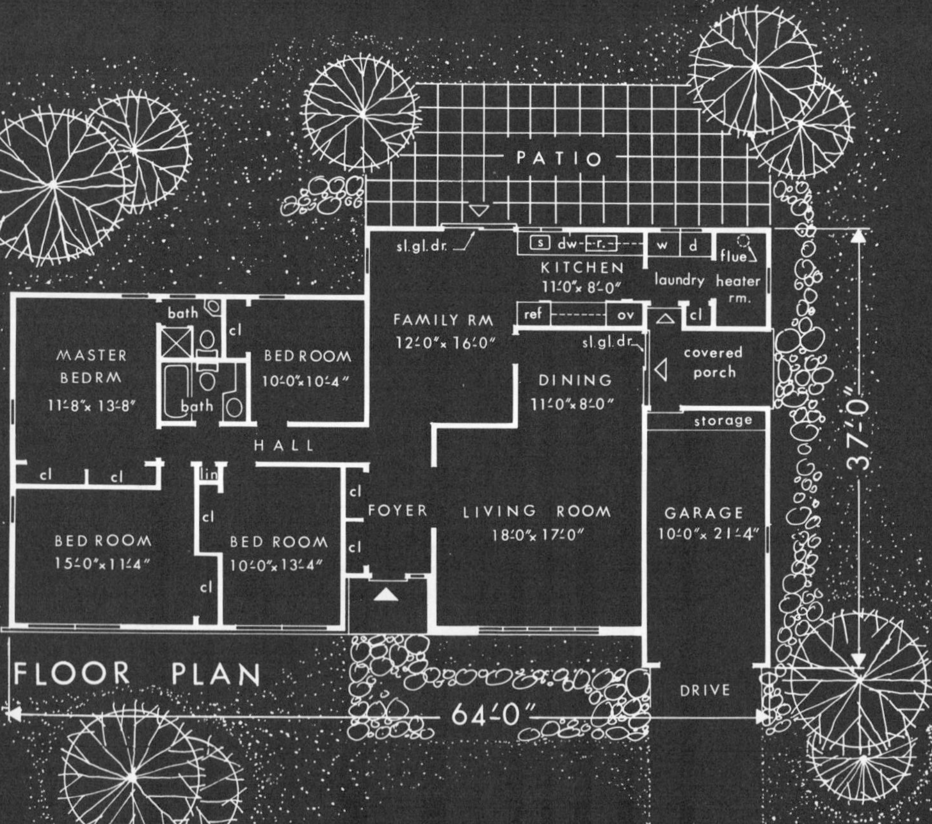


FLOOR PLAN

## AREA STATISTICS :

| | |
|---|---|
| basic house.......... | 1,594 sq. ft. |
| laundry, heater rm.... | 80 sq. ft. |
| covered porch........ | 63 sq. ft. |
| garage, storage.... | 213 sq. ft. |

DESIGN S-79

# Contemporary Ranch in Stone and Redwood

The contemporary lines of this home are achieved through stone and redwood siding and the graceful roof line, accenting the stone piers of the living room windows.

A pleasing angle is accomplished by the downward pitch of the roof, which continues over the garage to the rear. The result is a beautiful exterior which is carried through to the interior. A natural cathedral ceiling is produced by the design of the roof line in the main living areas of the house.

One enters the central foyer, to the left of which is a large guest closet. Straight ahead is the family room with its cathedral ceiling and sliding glass doors leading to the patio. Off the family room is the kitchen with a dropped luminous ceiling. The different ceiling heights in these two rooms produce a dramatic and interesting effect. Adjacent to the kitchen is the laundry and heater room with entrances to the patio and to the covered porch.

To the right of the foyer is the living and dining room shaped in an ell. A sliding glass door in the dining room leads to the covered porch and the outdoors.

The bedroom wing to the left of the foyer is private with its own center hall leading to all four rooms and the main bath.

Costs have been kept down without sacrificing either aesthetics or livability.

## Material List

**CONCRETE WORK**

| | |
|---|---|
| Foundations, footings, slabs, etc. | 69 cu. yds. |

**MASONRY**

| | |
|---|---|
| Stone Walls | 109 cu. ft. |

**FRAMING LUMBER**

| | |
|---|---|
| Total Sills, Joists, Rafters, Studs, Plates, etc. | 5856 B.F.M. |

**SHEATHING, INSULATION**

| | |
|---|---|
| Wall Sheathing | 1745 sq. ft. |
| Roof Sheathing | 2972 sq. ft. |
| Wall Insulation | 1685 sq. ft. |
| Ceiling Insulation | 1894 sq. ft. |

**FINISHES, INTERIOR**

| | |
|---|---|
| Vinyl Tile | 447 sq. ft. |
| Oak Flooring | 1100 sq. ft. |
| Ceramic Tile Floors | 50 sq. ft. |
| Ceramic Tile Walls | 180 sq. ft. |
| Gypsum Board - house walls | 2864 sq. ft. |
| Gypsum Board - house ceilings | 1674 sq. ft. |
| Gypsum Board - garage | 615 sq. ft. |

**FINISHES, EXTERIOR (other than masonry)**

| | |
|---|---|
| Vertical Siding | 1393 sq. ft. |
| Asphalt Shingle Roofing | 2972 sq. ft. |
| Plywood Eave & Porch Soffits | 425 sq. ft. |

**WINDOW SCHEDULE**

| | |
|---|---|
| Awning | 2 units |
| Wood casement | 8 units |
| Casement Bow Window | 1 unit |
| Gliding windows | 2 units |

**DOOR SCHEDULE**

| | |
|---|---|
| Ext. Hardwood, paneled & glazed side light | 1 unit |
| Ext. Hardwood, paneled & glazed | 1 unit |
| Aluminum Sliding | 2 units |
| Wood Overhead garage, paneled | 1 unit |
| Int. Hardwood, flush, staingrade | 6 units |
| Int. Hardwood, bi-folding | 10 units |

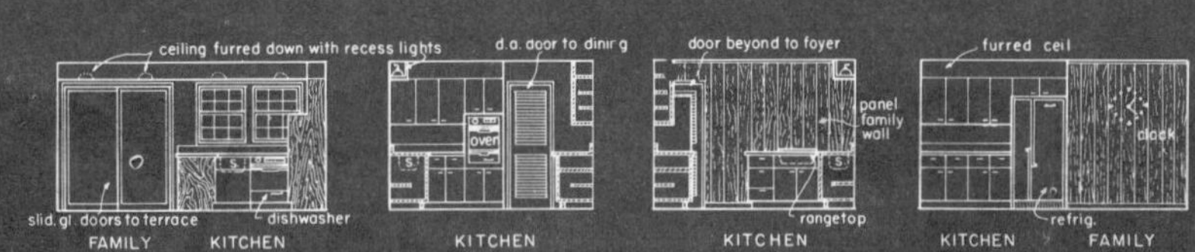


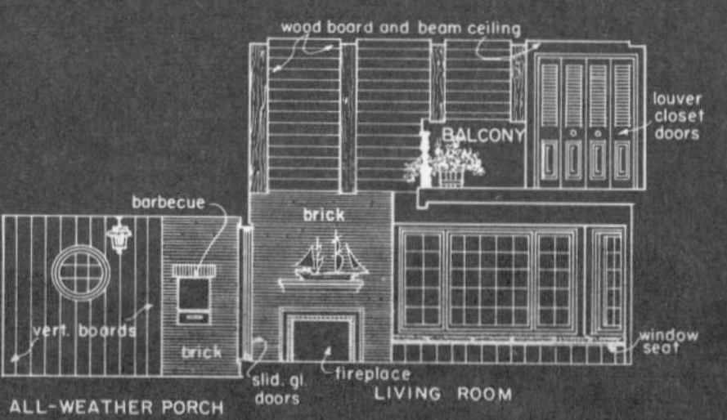


interior elevations

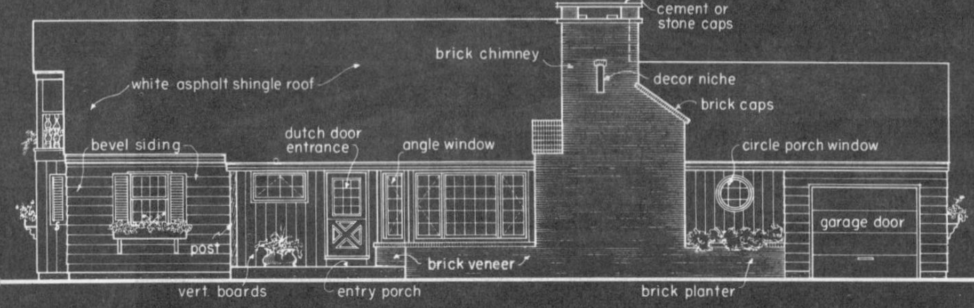


front elevation

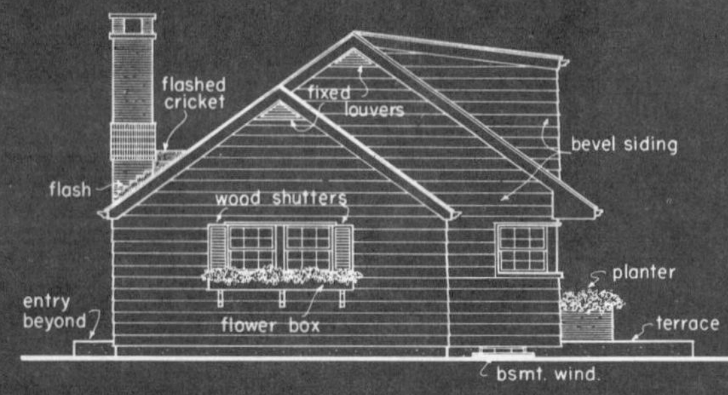


right side elevation

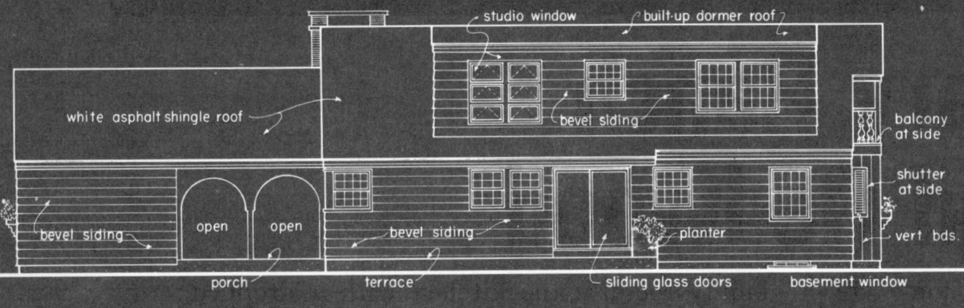


rear elevation

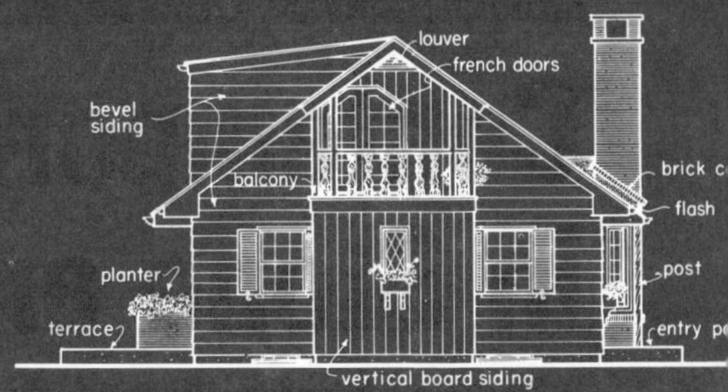


left side elevation

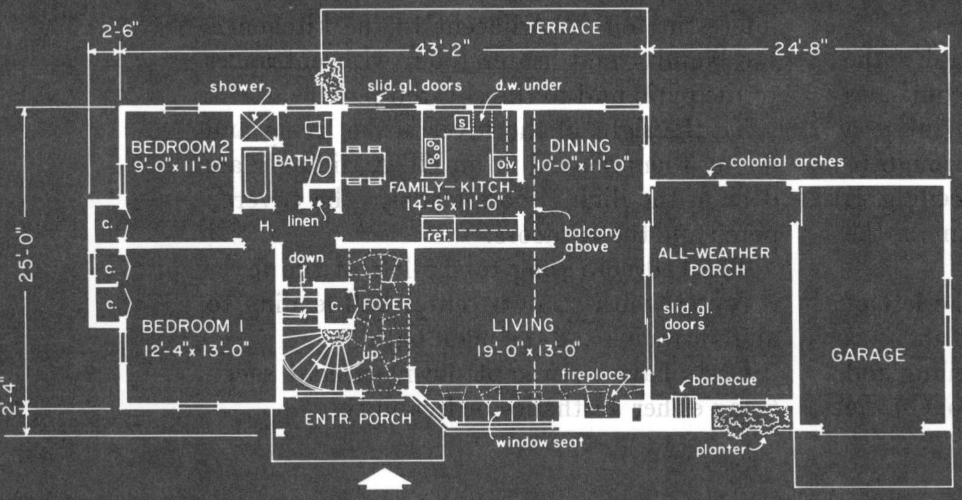


first floor

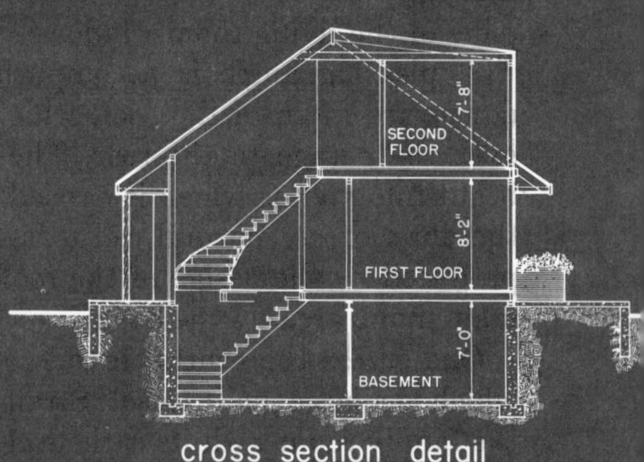


cross section detail

## SIZE DATA

1141 square feet, first floor excluding porches, garage and terrace.

525 square feet, second floor excluding porch, storage and upper area at the living and dining looms.

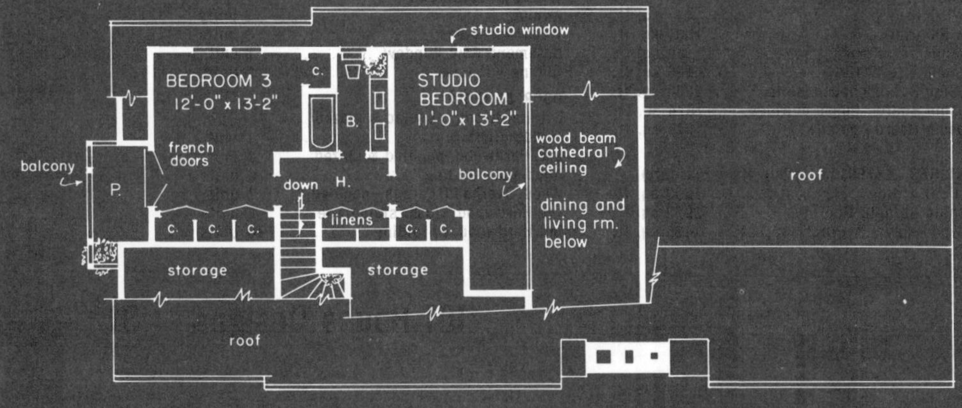


second floor

DESIGN L-43

# Desired Features in Modest Traditional

Maximum living space at minimum cost and desirable features keynote this 1½ story house.

A modest traditional exterior is employed because of its wide acceptability. A log-burning fireplace rates very high. Here it is incorporated into a wide chimney with a barbecue, and the large chimney located in front becomes a strong design feature, again a most popular item.

Circular stairs are a symbol of lavish living. A compact foyer provides sufficient room to install a semicircular stair. Another luxury feature is a bath containing both tub and stall shower. The first floor bath has this.

Direct-food-serving access to outdoor terrace and U-shaped, step-saving efficiency kitchen are two more "wanted" items. The all weather porch between the house and garage has great appeal.

Bay windows have always been in vogue as well as window seats. In this home there is an 11-foot window in the living room with seat to match. They are an extension of the fireplace wall, creating a beautifully designed entire front living room wall.

Balconies are desirable but seldom provided in smaller homes. Here we have no less than two — one over the living and dining rooms creating a 1½ story high cathedral ceiling and the other on the second floor, projecting outdoors.

## Material List

**CONCRETE**
8" 10" Wall Chimney Piers — Footings .................. 1700 cu. ft.
4" Basement, Garage & Porch Fl. . 1820 sq. ft.
**MASONRY**
4" Chimney Inside & Outside Veneer Planter .............. 484 sq. ft.
Common Brick Backup ..... 4800 Com. Brick
130 Fire Brick — Damper & Cleanouts
48 Lin. Ft. of T. C. Flue Avg. 8 x 13
1 Barbecue Outfit 8 Lin. Angle Iron
60 Sq. Ft. of Stone & Flag — Hearth & Caps
**CARPENTRY**
All Framing Lumber ............13,342 BM
All Plywood Sheathing, Floor Etc. Avg. ½" .................. 3600 sq. ft.
All Finishing Lumber, Siding, Panels, Trim ................ 3836 BM
Exterior Plywood ............. 464 sq. ft.
130 Lin. Ft. 4 x 6 Wood Gutter OG
1830 Lin. Ft. Moldings .........
Flooring 13/16 x 2¼ Clear R. Oak 1470 BM
**ROOFING**
19 Sq. 235 lb. White Asphalt Shingles
4½ Sq. 4 Ply Tar & Gravel
15 Rolls 15 lb. Felt — Roof & Walls
**PLASTERBOARD**
5750 Sq. Ft. All Walls & Ceilings — Avg. ½"
**MILLWORK**
5 O.S. Doors, Frames & Trim Avg. 2'6
2 Aluminum Sliding Door Units
8 Inside Doors 2'6 Jambs Trim 2 Sides
20 2 Panel Louver Doors Avg. 1'8 Jambs Trim 2 sides
1 Garage Door 8'0 x 7'0 Flush Panel — Overhead type
22 DH Windows — Frame & Sash Avg. 3'0 x 4'2 Trim 1 side
1 Circle Sash & Frame 2'6 9 Lt. Trim 1 side

FRONT ELEVATION

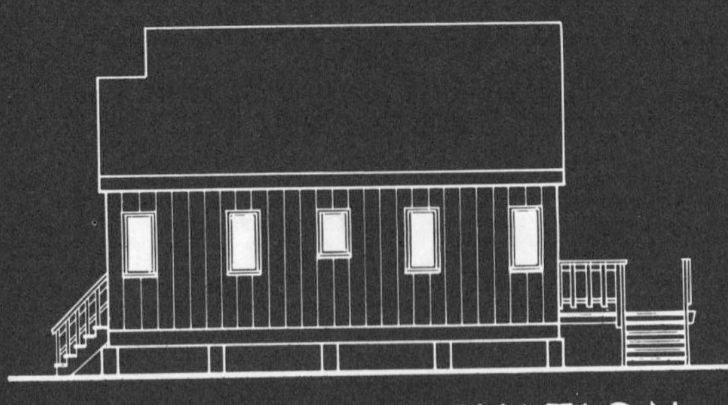

RIGHT SIDE ELEVATION

REAR ELEVATION

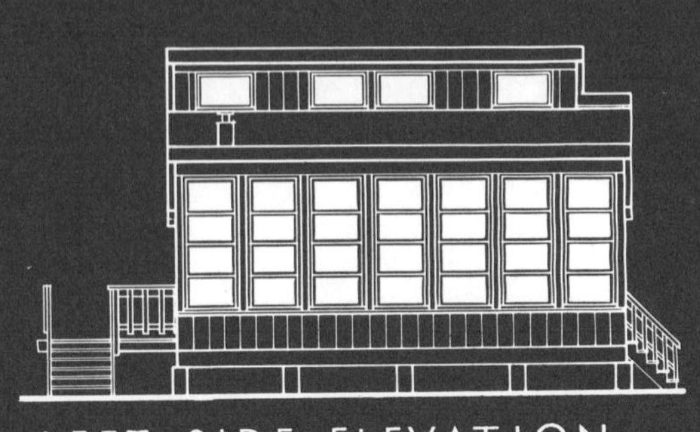

LEFT SIDE ELEVATION

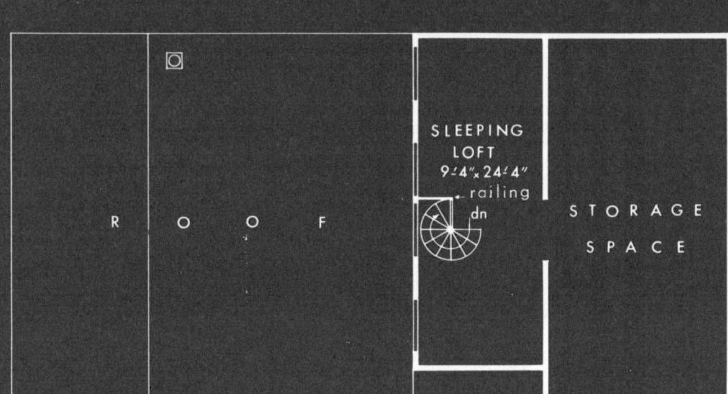


SECOND FLOOR PLAN

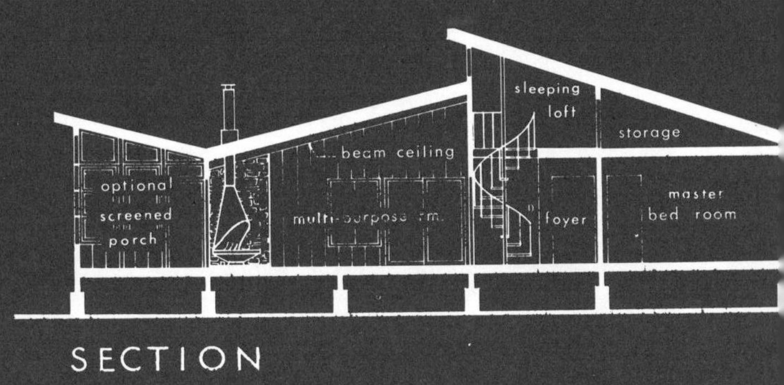


SECTION

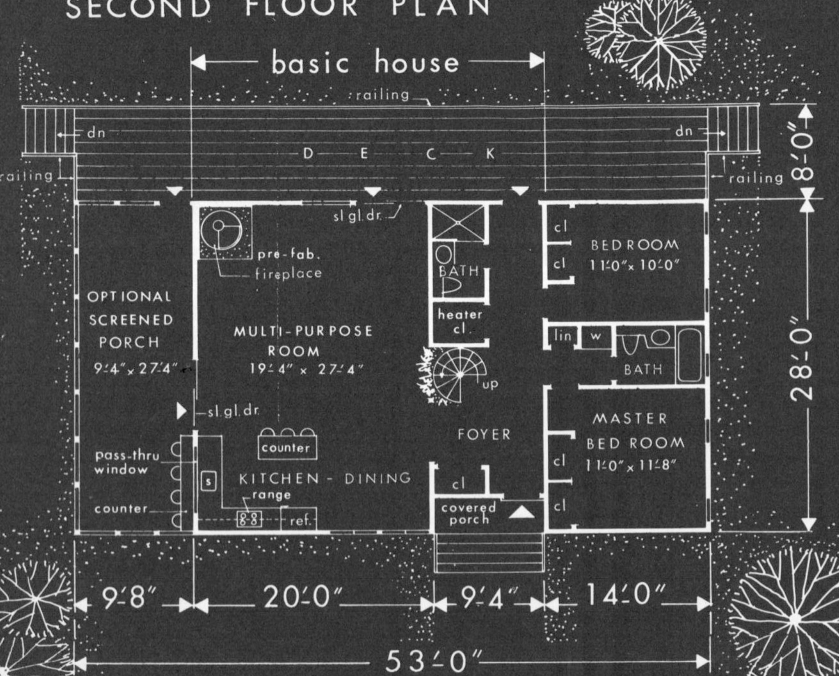


FIRST FLOOR PLAN

## AREA STATISTICS

| | |
|---|---|
| basic house : 1st floor . . . . | 800 sc |
| " " sleeping loft . | 225 |
| bed room wing . . . . . . . . | 392 |
| screened porch optional . | 270 |
| deck . . . . . . . . . . . . . . . | 424 |

DESIGN S-87

# Unusual Flexible Vacation House

How big a second home is depends on the size of the family and the amount of money it can spend. This house takes into account both of those factors. Entirely aside from its interesting contemporary exterior, it can be a compact little house with a total livable area of 800 square feet on the first floor and 225 square feet upstairs — or, with the addition of a wing, a medium-size house with an additional 392 square feet — or, with the addition of a screened porch and an extra 270 square feet, a house measuring 53′ by 36′.

Most important, the basic design remains the same during each phase of construction so that there is never a tacked-on look. Whether it be the countryside or the seashore, the house can be placed on a lot to take advantage of surrounding scenery.

The virtually maintenance-free exterior is a combination of redwood siding and glass. Its contemporary appearance is softened by a deck. From the deck three entrances lead to the optional screened porch, the multipurpose room and the central foyer.

Straight ahead through the central foyer is a spiral stairway which leads to the sleeping loft upstairs. Directly to the left of the foyer is the large multipurpose room which takes up the entire width of the house. This room contains a full kitchen, dining area, and living area. Sliding glass doors lead from the living room to the deck and porch.

## Material List

| CONCRETE WORK | |
|---|---|
| Foundations, footings, slabs, etc. | 20 cu. yds. |
| Reinforcing Mesh | 36 sq. ft. |
| **FRAMING LUMBER** | |
| Total Sills, Joists, Rafters, Studs, Plates, etc. | 3878 BFM |
| **SHEATHING, INSULATION** | |
| Sub Flooring | 1405 sq. ft. |
| Roof Sheathing | 1690 sq. ft. |
| Wall Insulation | 1240 sq. ft. |
| Ceiling Insulation | 400 sq. ft. |

| FINISHES, INTERIOR | |
|---|---|
| Resilient Floor Tile | 1325 sq. ft. |
| Ceramic Tile Floors | 25 sq. ft. |
| Ceramic Tile Walls | 70 sq. ft. |
| Gypsum Board — house walls | 1625 sq. ft. |
| Gypsum Board — house ceilings | 1707 sq. ft. |
| **FINISHES, EXTERIOR (other than masonry)** | |
| Vertical Siding (Texture 1-11) | 1240 sq. ft. |
| Asphalt Shingle Roofing | 1690 sq. ft. |
| Plywood Eave & Porch Soffits | 185 sq. ft. |

| WINDOW SCHEDULE | |
|---|---|
| Wood Casement | 8 units |
| Fixed Sash | 15 units |
| Awning | 16 units |
| Gliding | 1 unit |
| **DOOR SCHEDULE** | |
| Ext. Hardwood, Flush | 3 units |
| Aluminum Sliding | 2 units |
| Int. Hardwood, flush, staingrade | 4 units |
| Int. Hardwood, louvered, bifolding | 7 units |

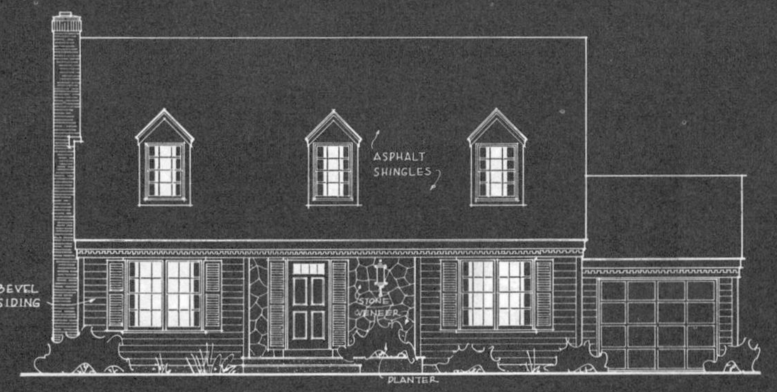

FRONT ELEVATION

RIGHT SIDE ELEVATION

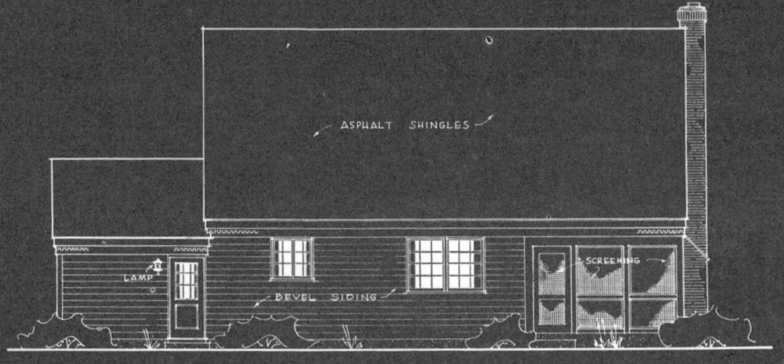

REAR ELEVATION

LEFT SIDE ELEVATION

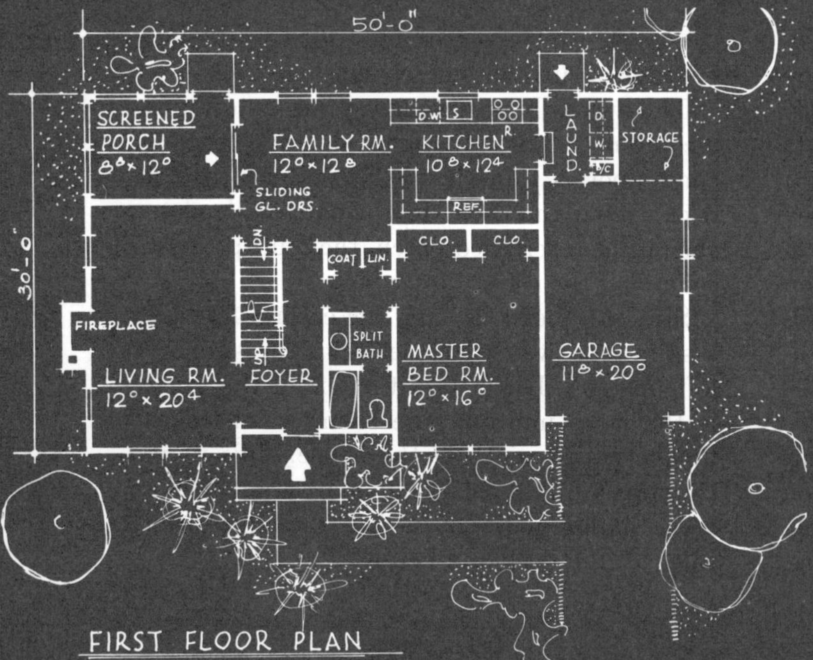

FIRST FLOOR PLAN

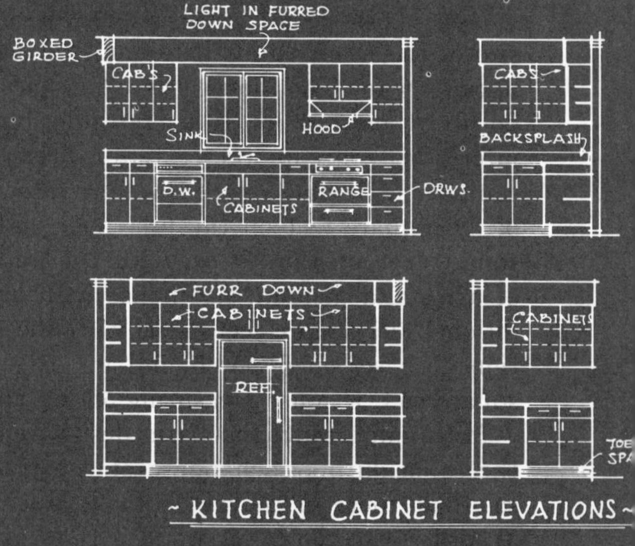

~ KITCHEN CABINET ELEVATIONS ~

## HOUSE DATA

| | |
|---|---|
| FIRST FLOOR | 1004 SQ. FT. |
| SCREENED PORCH | 111 SQ. FT. |
| LAUNDRY AREA | 46 SQ. FT. |
| SECOND FLOOR | 585 SQ. FT. |
| GARAGE & STOR. | 285 SQ. FT. |

DESIGN L-42

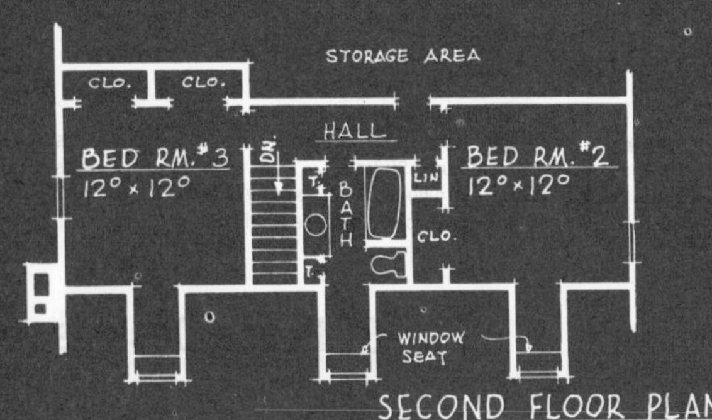

SECOND FLOOR PLAN

# Unusual Floor Arrangement in Cape Cod

This is a Cape Cod with a "twist."

The customary layout of a Cape Cod has a living room, dining room, kitchen and two bedrooms on the first floor, with two additional sleeping rooms on the attic floor. This one has the two upstairs bedrooms, but only one — the master bedroom — on the main floor. Also on the first floor are a family room and a screened porch. The owners' bedroom has a split bath, serving as a powder room from the foyer, with a door placed to cut off a view from the hall into the lavatory area.

The front entrance to the house has been recessed in stone, but it can be done as attractively in brick. A front entrance light on the stone wall helps give additional "curb appeal" to an elevation which can boast of well-designed simplicity.

Immediately on entering the house, one gets a view of the living room fireplace. Although the exterior indicates a formal, almost symmetrical facade, the interior is designed for casual living. The family room replaces the customary dining room, and adjacent to this one can see through sliding glass doors the screened porch.

The kitchen layout combines a straight in-line arrangement with a U-shaped counter nearby. The laundry is convenient for drying either in the garage or in the sunny rear yard.

## Material List

**CONCRETE WORK**

|  |  |
|---|---|
| Concrete Walls | 838 cu. ft. |
| Slabs | 485 cu. ft. |
| Foundation Footings | 187 cu. ft. |
| Misc. Concrete | 187 cu. ft. |

**STRUCTURAL STEEL**

|  |  |
|---|---|
| Lally Columns | 3 pieces |
| Girder ......S 7 15.3# | 47 lin. ft. |

**BRICK WORK**

|  |  |
|---|---|
| Veneer ............. Stone | 96 sq. ft. |
| Chimney & Fireplace .. Brick | 164 cu. ft. |
| Flue Lining | 54 lin. ft. |

**CARPENTRY**

|  |  |
|---|---|
| Framing Lumber | 7006 B.F. |
| Studs | 4000 B.F. |
| Plates | 1000 B.F. |
| Roof Sheathing | 2100 sq. ft. |
| Sub Flooring | 1560 sq. ft. |
| Side Wall Sheathing | 1740 sq. ft. |
| Wall Insulation | 982 sq. ft. |
| Ceiling Insulation | 1370 sq. ft. |
| Oak Flooring | 1320 sq. ft. |
| Kitchen Plywood | 132 sq. ft. |

**MILL WORK**

|  |  |
|---|---|
| Exterior Doors & Frames Complete | 3 pieces |
| Garage Door Complete Set | 1 unit |
| Sliding Glass Door Units | 1 unit |
| Interior Doors & Frames Complete | 12 pieces |
| Bi Fold Doors | 1 unit |
| Sliding Doors | 12 pieces |
| Window Units | 16 units |
| Fascia | 224 lin. ft. |
| Base | 590 lin. ft. |
| Pole Sockets | 6 pair |

**ROOFING**

|  |  |
|---|---|
| Shingles ...... 235# asphalt | 2100 sq. ft. |
| Roofing Paper 15# felt | 2100 sq. ft. |

FRONT ELEVATION

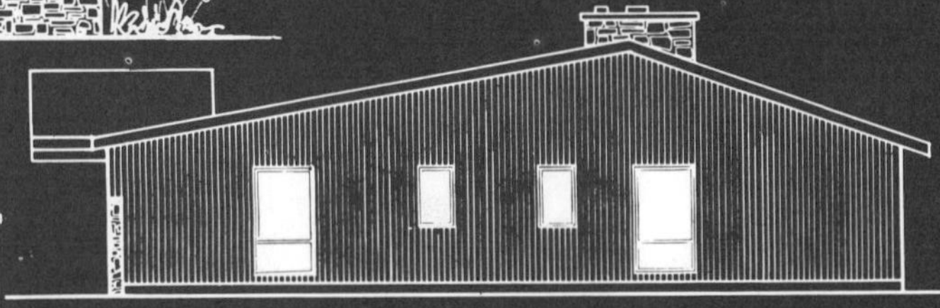

RIGHT SIDE ELEVATION

REAR ELEVATION

LEFT SIDE ELEVATION

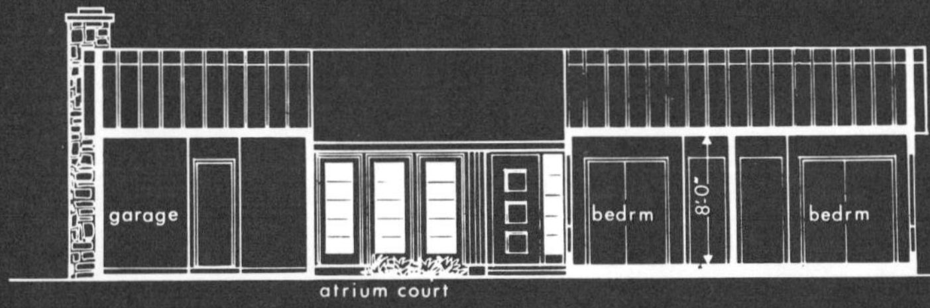

garage

bedrm

8'-0"

bedrm

atrium court

SECTION

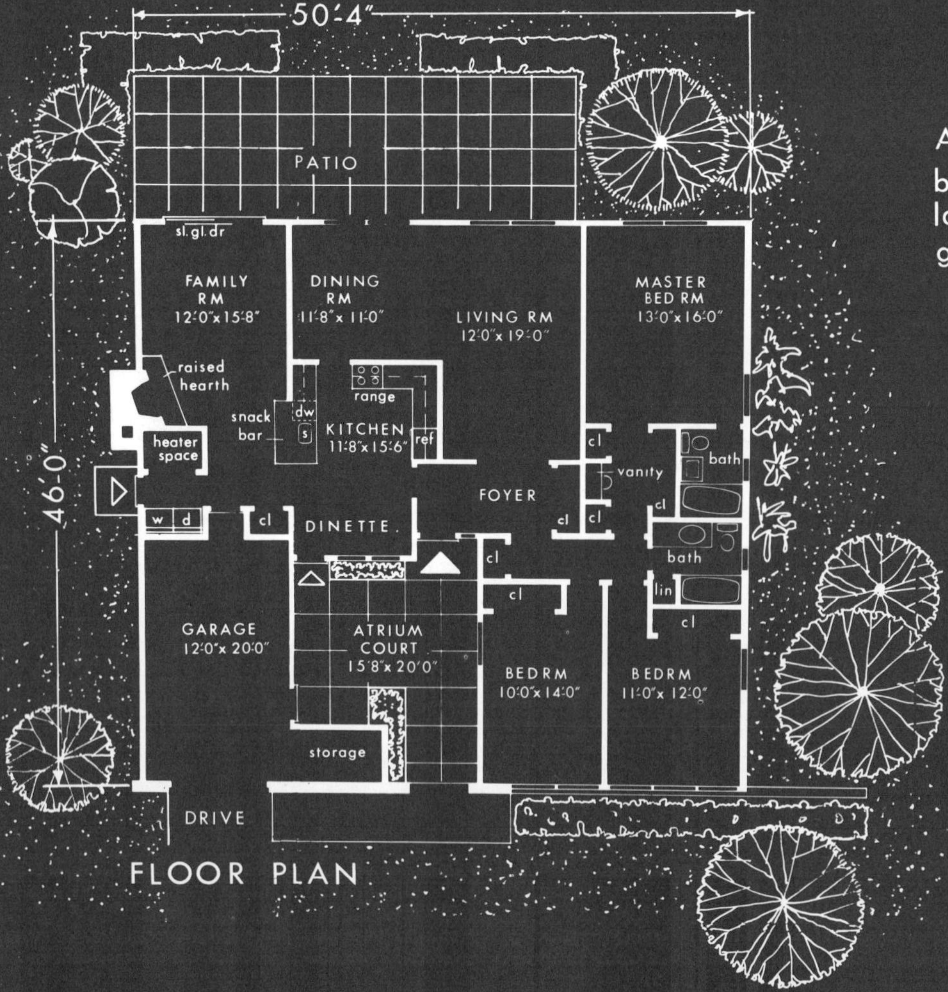

50'-4"

46'-0"

PATIO

sl. gl. dr

FAMILY RM 12'-0"x15'-8"

DINING RM 11'-8"x 11'-0"

LIVING RM 12'-0"x 19'-0"

MASTER BED RM 13'-0"x16'-0"

raised hearth

range

snack bar

dw

s

KITCHEN 11'-8"x15'-6"

ref

heater space

cl

vanity

bath

FOYER

w d

cl

DINETTE

cl

cl

cl

bath

lin

cl

cl

GARAGE 12'-0"x 20'-0"

ATRIUM COURT 15'-8"x 20'-0"

BEDRM 10'-0"x14'-0"

BEDRM 11'-0"x 12'-0"

storage

DRIVE

FLOOR PLAN

AREA STATISTICS

basic house........ 1745 sq. ft.
laundry, heater rm.... 40 sq. ft.
garage.............. 240 sq. ft.

DESIGN S-48

# Enclosed Courtyard Highlights Ranch

The atrium concept of this three-bedroom ranch combines economy with utility and aesthetics. The atrium, an outdoor enclosed courtyard, provides sunshine to adjacent rooms and serves as a desirable private outdoor relaxation area.

The atrium court is in the front center, and is a delightful entrance to the home. Metal gates behind a covered portico enclose the court from the street.

The front door of the house itself is 20′ inside the courtyard. It is flanked by a large sidelight and leads to a spacious reception foyer. Straight ahead is an L-shaped living room and dining room arrangement, both with large windows overlooking the rear patio and garden.

A sizable eat-in kitchen is off to the left of the foyer. Its dinette area features a large window wall and door overlooking and connecting to the atrium.

An indoor breakfast bar separates the kitchen from the adjoining family room. Next to the family room is the laundry-mud room, the side service door, a service closet and the inside entrance to the one-car garage.

The bedroom wing at the right side of the house has three bedrooms, two bathrooms and excellent closet space.

Off the garage is a large alcove, which is adjacent to the atrium court, thereby providing additional space for storage.

## Material List

|  |  |
|---|---|
| **CONCRETE WORK** | |
| Foundations, footings, slabs, etc. | 58 cu. yds. |
| **MASONRY** | |
| Stone Veneer | 100 cu. ft. |
| Stone Fireplace & Chimney | 240 cu. ft. |
| Stone Patio Walls | 40 cu. ft. |
| **FRAMING LUMBER** | |
| Total Sills, Joists, Rafters, Studs, Plates, etc. | 12,900 B.F.M. |
| **SHEATHING, INSULATION** | |
| Wall Sheathing | 1822 sq. ft. |
| Roof Sheathing | 2304 sq. ft. |
| Wall Insulation | 1600 sq. ft. |
| Ceiling Insulation | 1785 sq. ft. |
| **FINISHES, INTERIOR** | |
| Vinyl Tile | 560 sq. ft. |
| Oak Block Flooring | 644 sq. ft. |
| Ceramic Tile Floors | 52 sq. ft. |
| Ceramic Tile Walls | 266 sq. ft. |
| Gypsum Board — house walls | 2760 sq. ft. |
| Gypsum Board — house ceilings | 1785 sq. ft. |
| Gypsum Board — garage | 2300 sq. ft. |
| **FINISHES, EXTERIOR (other than masonry)** | |
| Vertical Siding | 1585 sq. ft. |
| Asphalt Shingle Roofing | 2304 sq. ft. |
| Plywood Eave & Porch Soffits | 597 sq. ft. |
| **WINDOW SCHEDULE** | |
| Wood Casement | 2 units |
| Flexivent Awning | 14 units |
| **DOOR SCHEDULE** | |
| Ext. Hardwood, paneled & glazed | 5 units |
| Ext. Hardwood, paneled | 1 unit |
| Aluminum Sliding | 1 unit |
| Wood Overhead Garage, paneled | 1 unit |
| Int. Hardwood, flush, staingrade | 5 units |
| Int. Hardwood, louvered, bi-folding | 10 units |

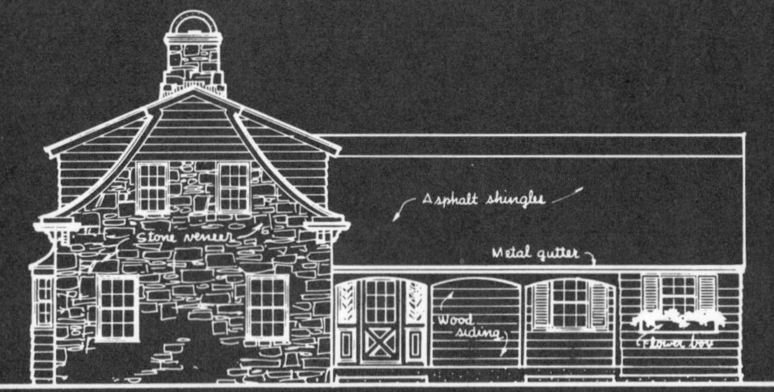


Front elevation

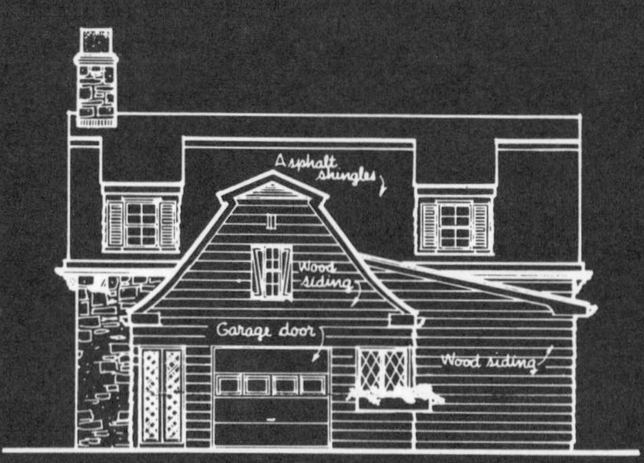


Right side elevation

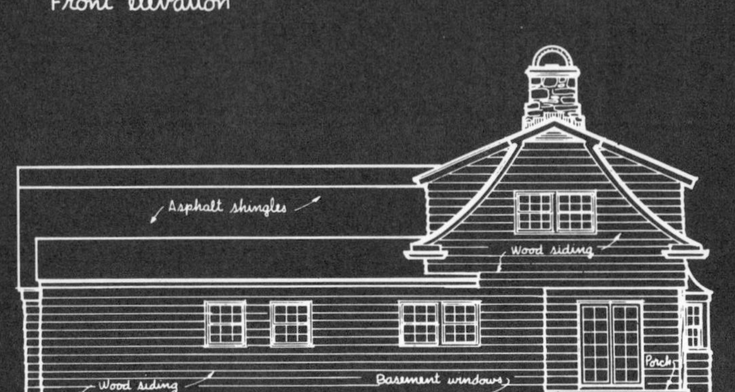


Rear elevation

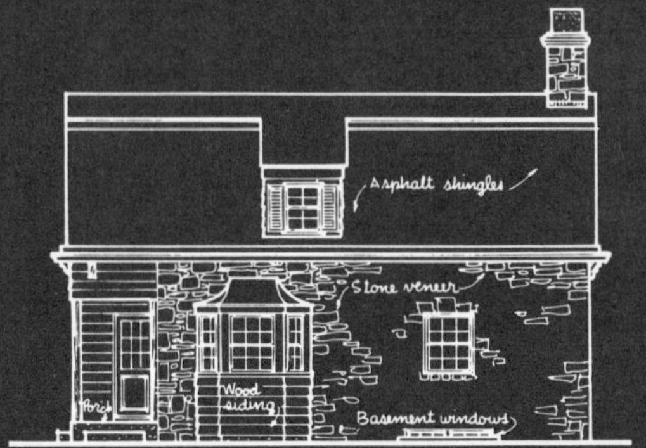


Left side elevation

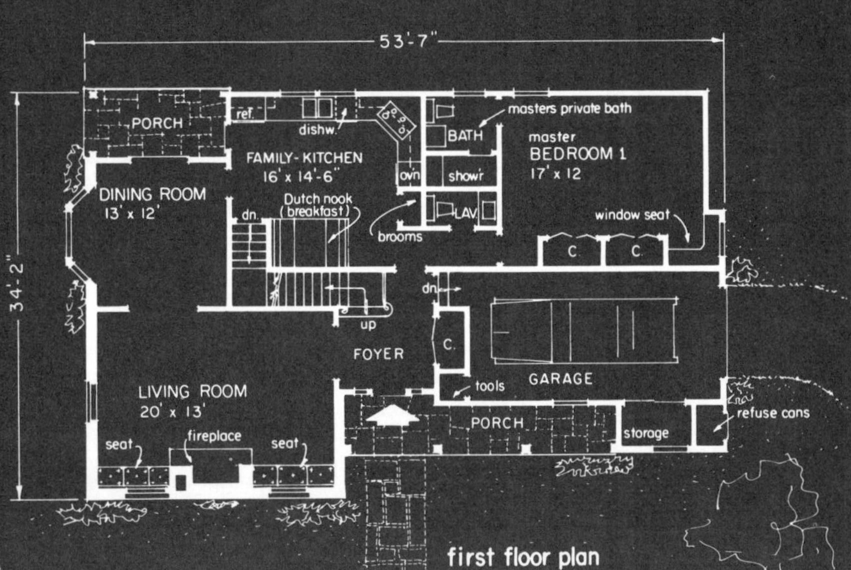


first floor plan

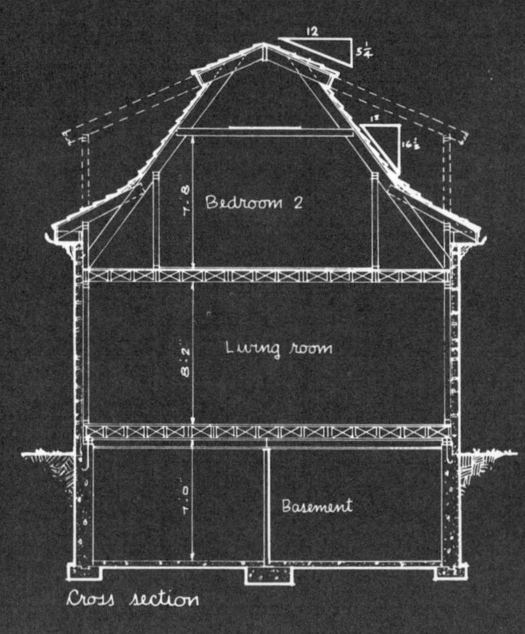


Cross section

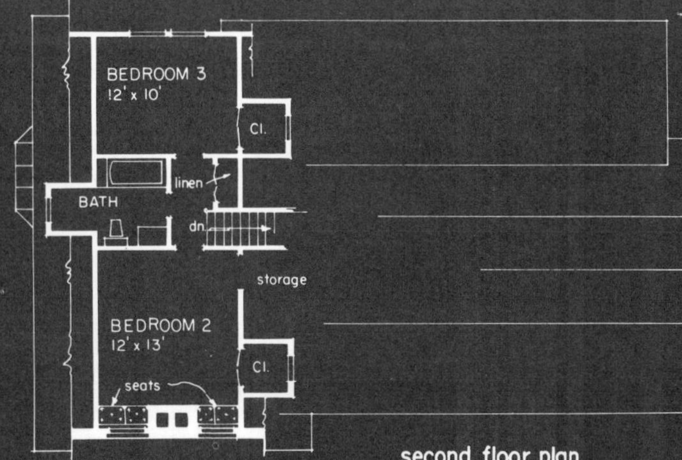


second floor plan

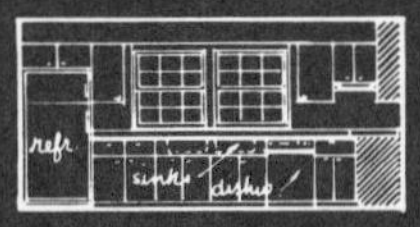


Kitchen elevations

DESIGN Z-84

## Size data.

1,280 sq. ft. on first floor not including garage and porches.

489 sq. ft. on second floor

# Dutch Colonial with Family Kitchen

Anyone who has ever visited The Netherlands knows that the houses there combine quaintness and charm with a kind of neat and orderly look, as though every detail had been planned with considerable thought.

One is likely to get the same impression from this delightful Dutch Colonial that seems larger than it actually is. Much of the reason for its spacious appearance is the characteristic gambrel roof, introduced to America hundreds of years ago by Dutch colonists.

At the back half of the house is what the family is likely to consider the main interior feature a sizable farm-type kitchen. It has a large Dutch-styled breakfast nook, splayed counter corner with the cooking unit set in, a brick-patterned floor and broom closet. Wood paneling is on the walls around the breakfast nook. There is an entrance from the kitchen to a porch at the left rear of the house.

Adjacent to the kitchen is the dining room, which also has an entrance to the porch. A huge bay window increases the size of the dining room and enhances its appearance.

The living room is accessible from the front foyer and the dining room. The master bedroom on the first floor has two closets, a corner window seat and a bathroom with shower.

With the master bedroom on the first floor, Design Z-84 could be used by a couple who might wish to reserve the two upstairs bedrooms for guests.

## Material List

**CONCRETE**
8" - 10" - 12" walls ........... 1814 cu. ft.
Conc. floors 4" ............... 1542 sq. ft.

**STEEL**
3-3½" — 6'8 lally columns
34 Lin. 8" I — wide flange 24 lb.
20 Lin. 4 x 3-5/16 Angle Iron ..

**MASONRY**
920 ft. 4" Stone veneer incl. chimney
Common brick back up chimney.. 1300 brick
54 Lin. 12 x 12 flue
100 fire brick

**CARPENTRY**
Floor framing — 2 floors —
plates, etc. .................. 4000 BM
⅝ — sub floor plyscore ........ 2200 sq. ft.
All stud headers etc. beams .... 4630 BM
Ceil. Joists, etc. .............. 840 BM
Rafters, etc. .................. 3400 BM
½ plyscore roof bds, — catwalk. 3680 sq. ft.
⅜ plyscore sheathing .......... 2500 sq. ft.
1 x 10 bevel siding ........... 1650 BM
Oak floor 13/16 x 2¼ ......... 1570 BM
10 Rolls — 15 lb. felt
⅜ Exterior plywood ........... 470 sq. ft.

**INSULATION**
1750 sq. ft. Semi-thick batts
1400 sq. ft. full thick batts

**ROOFING**
6 sq. Asphalt shingles — 340 lb.
20 sq. Asphalt shingles — 240 lb.
240 Indiv. shingles ridges — etc.
5 Rolls — 15 lb. felt
3 Rolls — 30 lb. felt

**SHEETROCK**
½ for walls tapered ........... 5200 sq. ft.
⅜ for ceiling tapered ......... 1800 sq. ft.

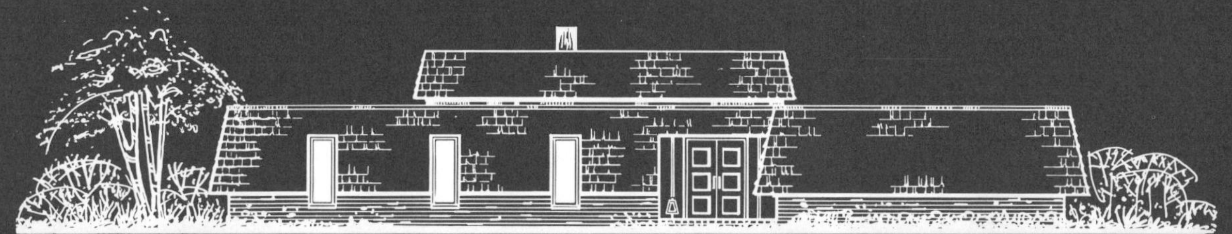

FRONT ELEVATION

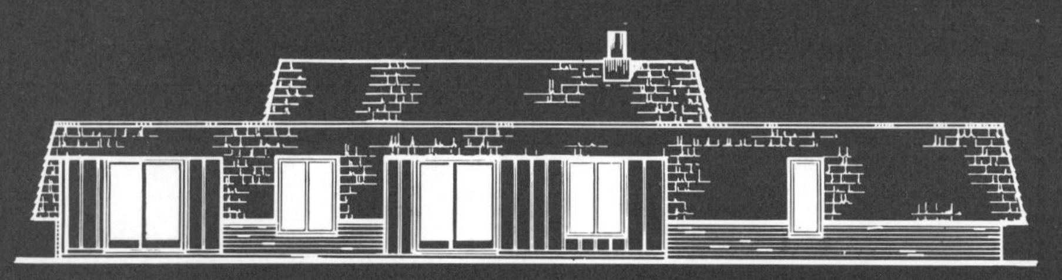

REAR ELEVATION

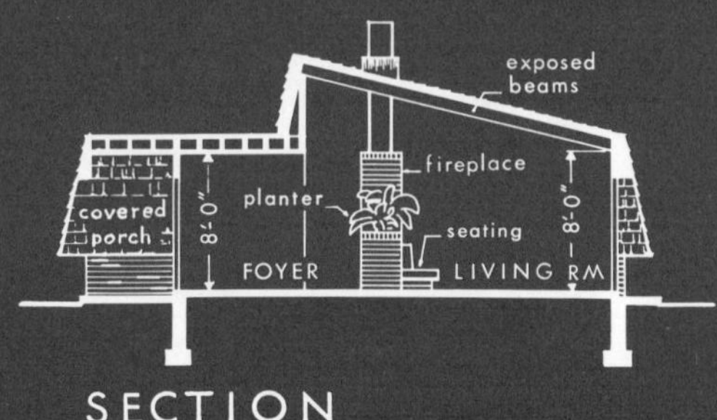


SECTION

RIGHT SIDE ELEVATION

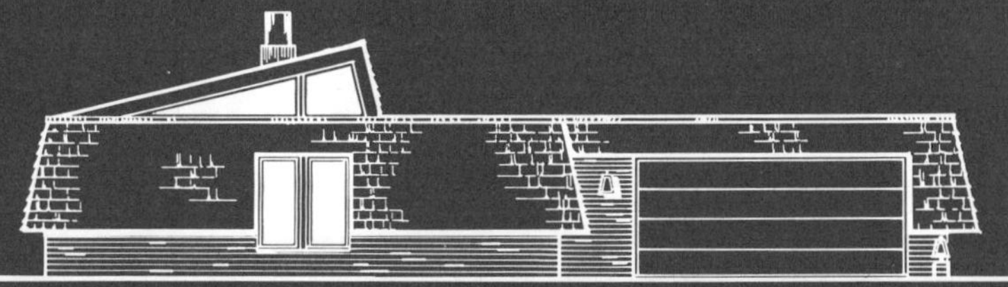

LEFT SIDE ELEVATION

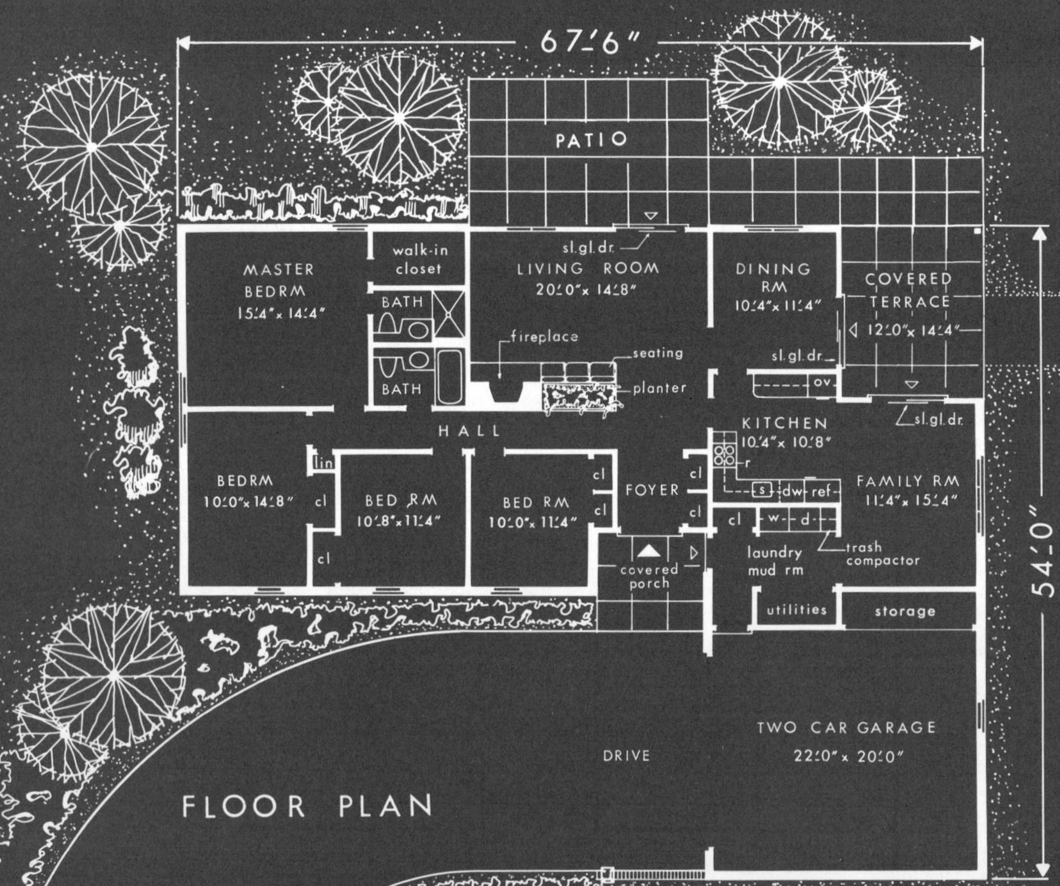


AREA STATISTICS :

basic house, including
laundry-mud rm ......1778 sq
garage & storage ... 510

Design L-19

# Contemporary Ranch Has Own Character

One of the advantages of contemporary design is the flexibility afforded the architect in expressing himself through the use of form and materials of his own choice.

In this contemporary ranch, hand-split shakes are used on a mansard-type roof. The continuation of the same material on most of the exterior is dramatic, yet soft.

There is no upstairs, no downstairs, no basement, but the house is so well designed that the active, formal and private areas remain intact.

The entry to the house is off the main drive from a covered porch into the entrance foyer. Straight ahead is the living room, definitely a highlight of this home. Featured at the entrance is a planter which acts as a separation between the living room and the foyer. As one looks beyond the living room, he sees the outdoor patio and the rear garden. Extended to one side of the fireplace, and placed on a raised hearth, is a cozy seating arrangement shielded by the planter in the foyer.

The use of a round metal flue extending from the fireplace up through the exposed beamed ceiling, together with the clerestory above the living room, adds to the contemporary character and unusually dramatic result.

To the left of the foyer are the four bedrooms. To the right of the foyer is the active area consisting of the kitchen-family room, laundry-mud room and dining room.

## Material List

**CONCRETE WORK**

| | |
|---|---|
| Foundations, footings, slabs, etc. | 83 cu. yds. |

**STEEL**

| | |
|---|---|
| Flitch Plates | 26 lin. ft. |
| Lally Cols. | 1 pc. |
| Reinforcing Mesh | 1788 sq. ft. |

**MASONRY**

| | |
|---|---|
| Brick Walls | 538 sq. ft. |
| Brick Fireplace | 180 sq. ft. |

**FRAMING LUMBER**

| | |
|---|---|
| Total Sills, Joists, Rafters, Studs, Plates, etc. | 7287 B.F.M. |

**SHEATHING, INSULATION**

| | |
|---|---|
| Wall Sheathing | 1932 sq. ft. |
| Roof Sheathing | 2716 sq. ft. |
| Wall Insulation | 1624 sq. ft. |
| Ceiling Insulation | 1228 sq. ft. |
| Roof Insulation | 2288 sq. ft. |

**FINISHES, INTERIOR**

| | |
|---|---|
| Vinyl Tile or Oak Flooring (option) | 1778 sq. ft. |
| Ceramic Tile Floors | 50 sq. ft. |
| Ceramic Tile Walls | 216 sq. ft. |
| Gypsum Board - house walls | 2992 sq. ft. |
| Gypsum Board - house ceilings | 1778 sq. ft. |
| Gypsum Board - garage | 1024 sq. ft. |

**FINISHES, EXTERIOR (other than masonry)**

| | |
|---|---|
| Cedar Shingles | 1932 sq. ft. |
| Board & Batten Siding | 160 sq. ft. |
| Built-up Roofing | 2716 sq. ft. |
| Plywood Eave & Porch Soffits | 450 sq. ft. |

**DOOR SCHEDULE**

| | |
|---|---|
| Ext. Hardwood, Paneled | 2 units |
| Ext. Hardwood, Paneled | 1 unit |
| Aluminum Sliding | 3 units |
| Wood Overhead Garage, paneled | 1 unit |
| Int. Hardwood, flush, staingrade | 10 units |
| Int. Hardwood, louvered, bi-folding | 9 units |

FRONT ELEVATION

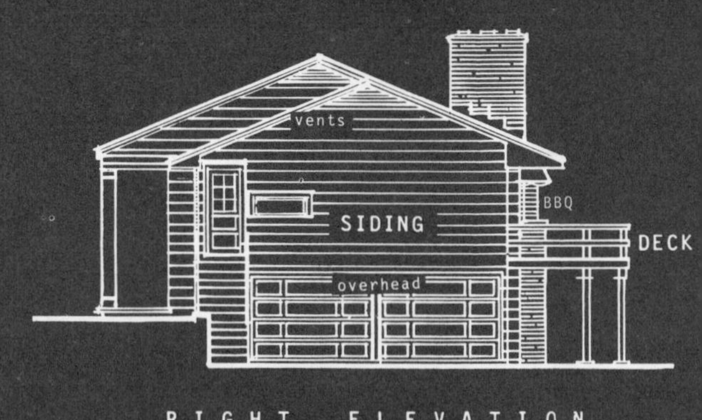

RIGHT ELEVATION

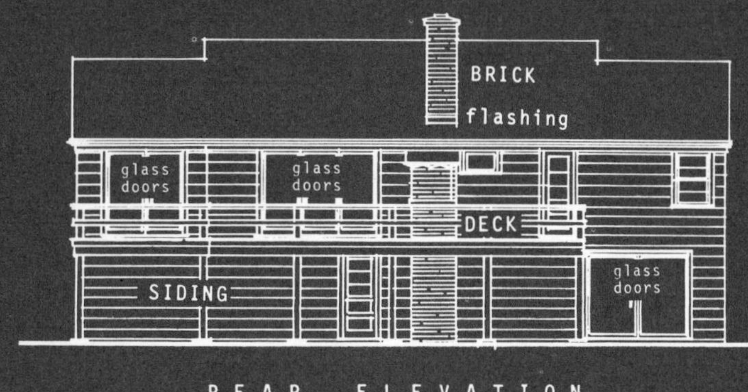

REAR ELEVATION

LEFT ELEVATION

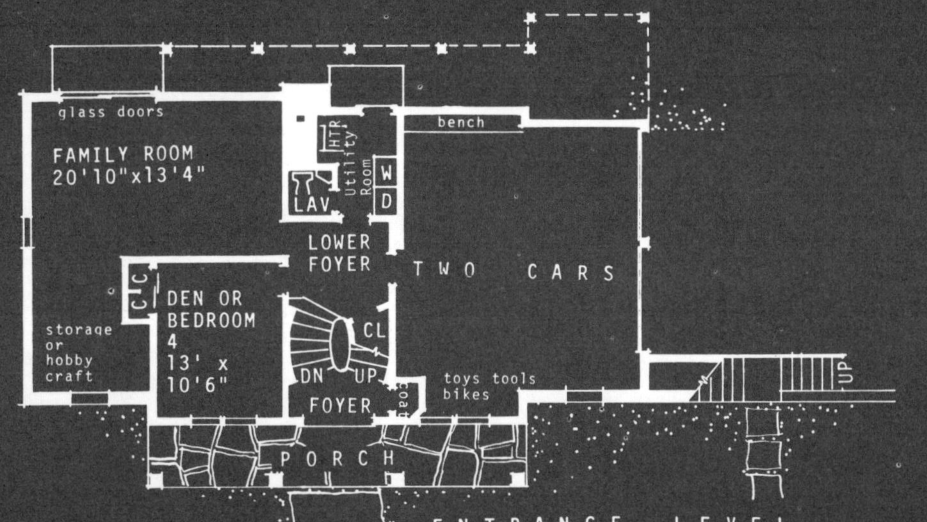

ENTRANCE LEVEL

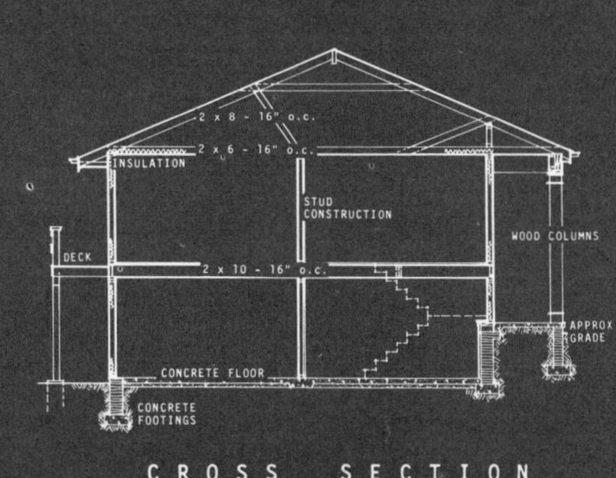

CROSS SECTION

51' 10"

DINING TERRACE

DECK

BBQ

MASTER BEDROOM 14'6" x 13'4"

BATH 2

glass doors

glass doors

CL CL

BATH 1

LIVING ROOM 19'6" x 12'

DINING ROOM 11'x10'

31' 10"

HALL

REF

RG

BEDROOM 2 11' x10'

L

CL

BEDRM 3 10' x 8'

rail

21' x 13'4" overall

LS

DN UP

plant

FAMILY ROOM

snacks

KITCHEN

DOWN

CL

5'

UPPER LEVEL

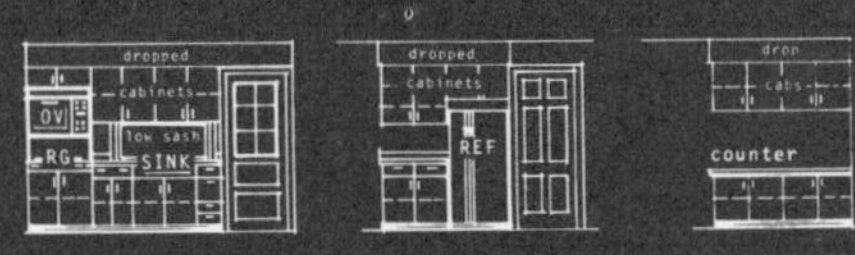

KITCHEN CABINETS

Size data
1,786 sq. ft.

DESIGN L-8

# High Ranch Gets Colonial Styling

A modern plan in traditional Colonial styling is one answer for many home-building families. Here's one that offers a symmetrical exterior, dignified and yet not too formal, with a solid hospitable look.

The plan, a bi-level or high ranch, offers the convenience of one-floor living and, at the option of the owner, a lower level that can be finished as suggested or in any number of ways. The level might even be put to use as an income apartment.

The main level includes seven rooms, with the living and dining rooms at the back. Both have sliding glass doors to a rear deck. An outdoor barbecue and the corner fireplace in the living room are fine adjuncts to family use or entertaining. In the bedroom wing to the left, the master bedroom enjoys privacy with its own bath, a closet wall and a door to the deck. The front bedrooms share a closet wall and use the main bathroom.

For housekeeping, the kitchen is a modern adaptation of the kitchen of Colonial times. The area designated on the plan as "family room" is big enough to use as a multi-activity area.

Although its somewhat imposing exterior makes it seem like a very large house, Design L-8 actually has modest overall dimensions of 51′10″ by 32′3″, including the front porch.

## Material List

|  |  |
|---|---|
| **CONCRETE WORK** | |
| Footings | 8½ cu. yds. |
| Floors (basement & platforms) | 30 cu. yds. |
| **MASONRY** | |
| 4 x 8 x 18 CB | 160 lin. ft. |
| 8 x 8 x 18 CB | 200 sq. ft. |
| 10 x 8 x 18 CB | 440 sq. ft. |
| 12 x 8 x 18 CB | 95 sq. ft. |
| Brick veneer - chimney fill | 450 sq. ft. |
| Flagstone | 160 sq. ft. |
| **CARPENTRY** | |
| 4″ Lally columns | 3 pieces |
| Framing lumber | 6607 B.F. |
| Stud & plates | 4257 B.F. |
| Roof sheathing 1 x 6 | 2540 sq. ft. |
| Plywood | 2100 |
| Fascia #1 pine 1 x 6 | 105 lin. ft. |
| Soffit ⅜″ x 2′ plywood | 105 lin. ft. |
| Shingle mould | 105 lin. ft. |
| Wall sheathing 1 x 6 | 3460 sq. ft. |
| Plywood | 2855 |
| Siding | 3240 sq. ft. |
| Roofing 210# asphalt | 2100 sq. ft. |
| Insulation - walls | 2660 sq. ft. |
| Ceiling | 1500 sq. ft. |
| Flooring | 530 sq. ft. |
| Finish wood flooring | 1535 sq. ft. |
| **DRY WALL** | |
| Ceilings ⅜″ | 2865 sq. ft. |
| Walls ½″ | 6505 sq. ft. |
| **MILLWORK** | |
| Base | 700 lin. ft. |
| Shelving 12″ pine | 35 lin. ft. |
| Windows | 16 pieces |
| Exterior doors | 9 pieces |
| Interior doors | 17 pieces |
| Louvers | 4 pieces |

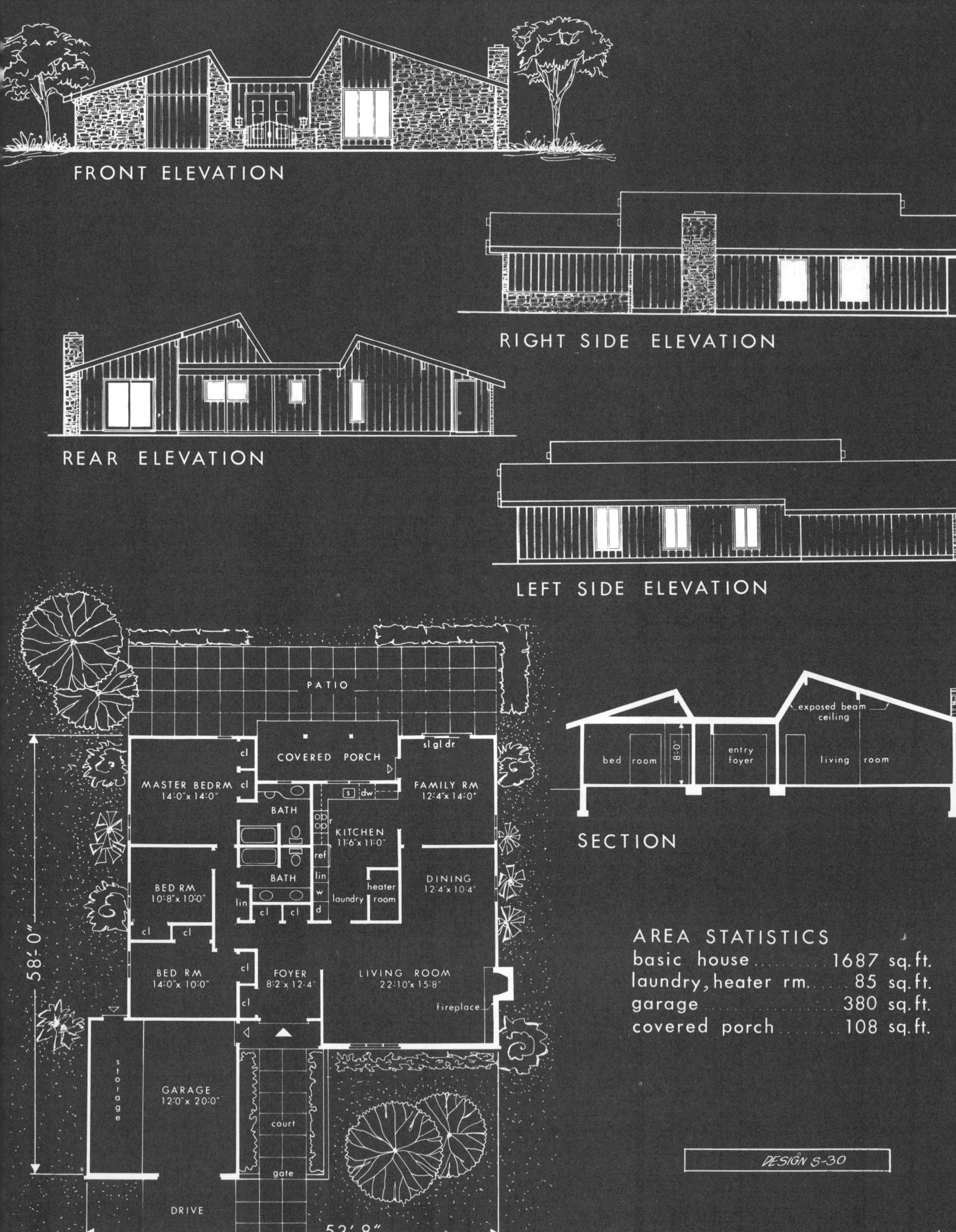

FRONT ELEVATION
RIGHT SIDE ELEVATION
REAR ELEVATION
LEFT SIDE ELEVATION
PATIO
COVERED PORCH
sl gl dr
cl
MASTER BEDRM
14'0" x 14'0"
FAMILY RM
12'4" x 14'0"
BATH
s
dw
ob
opr
KITCHEN
11'6" x 11'0"
ref
lin
w
d
BATH
BED RM
10'8" x 10'0"
lin
laundry
heater room
DINING
12'4" x 10'4"
BED RM
14'0" x 10'0"
FOYER
8'2" x 12'4"
LIVING ROOM
22'10" x 15'8"
fireplace
58'0"
storage
GARAGE
12'0" x 20'0"
court
gate
DRIVE
53' 8"
FLOOR PLAN
exposed beam ceiling
bed room
8'-0"
entry foyer
living room
SECTION
AREA STATISTICS
basic house ........ 1687 sq. ft.
laundry, heater rm. ..... 85 sq. ft.
garage .......... 380 sq. ft.
covered porch ........ 108 sq. ft.
DESIGN S-30

# Happy Ranch Home on Concrete Slab

This inviting and distinctive contemporary ranch home has a happy look.

Whether it's the roof line or the overall shape or the materials that create this appearance, it's there.

There's a nice blending of stone, vertical siding and glass. The entrance, which is undoubtedly the focal point, is reached through a welcoming courtyard flanked on one side by a low stone wall. The courtyard entrance gates, the outside decorative lights, the planting that escorts one to the front door, the vertical battens, the protruding beams — all contribute to the happy feeling.

The house is functional in its plan and economical to build. It can fit on a modest-sized lot. The one-story structure is strictly one-story living, since it has neither a basement nor an attic.

As one enters the living room, he finds an impressive beamed cathedral ceiling following the contour of the roof above and sloping down toward the fireplace. The front wall is highlighted by large windowed panels.

The kitchen, laundry and family room toward the rear of the house merge into each other, but are individually defined. The family room is offset from the kitchen and opens out onto a large patio by means of glass sliding doors and also leads to a covered porch through a Dutch door arrangement. Off the kitchen is a compact laundry room.

## Material List

**CONCRETE WORK**

| | |
|---|---|
| Foundations, footings, slabs, etc. | 72 cu. yds. |

**MASONRY**

| | |
|---|---|
| Stone Veneer | 103 cu. ft. |
| Stone Fireplace & Chimney | 126 cu. ft. |
| Stone Walls (entry court) | 49 cu. ft. |

**FRAMING LUMBER**

| | |
|---|---|
| Total Sills, Joists, Rafters, Studs, Plates, etc. | 9360 B.F.M. |

**SHEATHING, INSULATION**

| | |
|---|---|
| Wall Sheathing | 1356 sq. ft. |
| Roof Sheathing | 3252 sq. ft. |
| Wall Insulation | 1100 sq. ft. |
| Ceiling Insulation | 1687 sq. ft. |

**FINISHES, INTERIOR**

| | |
|---|---|
| Vinyl Tile | 366 sq. ft. |
| Oak Block Flooring | 1406 sq. ft. |
| Ceramic Tile Floors | 100 sq. ft. |
| Ceramic Tile Walls | 120 sq. ft. |
| Gypsum Board | 5597 sq. ft. |

**FINISHES, EXTERIOR (other than masonry)**

| | |
|---|---|
| Vertical Siding | 925 sq. ft. |
| Asphalt Shingle Roofing | 2724 sq. ft. |
| 4 Ply Built-up Roofing | 504 sq. ft. |
| Plywood Eave & Porch Soffits | 410 sq. ft. |

**WINDOW SCHEDULE**

| | |
|---|---|
| Wood Casement | 10 units |
| Gliding | 1 unit |

**DOOR SCHEDULE**

| | |
|---|---|
| Ext. Hardwood, paneled & sidelite | 1 unit |
| Ext. Hardwood, paneled | 3 units |
| Aluminum Sliding | 1 unit |
| Wood Overhead garage paneled | 1 unit |
| Int. Hardwood, flush, staingrade | 8 units |
| Int. Hardwood, louvered, bi-folding | 9 units |

FRONT ELEVATION

RIGHT SIDE ELEVATION

REAR ELEVATION

LEFT SIDE ELEVATION

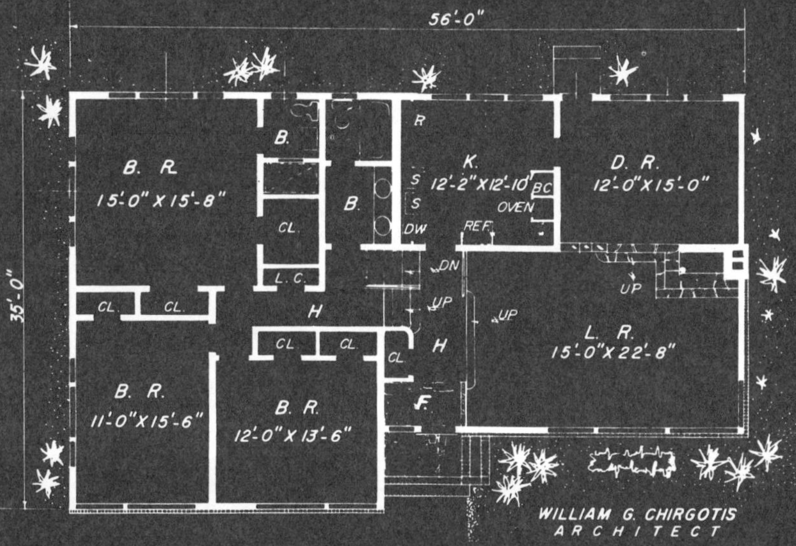


FLOOR PLAN · FIRST & SECOND LEVELS

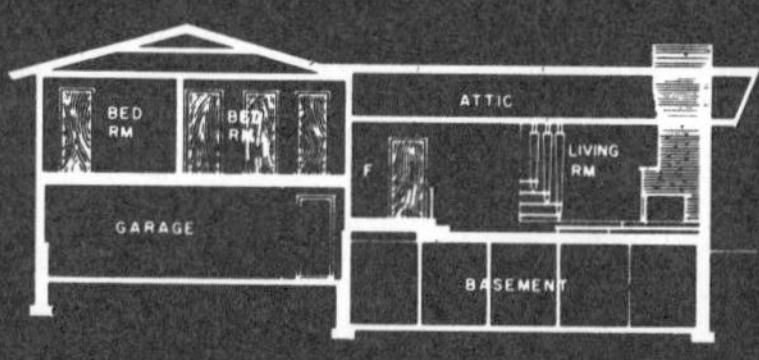


LONGITUDINAL SECTION

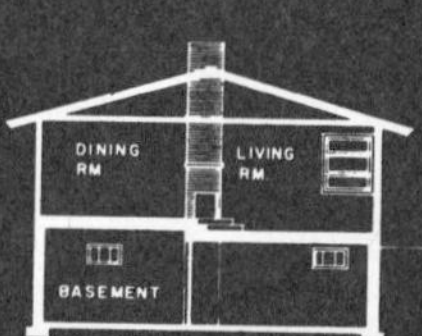


CROSS SECTION

AREA STATISTICS

| | |
|---|---|
| BASEMENT | 1800 SQ FT |
| FIRST FLOOR | 1800 SQ FT |

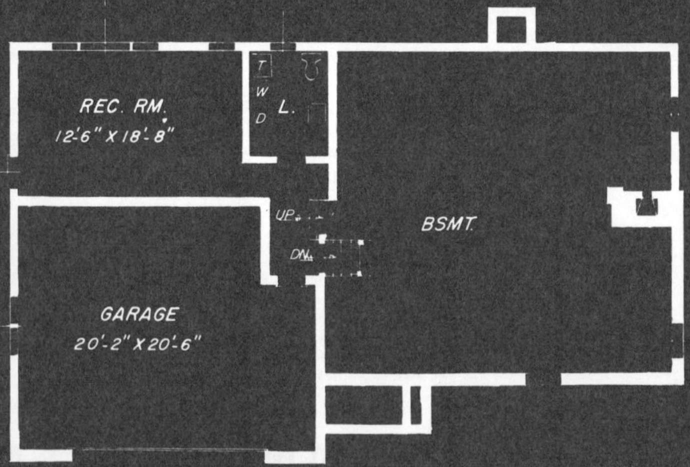


BASEMENT & GARAGE PLAN

DESIGN S-55

# Split-Level with Sunken Living Room

Design S-55 is a transitional split; it blends contemporary and traditional elements.

There are the usual first and second levels, with a recreation room on the garage level and a basement a few steps below that. The eye-catcher is a spacious living room, slightly sunken below the adjoining dining room and separated from it by, on one side, a tapered slat divider and, on the other, by a stone corner fireplace. Adding to the attractiveness of the room are a triple-window unit at the front of the house and a single side window meeting it at the corner.

The dining room looks out on the rear and is adjacent to an almost-square kitchen with sufficient space for a dinette set.

To the left of the living area, up a few steps, is the bedroom level with plenty of wall and closet space. The master bedroom has a private bath with glass-enclosed stall shower and two closets, one a walk-in.

On the lower level, a wood-paneled recreation room, lavatory and laundry room are located behind the two-car garage.

Because the exterior of a split is necessarily "broken," it can be made attractive without too much trouble. In this transitional design, the architect has effectively combined red cedar siding, vertical V-joint boarding, brick veneer, asphalt shingled roofing and a disciplined repetition of awning-type and fixed ribbon windows.

## Material List

**CONCRETE WORK**

| | |
|---|---|
| Footings, floors, etc. | 36 cu. yds. |

**STEEL**

| | |
|---|---|
| 4" lally columns | 6 pieces |
| Metal areaways | 3 pieces |

**MASONRY**

| | |
|---|---|
| Flagstone | 60 sq. ft. |
| 8" Concrete Block | 600 blocks |
| 12" Concrete Block | 1500 blocks |
| Brick | 400 sq. ft. |
| 4" Pancake Block | 160 blocks |

**FRAMING LUMBER**

| | |
|---|---|
| Sills, joists, rafters, studs, plates | 16,000 BFM |

**EXTERIOR SHEATHING**

| | |
|---|---|
| ½" Plyscore wall sheathing | 1600 sq. ft. |
| ½" Plyscore roof sheathing | 2000 sq. ft. |

**SUBFLOORING**

| | |
|---|---|
| ⅝" Plyscore | 1800 sq. ft. |

**DOOR SCHEDULE**

(1) 16′ x 6-6 wp overhead
(1) 3-0 x 6-8 x 1¾ flush front entrance doors
(2) 2-8 x 6-8 x 1¾ wp rear sash door
(1 )2-6 x 6-8 x 1¾ scfp. door
(15) Flush interior veneered doors

**WINDOW SCHEDULE**

| | |
|---|---|
| (4) 4-0 x 4-4 Awn | (6) 3-1 x 4-2 Fixed |
| (8) 2-2 x 3-2 Jalousie | (1) 4-4 x 4-2 Fixed |
| (6) 3-1 x 2-2 Jalousie | (2) 4-4 x 3-2 Fixed |
| (2) 2-2 x 4-2 Jalousie | (4) 5-8 x 4-4 Fixed |
| (2) 2-2 x 2-2 Jalousie | (3) 2-8 x 1-8 Basement |

**EXTERIOR FINISHES**

| | |
|---|---|
| Asphalt shingles | 20 squares |
| Wood shingles siding | 1300 sq. ft. |
| Vertical "V" Joint Boarding | 100 sq. ft. |

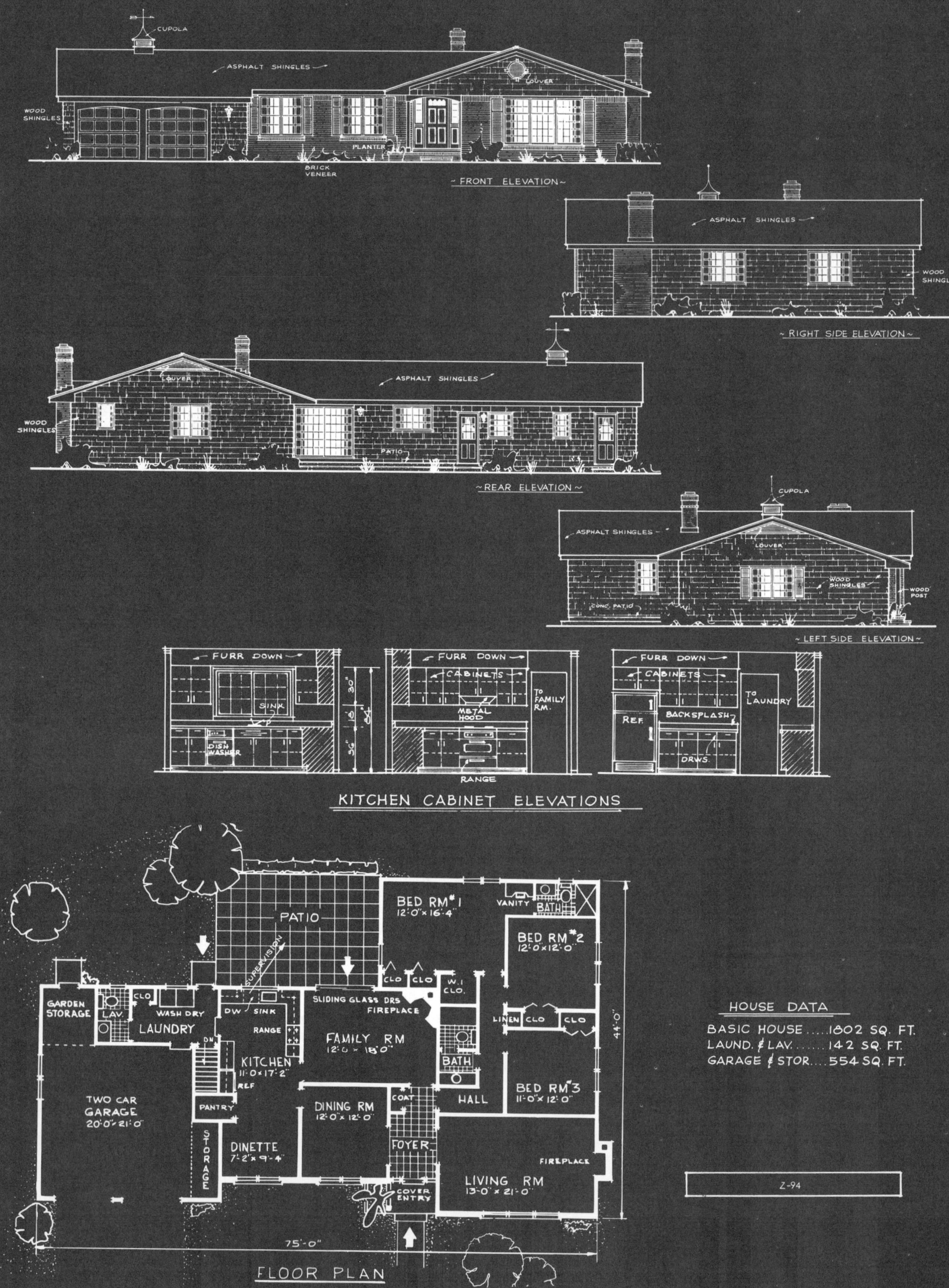
CUPOLA
ASPHALT SHINGLES
LOUVER
WOOD SHINGLES
PLANTER
BRICK VENEER
~ FRONT ELEVATION ~
ASPHALT SHINGLES
WOOD SHINGLE
~ RIGHT SIDE ELEVATION ~
LOUVER
ASPHALT SHINGLES
WOOD SHINGLES
PATIO
~ REAR ELEVATION ~
CUPOLA
ASPHALT SHINGLES
LOUVER
WOOD SHINGLES
WOOD POST
CONC. PATIO
~ LEFT SIDE ELEVATION ~
FURR DOWN
SINK
DISH WASHER
30"
18"
36"
CABINETS
METAL HOOD
RANGE
TO FAMILY RM.
TO LAUNDRY
REF.
BACKSPLASH
DRWS.
KITCHEN CABINET ELEVATIONS
PATIO
SUPERVISION
BED RM #1
12'-0" x 16'-4"
VANITY
BATH
BED RM #2
12'-0" x 12'-0"
CLO
W.I. CLO.
GARDEN STORAGE
LAV.
WASH DRY
LAUNDRY
DW
SINK
RANGE
SLIDING GLASS DRS
FIREPLACE
FAMILY RM
12'-0" x 18'-0"
LINEN
KITCHEN
11'-0" x 17'-2"
REF
BATH
HALL
BED RM #3
11'-0" x 12'-0"
44'-0"
TWO CAR GARAGE
20'-0" x 21'-0"
PANTRY
DINING RM
12'-0" x 12'-0"
COAT
STORAGE
DINETTE
7'-2" x 9'-4"
FOYER
FIREPLACE
LIVING RM
13'-0" x 21'-0"
COVER ENTRY
75'-0"
FLOOR PLAN
HOUSE DATA
BASIC HOUSE.....1802 SQ. FT.
LAUND. & LAV.......142 SQ. FT.
GARAGE & STOR....554 SQ. FT.
Z-94

# Women Will Love This Large Kitchen

Few women could resist the kitchen designed for this center-hall ranch. It has everything the ladies have asked for in various nationwide housing surveys, including generous counter and cabinet space, a pantry, an adjacent laundry room and a truly adequate informal dining area. The kitchen is 24′4″ long, more than just a little unusual these days. The large and separate laundry room has a closet, a window, a lavatory with its own window, and two rear doors.

Immediately visible on entering the front foyer is a dramatic fireplace, located in the corner of the family room. On one side of the foyer is a large living room, also with a fireplace as the focal point. On the other is the dining room. Sliding glass doors in the family room lead to a rear patio. Three bedrooms are in the right wing of the house. The main bedroom has three closets, one a walk-in, and a bathroom with a stall shower. The hall bath is sectionalized, with a luminous ceiling and a mechanically ventilated exhaust.

Note the two large storage areas in the garage. The storage space at the rear is excellent for garden tools and includes an entrance door for power equipment.

The exterior design is traditional, with brick, wood shingles and double-hung, multi-paned windows. The roof lines accent the horizontal. Using a minimum of colors on the exterior will make the house appear larger.

## Material List

|  | |  |
|---|---|---|
| **CONCRETE WORK** | | |
| Concrete Walls | | 1068 cu. ft. |
| Slabs | | 937 cu. ft. |
| Foundation Footings | | 251 cu. ft. |
| Misc. Concrete | | 286 cu. ft. |
| **STRUCTURAL STEEL** | | |
| Lally Columns | 3½″ diam. | 7 pieces |
| Girder | 7″ I | 52 lin. ft. |
| **BRICK WORK** | | |
| Chimney | Brick | 240 cu. ft. |
| Flue Lining | T. C. | 44 lin. ft. |
| Veneer | Brick | 348 sq. ft. |
| **CARPENTRY** | | |
| Framing Lumber | | 9876 B.F. |
| Studs | | 3467 B.F. |
| Plates | | 1067 B.F. |
| Roof Sheathing | | 3488 sq. ft. |
| Sub Flooring | | 1840 sq. ft. |
| Side Wall Sheathing | | 2286 sq. ft. |
| Insulation Walls | | 1380 sq. ft. |
| Insulation Ceilings | | 2018 sq. ft. |
| Wood Flooring | | 1550 sq. ft. |
| Kitchen Underlayment | | 121 sq. ft. |
| **MILLWORK** | | |
| Exterior Doors & Frames Compl. | | 3 units |
| Interior Doors & Frames Compl. | | 14 units |
| Sliding Doors | | 1 unit |
| Bi Fold Doors | | 6 units |
| Fascia | | 280 lin. ft. |
| Base | | 680 lin. ft. |
| Windows | | 18 units |
| **KITCHEN CABINETS** | | |
| Base Cabinets | | 14 lin. ft. |
| Wall Cabinets | | 20 lin. ft. |
| **ROOFING** | | |
| Shingles | 235# asphalt | 3488 sq. ft. |
| Roofing Paper | 15# felt | 3488 sq. ft. |

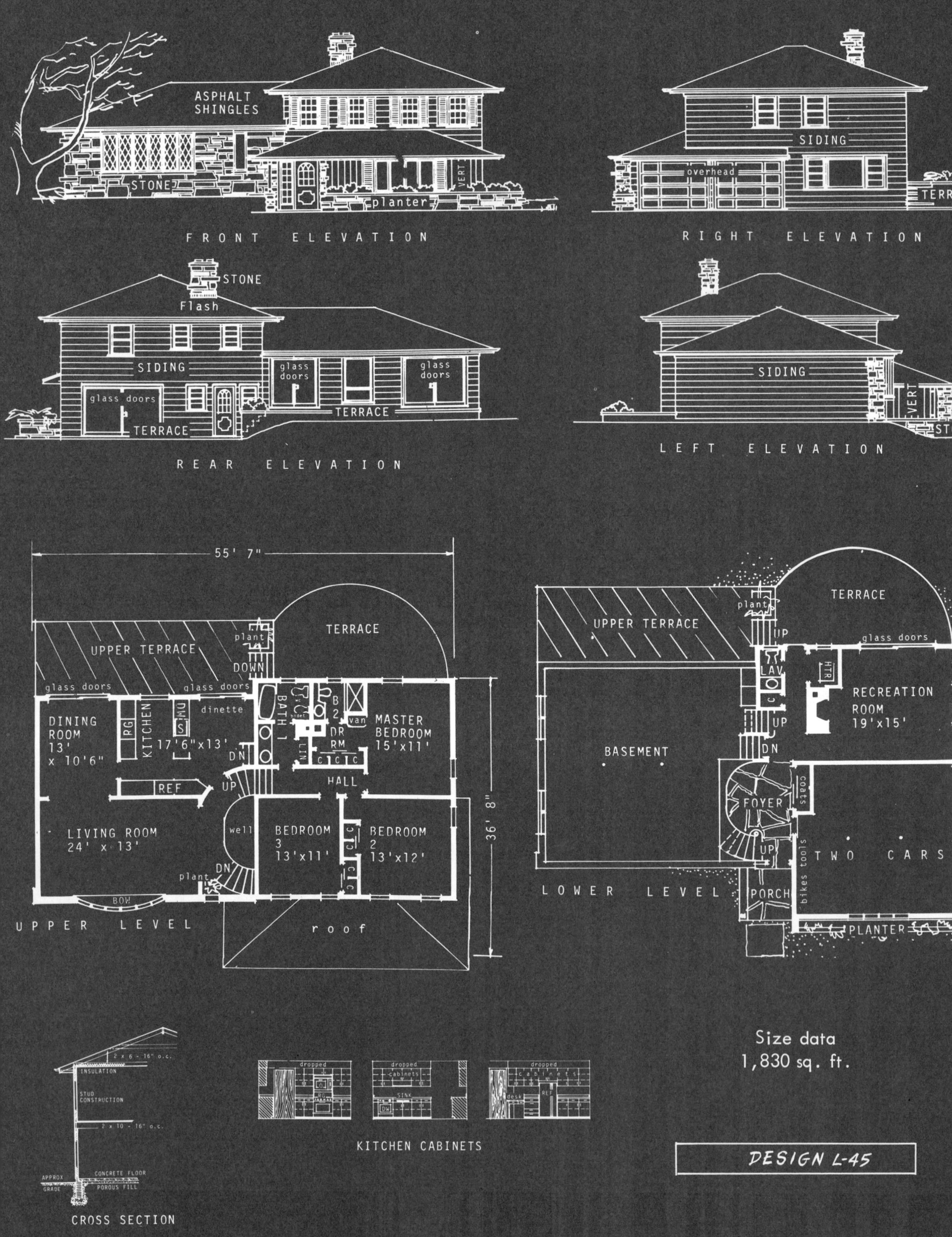


Size data
1,830 sq. ft.

DESIGN L-45

# Split-Level with Accent on Outdoors

Indoor-outdoor living gets special emphasis in this interesting split-level.

A spacious upper terrace on the main level is accessible through sliding glass doors from the dining room and the dinette section of the sizable kitchen. A ground level terrace off the recreation room can be reached through sliding glass doors, with the two terraces connected by steps.

The attractiveness of the rear elevation is matched by the charm of the front, with hip roofs on all levels, enhanced by stone, vertical and horizontal siding, planters and decorative doors and windows.

The interior plan is roomy and convenient with easy stairways between the areas. Of special interest is the repetition of curves in the foyer, upper hall and terrace.

On the lower level, a covered entry porch and interior access from the garage are provided near a large coat closet. On the same level are the access stairs to the basement area or to the ground-level recreation room.

The living room is a well-proportioned room with a wide, diamond-paned bow window. The kitchen is over 17′ wide, dividing neatly for family meals, formal ones or outdoor serving.

Up the last staircase curve are three bedrooms and two bathrooms. The master bedroom unit includes dressing room, stall-shower bathroom, closet area and a built-in vanity.

## Material List

|  |  |
|---|---|
| **CONCRETE WORK** | |
| Footings | 15 cu. yds. |
| Floors (basement & platforms) | 27 cu. yds. |
| **MASONRY** | |
| 4 x 8 x 18 CB | 185 lin. ft. |
| 8 x 8 x 18 CB | 325 sq. ft. |
| 10 x 8 x 18 CB | 755 sq. ft. |
| 12 x 8 x 18 CB | 335 sq. ft. |
| Chimney fill | 130 cu. ft. |
| Flagstone | 165 sq. ft. |
| Stone Veneer | 380 sq. ft. |
| **CARPENTRY** | |
| 4″ Lally Columns | 4 pieces |
| Framing Lumber | 7891 B.F. |
| Stud & Plates | 4113 B.F. |
| Roof Sheathing 1 x 6 | 3280 sq. ft. |
| Plywood | 2700 sq. ft. |
| Fascia #1 pine 1 x 6 | 380 lin. ft. |
| Soffit 3/8″ x 2′ plywood | 380 lin. ft. |
| Shingle Mould | 350 lin. ft. |
| Wall Sheathing 1 x 6 | 3130 sq. ft. |
| Plywood | 2580 sq. ft. |
| Siding | 2200 sq. ft. |
| Roofing 210# asphalt | 27 squares |
| Insulation — Walls | 2150 sq. ft. |
| Ceiling | 2040 sq. ft. |
| Finish Wood Flooring | 1665 sq. ft. |
| **DRY WALL** | |
| Ceilings 3/8″ | 2450 sq. ft. |
| Walls 1/2″ | 6080 sq. ft. |
| **MILLWORK** | |
| Base | 720 lin. ft. |
| Cove or Bed Moulding | 720 lin. ft. |
| Windows | 18 pieces |
| Exterior Doors | 7 pieces |
| Interior Doors | 18 pieces |
| Louvers | 16 pieces |

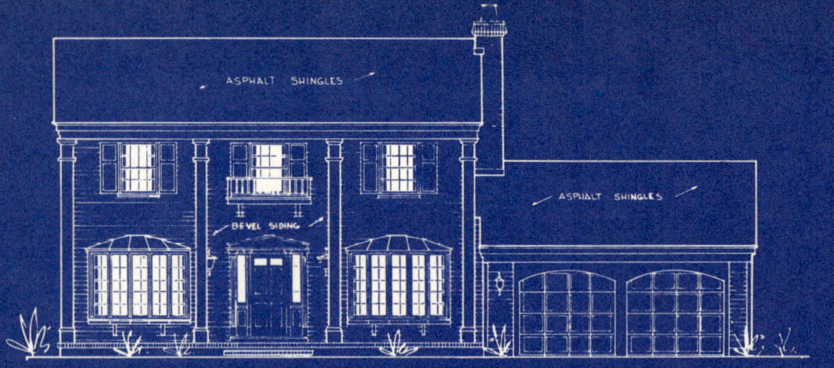


FRONT ELEVATION

RIGHT SIDE ELEVATION

REAR ELEVATION

LEFT SIDE ELEVATION

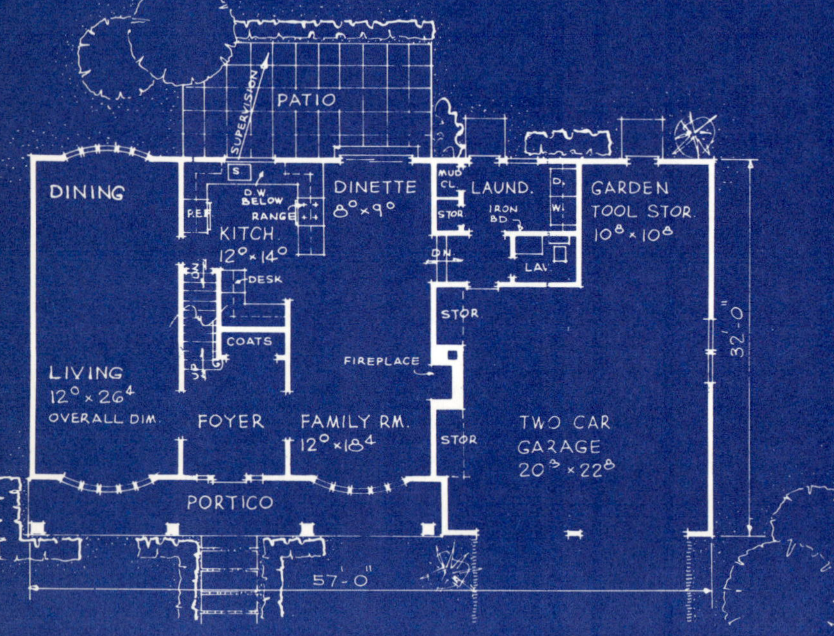


FIRST FLOOR PLAN

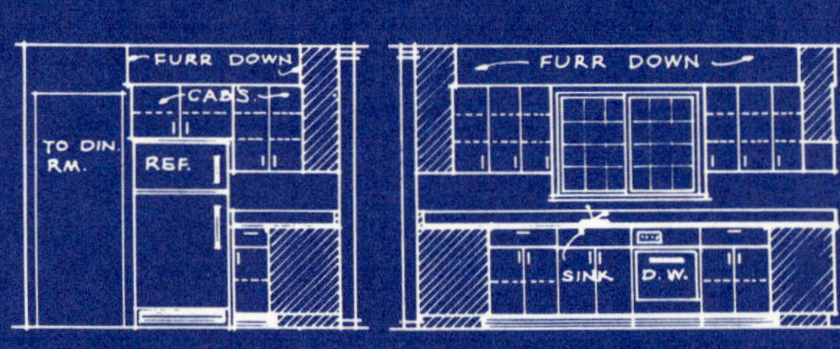


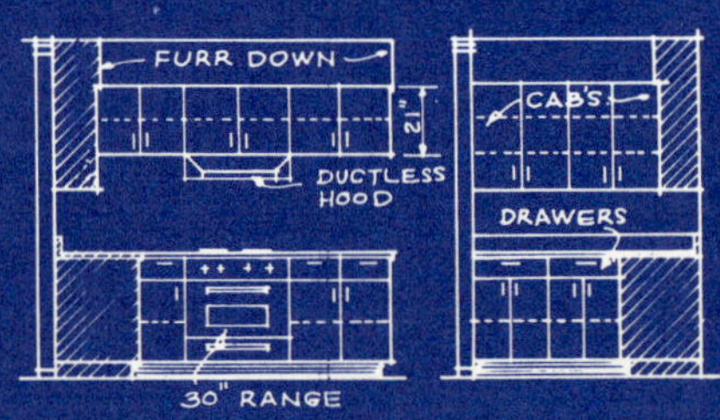

KITCHEN CABINET ELEVATIONS

BED RM #1 $12^0 \times 16^0$
SPLIT BATH
BED RM #3 $10^0 \times 11^4$
HALL
LIN
DN.
STOR
CLO
CLO
27'-0"
CLO
WALK IN CLO
CLO
BOOK SHELVES
BATH
BED RM #4 $9^0 \times 11^0$
BED RM #2 $12^4 \times 12^4$
34'-0"

SECOND FLOOR PLAN

HOUSE DATA

| | | |
|---|---|---|
| FIRST FLOOR | 918 | SQ. FT. |
| SECOND FLOOR | 918 | SQ. FT. |
| LAUND. & LAV. | 132 | SQ. FT. |
| GARAGE & STOR. | 598 | SQ. FT. |

Design S-36

# Elegance in Two-Story Southern Colonial

There are few houses these days which can boast of three bow windows. This one has just that. These symbols of Early American elegance have been added to a house which already possesses many details ordinarily found only in those of the luxury class.

For example, the front elevation (on which two of the bow windows are located) is framed by a two-story-high Southern Colonial portico with four well-proportioned wood posts arranged symmetrically about the center hall entrance.

The floor plan carries out the traditional layout of typical center hall design, with one notable exception. The family room is in the area normally reserved for the dining room. Accessible from the foyer, the family room is open-planned to the kitchen and dinette area, which in a sense is a throwback to the days of the huge family kitchen.

Another departure is the oversized foyer, designed in keeping with the desire for an impressive entrance hall. The fireplace is located to be seen from the foyer.

On the second floor, four bedrooms and two bathrooms, arranged in a rectangle to minimize costs, surround the center upstairs hall. Care has been taken to provide ample wall space for beds and furniture. The hall bathroom is in a split plan with two lavatories. The master bedroom has a private bath with a stall shower.

## Material List

| CONCRETE WORK | | |
|---|---|---|
| Foundation Walls | | 968 cu. ft. |
| Foundation Footing | | 222 cu. ft. |
| Slabs | | 654 cu. ft. |
| Misc. Concrete | | 173 cu. ft. |
| **BRICK WORK** | | |
| Chimney | Brick | 240 cu. ft. |
| Flue Lining | T.C. | 56 lin. ft. |
| **STRUCTURAL STEEL** | | |
| Lally Columns | 3½" diam. | 6 ea. |
| Girder | 7" I Beam | 40 lin. ft. |

| CARPENTRY | |
|---|---|
| Framing Lumber | 8997 B.F. |
| Studs | 3680 B.F. |
| Plates | 1100 B.F. |
| Roof Sheathing | 2228 sq. ft. |
| Sub Flooring | 1836 sq. ft. |
| Side Wall Sheathing | 2372 sq. ft. |
| Insulation Walls | 1680 sq. ft. |
| Insulation Ceilings | 918 sq. ft. |
| Wood Flooring | 1607 sq. ft. |
| Kitchen Underlayment | 140 sq. ft. |
| **MILLWORK** | |
| Ext. Doors & Frames Complete | 3 units |
| Garage Door Complete Set | 2 units |
| Int. Doors & Frames Complete | 16 units |
| Sliding Glass Doors | 1 unit |
| Bi Fold Doors | 6 units |
| Window Units | 18 units |
| Fascia | 120 lin. ft. |
| Base | 700 lin ft. |
| Stairs 12 risers | 1 set |
| Stairs 13 risers | 1 set |
| Sliding Doors Int. | 2 ea. |
| **ROOFING** | |
| Shingles | 2228 sq. ft. |
| Roofing Paper | 2228 sq. ft. |

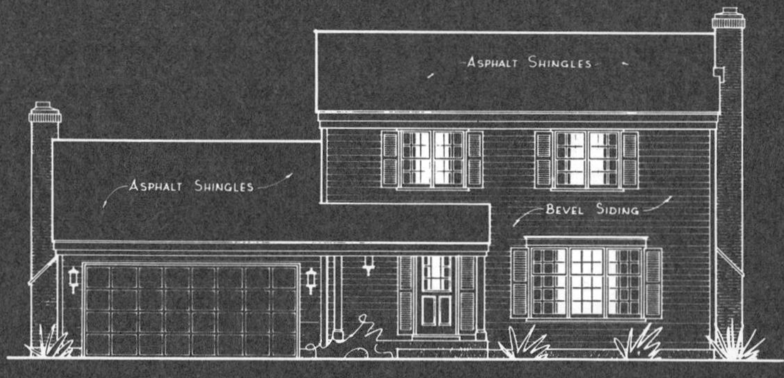


Front Elevation

Right Side Elevation

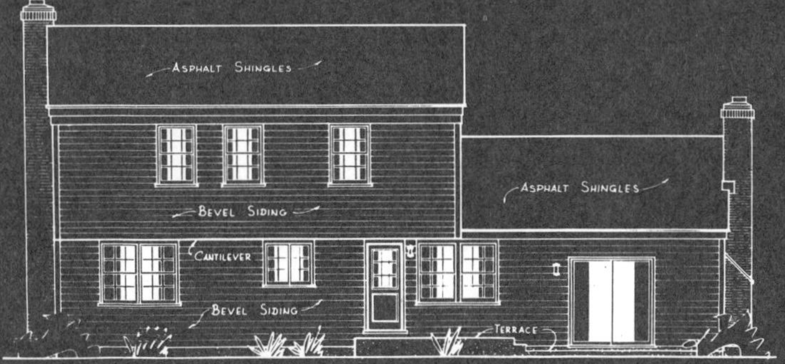


Rear Elevation

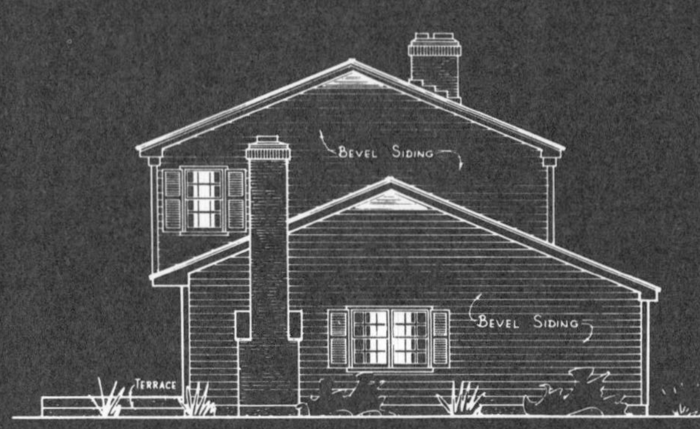


Left Side Elevation

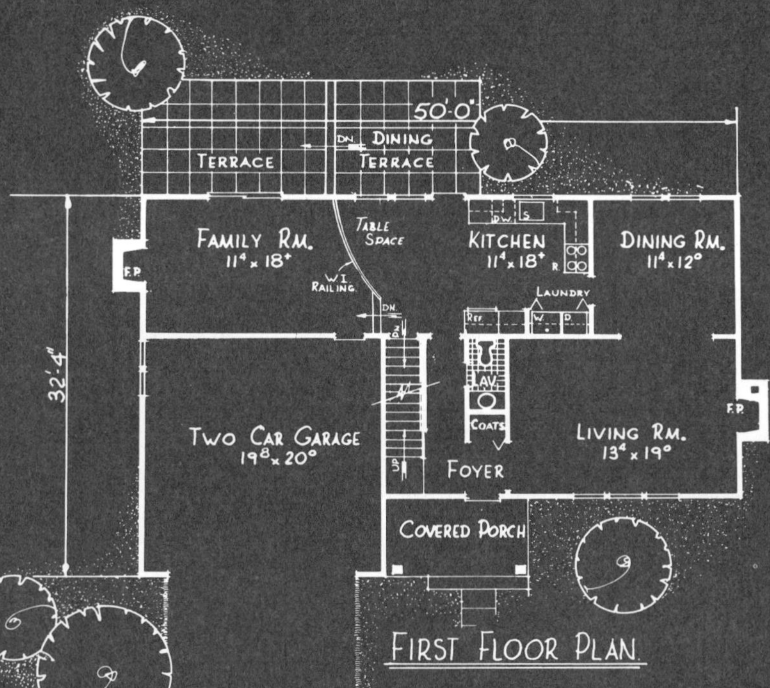


First Floor Plan

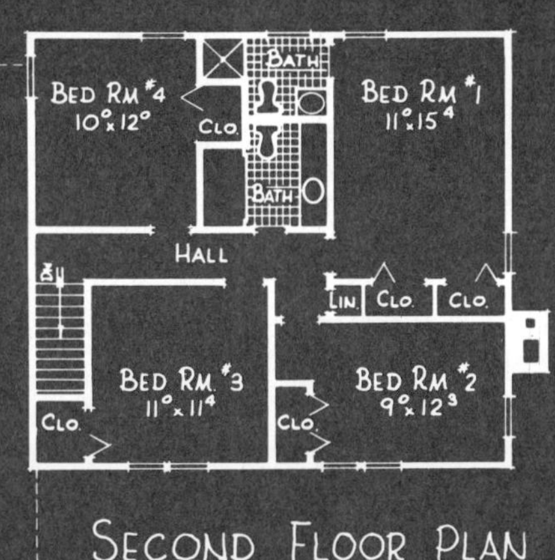


Second Floor Plan

## HOUSE DATA

FIRST FLOOR ..... 1011 SQ. FT.
SECOND FLOOR ..... 831 SQ. FT.
GARAGE ............ 406 SQ. FT.

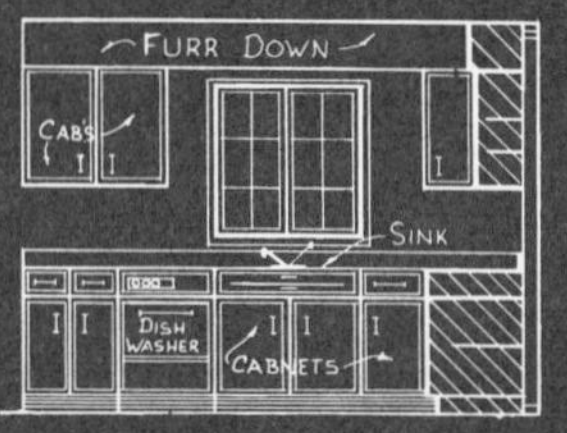


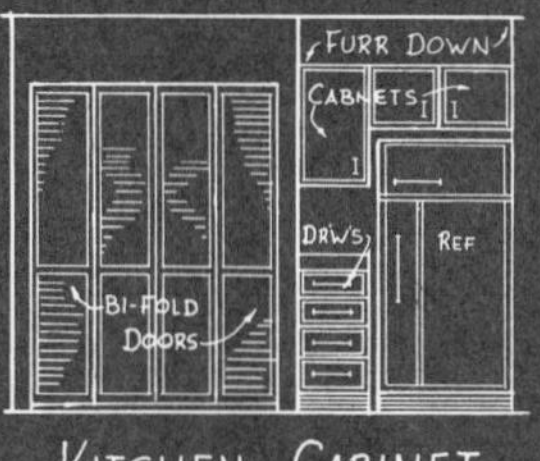


Kitchen Cabinet Elevations

DESIGN L-34

# Large Informal Area in Traditional Two-Story

This two-story house has a conventional four-bedroom, two-bath layout upstairs, but is distinguished by a first-floor plan with a large and distinctive informal area.

The section devoted to informal living is at the rear, the kitchen and family room combining in an open plan more than 36′ long. The two spaces are divided only by a curved, attractive railing which creates, on the kitchen side, a balcony for a dinette table. Direct access to rear terraces is possible from both rooms. At the far end of it all is a fireplace centered on the outside wall and a focal point whether you are in the kitchen or family room.

The balance of the first floor comprises a living room, dining room, foyer and lavatory, with a two-car garage tucked into the structure on the left side and in front of the family room.

For formal entertaining, the living room and dining room are in a convenient L shape at the right side of the house. The kitchen is a central control area between the dining room and the family room.

The four bedrooms upstairs are of medium size, designed for a family with young children.

In general proportion and detail, the house is traditional, with a simplicity reminiscent of that found in New England.

With an overall length of only 50′, the house will fit on a lot with moderate frontage.

## Material List

**CONCRETE WORK**

|  | |  |
|---|---|---|
| Concrete Walls | | 990 cu. ft. |
| Foundation Footings | | 165 cu. ft. |
| Slabs | | 497 cu. ft. |
| Misc. Concrete | | 238 cu. ft. |

**STRUCTURAL STEEL**

|  | |  |
|---|---|---|
| Lally Columns | 3½″ diam. | 5 pieces |
| Girders | 6″ I 12.5# | 22 lin. ft. |
| Girder | 10 B 19# | 17 lin. ft. |

**BRICK WORK**

|  | |  |
|---|---|---|
| Chimney | | 245 cu. ft. |
| Flue Lining | T. C. | 66 lin. ft. |

**CARPENTRY**

|  |  |
|---|---|
| Framing Lumber | 5999 B.F. |
| Studs | 3573 B.F. |
| Plates | 1067 B.F. |
| Roof Sheathing | 1820 sq. ft. |
| Sub Flooring | 1601 sq. ft. |
| Side Wall Sheathing | 2275 sq. ft. |
| Wall Insulation | 1750 sq. ft. |
| Ceiling Insulation | 1063 sq. ft. |
| Wood Flooring | 1512 sq. ft. |
| Kitchen Plywood | 197 sq. ft. |

**MILLWORK**

|  |  |
|---|---|
| Exterior Doors & Frames Compl. | 2 units |
| Sliding Glass Door Unit | 1 unit |
| Interior Doors & Frames Compl. | 10 units |
| Bi Fold Doors | 7 units |
| Window Units | 17 units |
| Fascia & Rake | 240 lin. ft. |
| Base | 630 lin. ft. |

**KITCHEN CABINETS**

|  |  |
|---|---|
| Base Cabinets | 7′10″ long |
| Wall Cabinets | 13′6″ long |

**ROOFING**

|  | |  |
|---|---|---|
| Shingles | 235# asphalt | 1820 sq. ft. |
| Roofing Paper | 15# felt | 1820 sq. ft. |

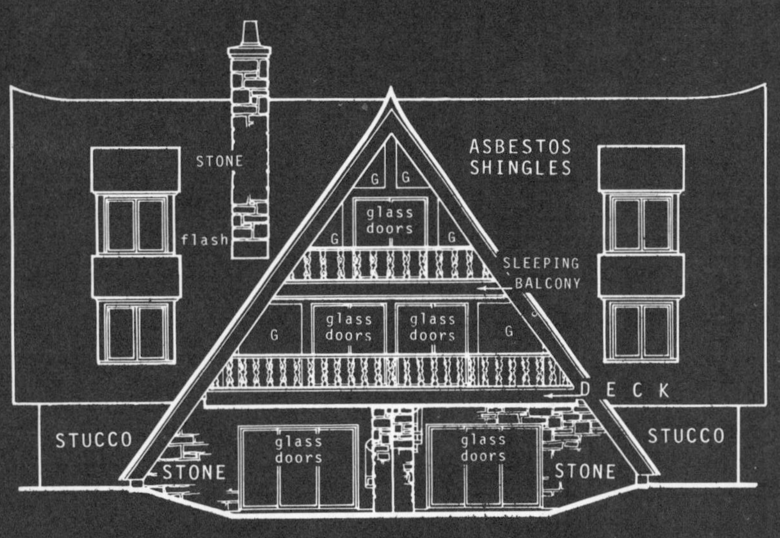


FRONT ELEVATION

SLEEPING BALCONY
DECK
STONE
PRE-FAB STAIRS
vent
REDWOOD SIDING
STUCCO

RIGHT ELEVATION

REAR ELEVATION

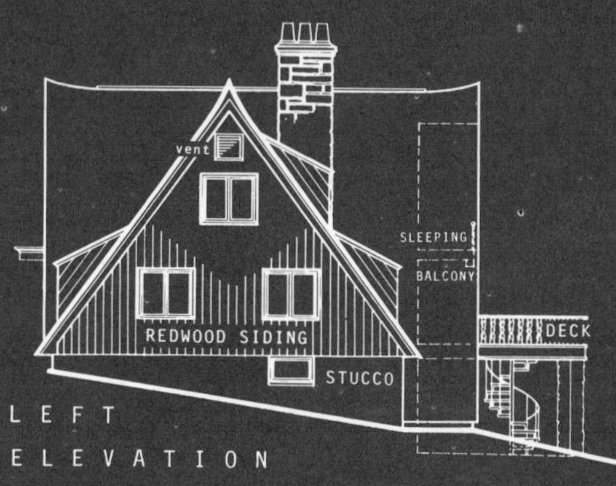


LEFT ELEVATION

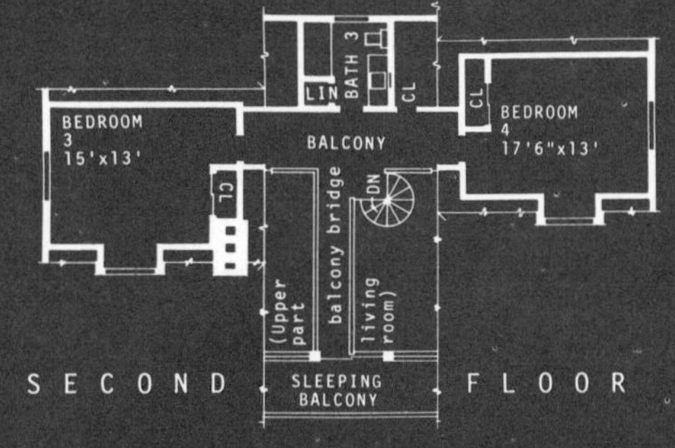


SECOND FLOOR

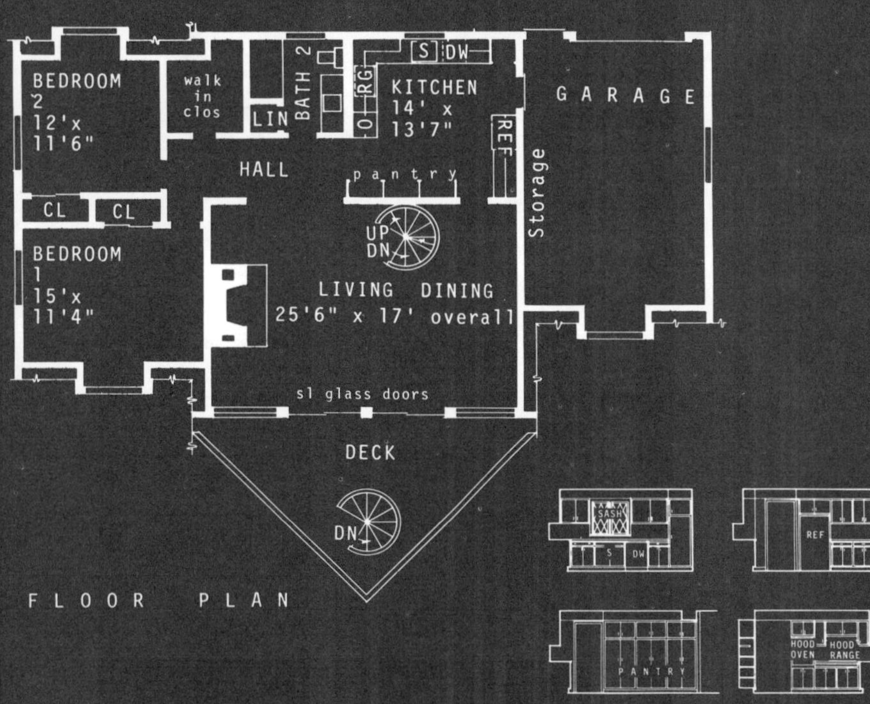


FLOOR PLAN

KITCHEN CABINET DETAILS

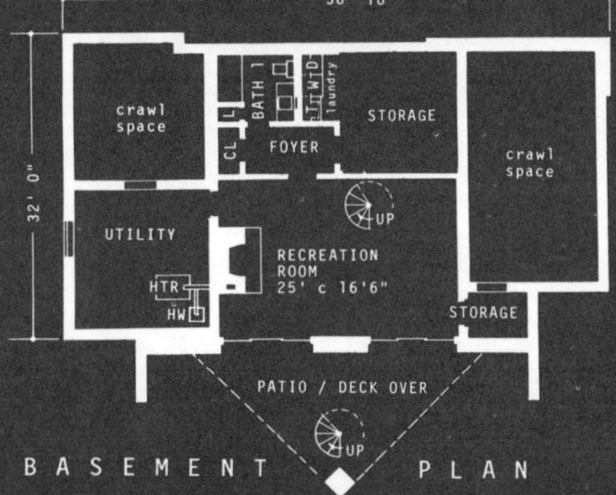


BASEMENT PLAN

**DESIGN L-41**

Size data
1,867 sq. ft.

# Exciting Combination A-Frame and Chalet

This exciting design combines the most striking features of the A-frame and the chalet in a house that is not a second or vacation home.

The front and rear elevations of Design L-41 have high A-peak additions and dormers add a most interesting contrast. Chalet balconies at the front provide sleeping on the upper floor, a deck on the main floor and a patio under the deck with a circular stairway connecting deck and patio.

For a lot that has a natural front slope, the basement level has a big recreation room opening out to the patio. On a flat lot this would be a basement recreation room, missing only the walk-out patio.

The main level incorporates a combined living-dining room, a big efficient kitchen, two bedrooms and a full bathroom, and excellent closet and storage area. The living room rises to the level above and is punctuated by the balcony bridge there, with the circular stair tying it all together. Off the back is a bedroom hall, insuring quiet and privacy for two dormered bedrooms that have fine exposures and share a closet wall.

The two bedrooms, third bath, closets and the sleeping balcony are fitted in to make a real upper floor. This can be completed at once or left for future finishing.

## Material List

**CONCRETE**

| | |
|---|---|
| Foundation Footings | 363 cu. ft. |
| Miscellaneous Footings | 44 cu. ft. |
| Slabs | 618 cu. ft. |

**STRUCTURAL STEEL**

| | |
|---|---|
| Columns | 8 pcs |
| Girders | 134 lin. ft. |
| Spiral Stairs | 2 units |

**BRICKWORK**

| | |
|---|---|
| Chimney and Fireplace | 590 cu. ft. |
| Block Walls ... 4" - 8" - 12" | 2274 S.F. |
| Stone Veneer | 260 S.F. |
| Flue Lining | 84 lin. ft. |

**CARPENTRY**

| | |
|---|---|
| Framing Lumber | 21,480 B.F. |
| Studs | 3200 B.F. |
| Plates | 1800 B.F. |
| Roof Sheathing | 5000 S.F. |
| Sub Flooring | 2350 S.F. |
| Side Wall Sheathing | 1800 S.F. |
| Insulation | 6800 S.F. |
| Wood Flooring | 1900 S.F. |
| Kitchen Plywood | 2000 S.F. |

**MILLWORK**

| | |
|---|---|
| Ext. Doors & Frames Complete | 1 unit |
| Int. Doors & Frames Complete | 17 pcs |
| Sliding Glass | 2 units |
| Sliding Doors | 8 pcs |
| Windows | 16 units |

**KITCHEN CABINETS**

| | |
|---|---|
| Base Cabinets | 13.50 lin. ft. |
| Wall Cabinets | 21.50 lin. ft. |

**ROOFING**

| | | |
|---|---|---|
| Shingles | Asbestos | 5000 S.F. |
| Roofing Paper | 15# felt | 5000 S.F. |

FRONT ELEVATION

LEFT SIDE ELEVATION

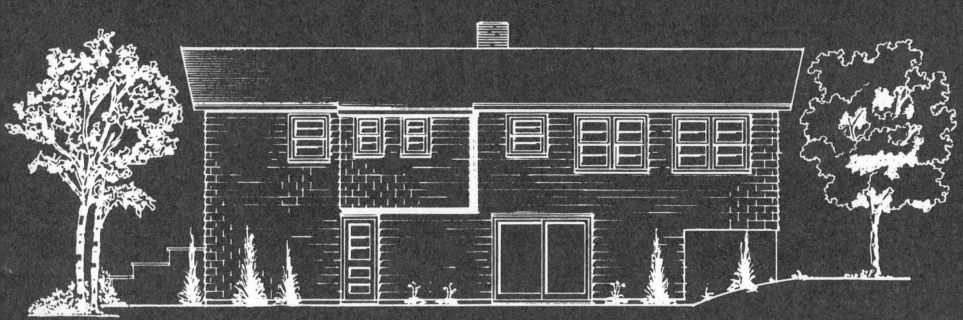

REAR ELEVATION

RIGHT SIDE ELEVATION

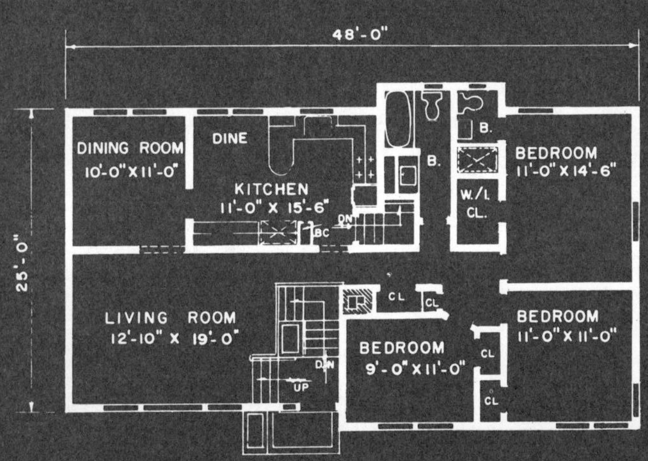


UPPER LEVEL

LIVING RM.
BED RM.
CL
BED RM.
SPARE RM.
LAV.
HTR.
GARAGE

LONGITUDINAL SECTION

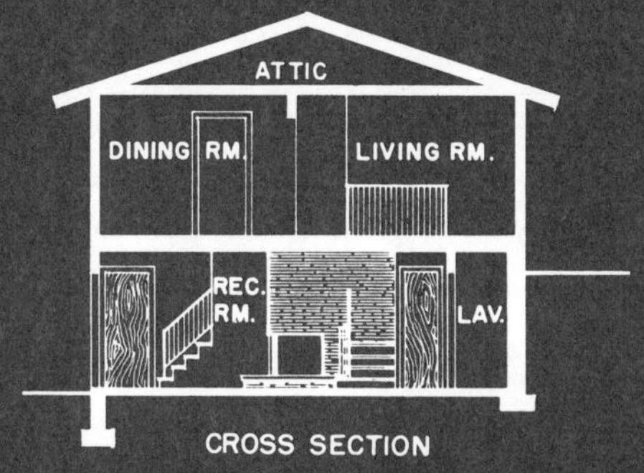


CROSS SECTION

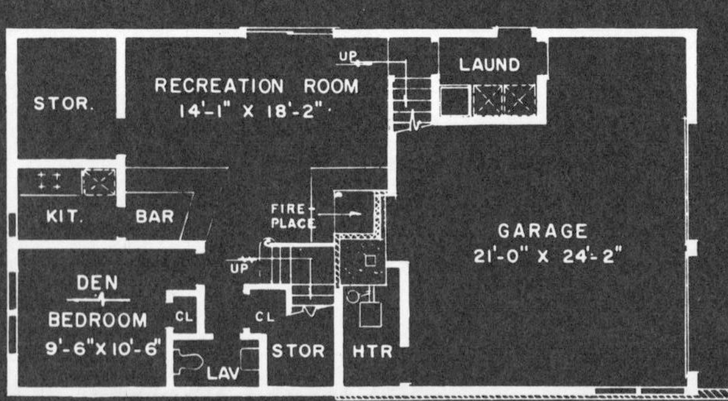


LOWER LEVEL

## AREA STATISTICS:

UPPER LEVEL . . . 1266 SQ. FT.
LOWER LEVEL . . . . 625 SQ. FT.
GARAGE . . . . . . . . . . 575 SQ. FT.

DESIGN L-15

# Modest Bi-Level Puts Emphasis On Value

A family that does not own a large lot, yet requires at least three bedrooms and a recreation area, will find this raised ranch ideal for its purposes.

Maximum living efficiency has been obtained in a design that combines hand-split wood shingles, vertical boards and battens, brick veneer and a gabled roof with outward sweeping angled surfaces.

With only 1266 well-utilized square feet on the upper living level and dimensions of 48′ by 26′8″, a large building lot is not needed.

Four steps up and to the left of the entrance is the formal living room, highlighted by the large picture window and the balconied effect created by the open stair, a dining room, kitchen, three bedrooms and two bathrooms. The view down from the foyer leads to the wood-paneled recreation room with corner fireplace and to the sliding glass doors that open to the patio. On this level are a den (or fourth bedroom); a laundry, heater room, lavatory, kitchenette, storage section and a two-car garage.

The bedrooms are located for maximum quiet and privacy. The master bedroom has two exposures, good wall space, a walk-in closet and a separate bathroom.

A mullion window in the dinette section and the window over the kitchen sink assure plenty of light and air as well as a clear view of the backyard.

## Material List

**CONCRETE WORK**

Foundations and footings ....... 8 cu. yds.
Floors, basement and porch 4″ slab ....... 15 cu. yds.

**STEEL**

Lally columns 4″ diameter ...... 4 pieces

**MASONRY**

Brick veneer ....... 350 sq. ft.
Concrete Block 8″ block ...... 400 blocks

**FRAMING LUMBER**

Sills, joists, rafters, studs, plates, etc. ....... 8300 B.F.M.

**EXTERIOR SHEATHING**

½″ Plyscore ext. wall sheathing . 2218 sq. ft.
⅝″ Plyscore roof sheathing ..... 1656 sq. ft.

**SUBFLOORING**

⅝″ Plyscore ....... 1190 sq. ft.
15#Saturated felt ....... 4 rolls

**DOOR SCHEDULE**

1 Sliding glass door units 8-0 x 6-8
1 Flush interior birch-veneer
2 Paneled & Glazed exterior hardwood 1¾″ x 6-8
1 2-6 x 6-8 S.C.F.P. door
1 2-4 x 6-8 Louvered door

**WINDOW SCHEDULE**

7 3-0 x 3-6 D.H.
1 3-0 x 3-2 D.H.
2 2-8 x 4-2 D.H.
1 1-8 x 1-8 Awning
2 2-8 x 1-8 Awning
2 2-0 x 3-2 D.H.
2 3-0 x 4-2 D.H.
2 3-0 x 2-0 Fixed
Picture Window Unit
5-0 x 5-6 Fixed
2-6 x 5-6 Flankers

**EXTERIOR FINISHES**

Exterior grade plywood ⅜″ thick. 50 sq. ft.
Low Pitch Asphalt Shingles ..... 17 squares
Vertical siding ....... 260 sq. ft.
Wood Shingles ....... 19 squares

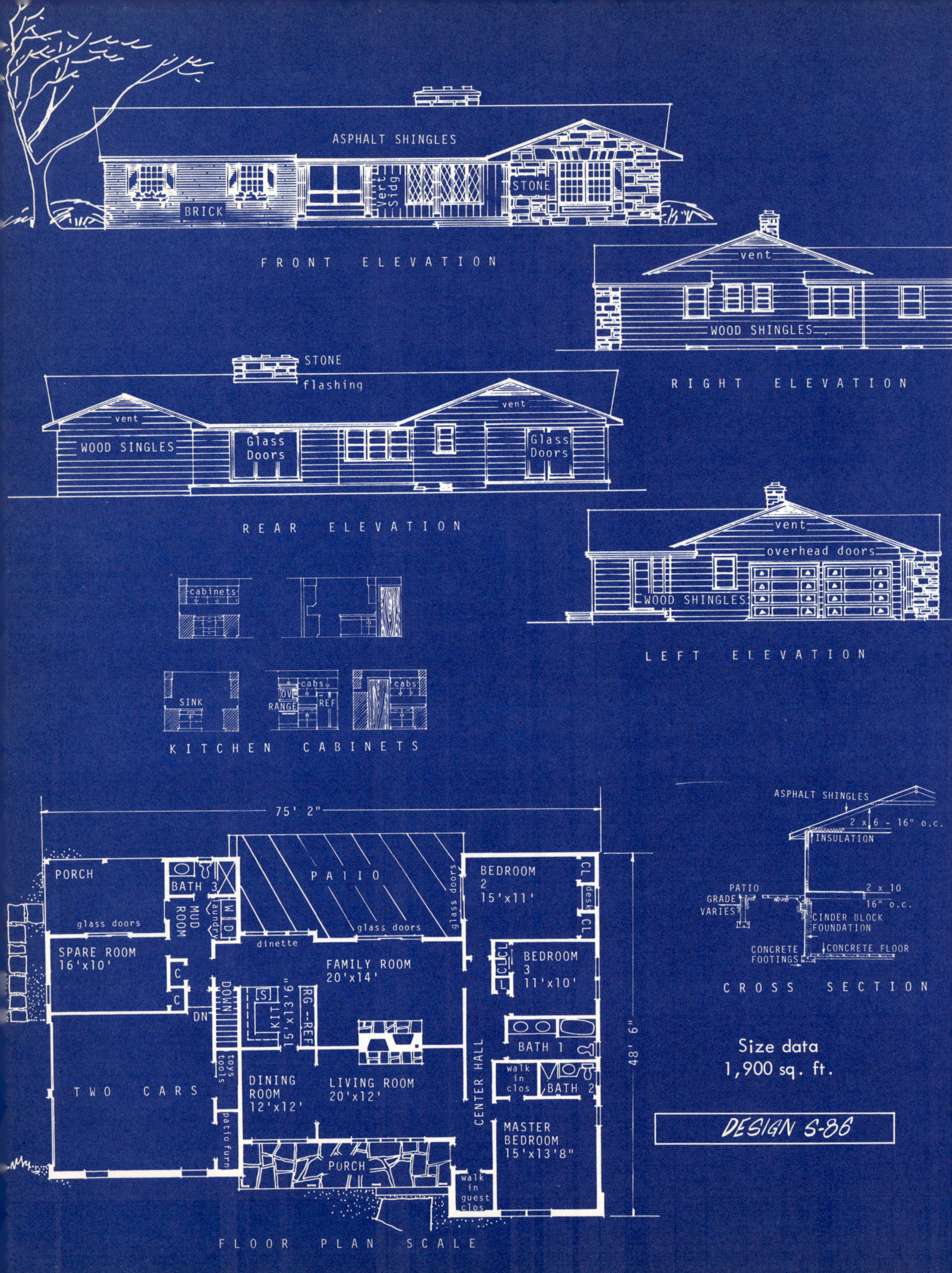
ASPHALT SHINGLES
BRICK
Vert Sidg
STONE
FRONT ELEVATION
vent
WOOD SHINGLES
RIGHT ELEVATION
STONE
flashing
vent
WOOD SINGLES
Glass Doors
Glass Doors
REAR ELEVATION
vent
overhead doors
WOOD SHINGLES
LEFT ELEVATION
cabinets
SINK
cabs
OV
RANGE
REF
cabs
KITCHEN CABINETS
75' 2"
PORCH
BATH 3
MUD ROOM
laundry
W D
PATIO
glass doors
glass doors
glass doors
BEDROOM 2 15'x11'
CL
desk
CL
SPARE ROOM 16'x10'
dinette
FAMILY ROOM 20'x14'
BEDROOM 3 11'x10'
C
C
DN
DOWN
KIT 15'x13'6"
RG--REF
BATH 1
48' 6"
TWO CARS
toys tools
DINING ROOM 12'x12'
LIVING ROOM 20'x12'
walk in clos
BATH 2
CENTER HALL
patio furn
MASTER BEDROOM 15'x13'8"
PORCH
walk in guest clos
FLOOR PLAN SCALE
ASPHALT SHINGLES
2 x 6 - 16" o.c.
INSULATION
PATIO
GRADE VARIES
2 x 10 16" o.c.
CINDER BLOCK FOUNDATION
CONCRETE FOOTINGS
CONCRETE FLOOR
CROSS SECTION
Size data
1,900 sq. ft.
DESIGN S-86

# H-Ranch with Bonus Area Behind Garage

Problems in today's family living might be solved by the area behind the garage in this H-shaped ranch.

While a three-bedroom plan may suit mother, father and the children, the need for a guest room often arises. An even more common need is a good, private area for an older member of the family — who should feel at home in the household yet have a place set away from the other rooms.

The spare room here answers either of these requirements in an efficient way. There's a stall-shower bathroom off a hall area with access from the spare room, permitting use without going through any part of the basic house. The porch next to the spare room could be built as a living room, giving a separate apartment for family use or as an income unit, adding a kitchenette if desired.

An entrance porch offers a degree of privacy. The living and dining rooms combine at the front in a 32′ expanse, the rooms defined by an arch, and the living room has diamond-paned windows. The spacious family room is conveniently located off the kitchen. There is direct access from the hall.

The three-bedroom wing is set off by a center hall. The back bedroom enjoys a patio view and access to it as well. Bedroom three isn't large, but it can offer privacy and comfort. At the front, the master bedroom has a stall-shower bathroom and walk-in closet.

## Material List

**CONCRETE WORK**

| | |
|---|---|
| Footings | 20 cu. ft. |
| Floors (Basement & platforms) | 51 cu. ft. |

**MASONRY**

| | |
|---|---|
| 4 x 8 x 18 CB | 335 lin. ft. |
| 8 x 8 x 18 CB | 330 sq. ft. |
| 10 x 8 x 18 CB | 1610 sq. ft. |
| 12 x 8 x 18 CB | 285 sq. ft. |
| Brick veneer — chimney fill | 210 sq. ft. |
| | 485 cu. ft. |
| Flagstone — stone veneer | 200 sq. ft. |
| | 295 sq. ft. |

**CARPENTRY**

| | |
|---|---|
| 4″ Lally columns | 8 pcs. |
| Framing lumber | 11,586 B.F. |
| Stud & plates | 4280 B.F. |
| Roof sheathing 1 x 6 | 5000 sq. ft. |
| Plywood | 4100 sq. ft. |
| Fascia #1 pine, 1 x 6 | 150 lin. ft. |
| Soffit ⅜″ x 2′ plywood | 150 lin. ft. |
| Shingle mould | 150 lin. ft. |
| Sheathing 1 x 6 | 4055 sq. ft. |
| Plywood | 3345 sq. ft. |
| Siding | 2910 sq. ft. |
| Roofing 210# asphalt | 40 squares |

| | |
|---|---|
| Insulation — walls | 2375 sq. ft. |
| ceiling | 2330 sq. ft. |
| floor | 350 sq. ft. |
| Finish wood flooring | 2310 sq. ft. |

**DRY WALL**

| | |
|---|---|
| Ceilings ⅜″ | 2840 sq. ft. |
| Walls ½″ | 5850 sq. ft. |

**MILLWORK**

| | |
|---|---|
| Base | 540 lin. ft. |
| Windows | 17 pcs. |
| Exterior doors | 7 pcs. |
| Interior doors | 23 pcs. |
| Louvers | 4 pcs. |

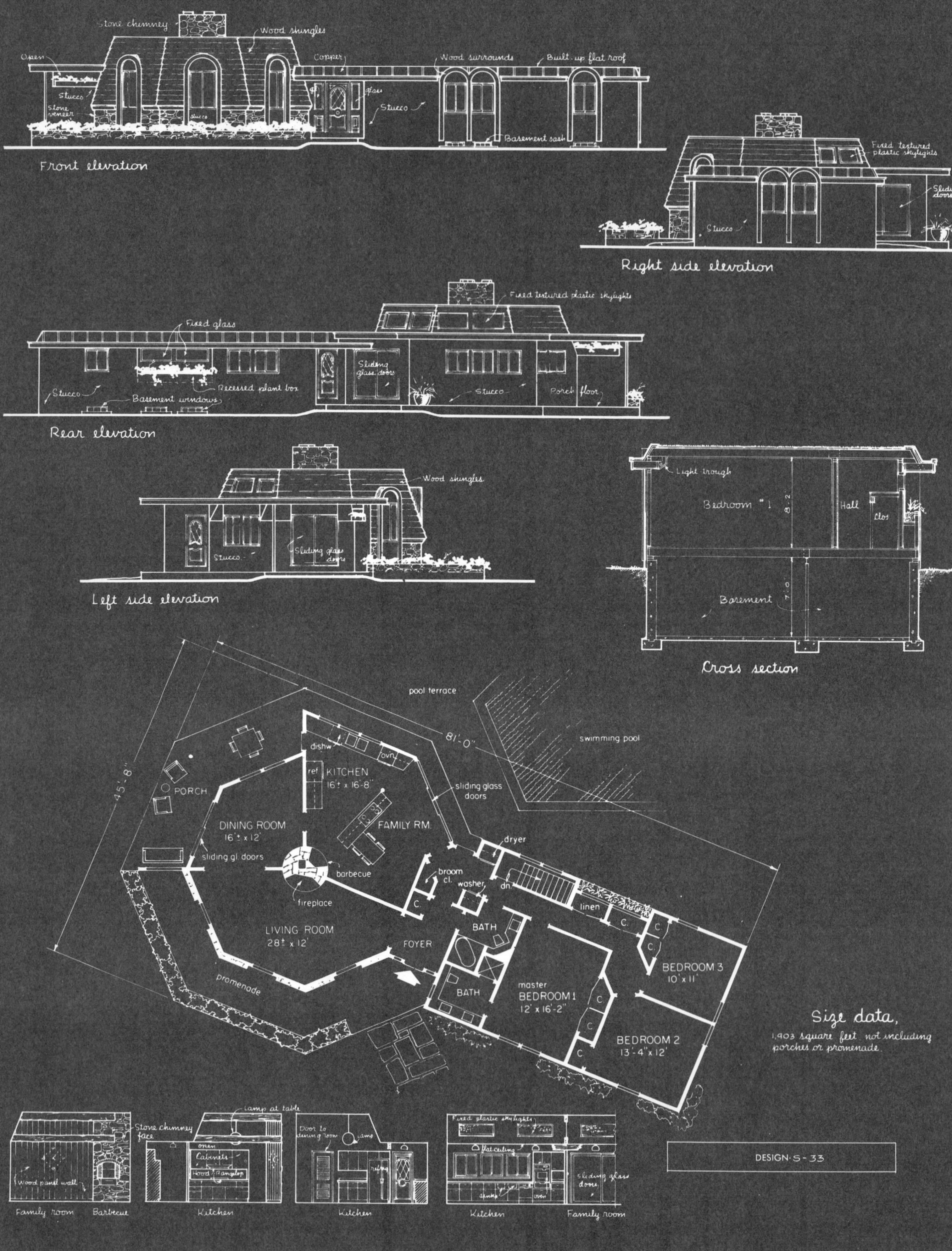

Stone chimney
Wood shingles
Open
Copper
Wood surrounds
Built-up flat roof
Stucco
Stone veneer
Stucco
Glass
Stucco
Basement sash
Front elevation
Fixed textured plastic skylights
Slidin doors
Stucco
Right side elevation
Fixed textured plastic skylights
Fixed glass
Sliding glass doors
Stucco
Recessed plant box
Basement windows
Stucco
Porch floor
Rear elevation
Wood shingles
Stucco
Sliding glass doors
Left side elevation
Light trough
Bedroom #1
8'-2"
Hall
Clos
Basement
7'-0"
Cross section
pool terrace
81'-0"
swimming pool
45'-8"
dishw
oven
ref
KITCHEN
16'± x 16'-8"
sliding glass doors
PORCH
DINING ROOM
16'± x 12'
FAMILY RM.
dryer
sliding gl. doors
barbecue
broom cl.
washer
dn.
fireplace
C.
linen
LIVING ROOM
28'± x 12'
BATH
FOYER
BEDROOM 3
10' x 11'
promenade
master
BEDROOM 1
12' x 16'-2"
BATH
BEDROOM 2
13'-4" x 12'
Size data,
1,903 square feet not including porches or promenade.
Lamp at table
Stone chimney face
Oven
Cabinets
Hood Rangetop
Wood panel wall
Door to dining room
Lamp
Refrig
Fixed plastic skylights
flat ceiling
Sinks
Oven
Sliding glass doors
Family room
Barbecue
Kitchen
Kitchen
Kitchen
Family room
DESIGN S-33

# Dramatic Three-Bedroom Contemporary

An octagon with an attached rectangular wing is the approach in this three-bedroom home offering unusual room shapes.

The exterior is contemporary, yet it has a traditional feeling. This is due to the mansard-shaped roof.

Inside the protected front entrance, a 54-square-foot foyer splits traffic to three different locations: the living room and dining room; the family room and kitchen; and the bedrooms and baths. A glance into the large living room reveals the sharp departure from the usual interior. The outside walls reflect the feeling of the exterior but in reverse. Arched windows and the sloping roof are dominant features of the design.

The circular stone chimney goes from the basement through the roof and is exposed in all four rooms of the octagon. It contains a fireplace in the living room, a barbecue grill in the kitchen-family room and the heater flue. All four rooms are spacious — which is especially necessary with angled walls — and all have large glass areas. Women will have a delightful time discovering the interesting ways in which furniture can be arranged. The ceiling in the octagon is laced with exposed beams, like wheel spokes, ranging from 8′ to 12′ high.

An island-type counter separates the kitchen and the family room. In all, there are 28 lineal feet of counter and appliances.

## Material List

**CONCRETE**

| | |
|---|---|
| 8″ 10″ Walls-Piers-Chimney | 2200 cu. ft. |
| 4″ Base Floor | 1800 sq. ft. |
| 4 Porch floor 6 x 6 mesh smooth finish | 870 sq. ft. |

**MASONRY**

4″ Veneer-stone-100 sq. ft.
4″Planter-stone-100 sq. ft.
4 Veneer chimney-300 sq. ft.
Back up-4800 common may vary
27 Lin. 13 x 13-flue lining
12 Lin. 8 x 13-flue lining
200 Fire Brick

**CARPENTRY**

| | |
|---|---|
| Plates-Joists etc. | 3760 BM |
| Sub floor 5/8 plyscore | 1860 sq. ft. |
| Studs-headers etc. | 3760 BM |
| Rafters | 4360 BM |
| Roof boards - 1/2 plyscore | 3000 sq. ft. |
| Finishing lumber | 280 BM |
| 3/8″ Exterior Plywood | 600 sq. ft. |
| Sheathing - 3/8 plyscore | 1700 sq. ft. |
| 4 x 10 - 4 x 12 Sel. fir | 924 BM |
| 13/16 x 2¼ Cl. Oak floor | 1200 BM |
| 5/8 plugged plyscore | 320 sq. ft. |

**INSULATION**

1250 ft. Semi-thick batts
1800 ft. Full-thick batts

**ROOFING**

8-sq. Cl. red cedar shingles 24″ - 10″ exposure
2 rolls - 15 lb. Saturated felt
22 sq. - 4 ply - tar & 400 lb. gravel per square

**PLASTERBOARD**

| | |
|---|---|
| 1/2 for walls | 6000 sq. ft. |
| 3/8 for ceilings | 2000 sq. ft. |

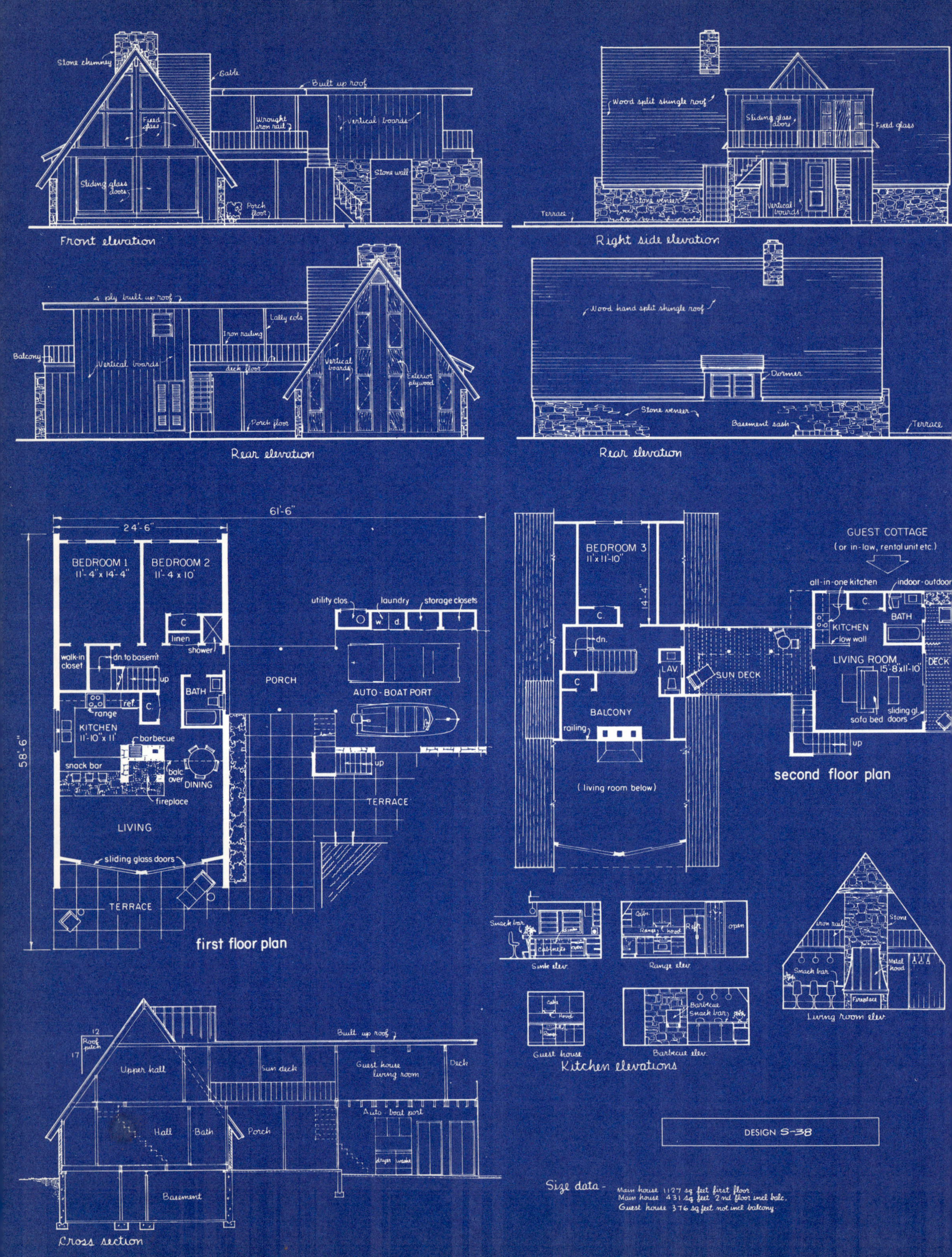

Stone chimney
Gable
Built up roof
Fixed glass
Wrought iron rail
Vertical boards
Sliding glass doors
Stone wall
Porch floor
Front elevation
Wood split shingle roof
Sliding glass doors
Fixed glass
Terrace
Stone veneer
Vertical boards
Right side elevation
4 ply built up roof
Latty cols
Iron railing
Balcony
Vertical boards
deck floor
Vertical boards
Exterior plywood
Porch floor
Rear elevation
Wood hand split shingle roof
Dormer
Stone veneer
Basement sash
Terrace
Rear elevation
61'-6"
24'-6"
BEDROOM 1
11'-4" x 14'-4"
BEDROOM 2
11'-4 x 10'
58'-6"
utility clos
laundry
storage closets
linen
shower
walk-in closet
dn. to basemt
up
BATH
PORCH
AUTO-BOAT PORT
range
ref.
KITCHEN
11'-10"x 11'
barbecue
snack bar
balc over
DINING
fireplace
LIVING
TERRACE
sliding glass doors
TERRACE
first floor plan
BEDROOM 3
11'x 11'-10"
14'-4"
dn.
LAV.
SUN DECK
BALCONY
railing
(living room below)
GUEST COTTAGE
(or in-law, rental unit etc.)
all-in-one kitchen
indoor-outdoor
KITCHEN
BATH
low wall
LIVING ROOM
15'-8"x11'-10"
DECK
sofa bed
sliding gl. doors
up
second floor plan
Snack bar
Cabinets
Sink elev.
Range elev.
Guest house
Barbecue elev.
Kitchen elevations
Stone
Iron rail
Snack bar
Metal hood
Fireplace
Living room elev.
Built up roof
Roof pitch
Upper hall
Sun deck
Guest house living room
Deck
Auto boat port
Hall
Bath
Porch
Basement
Cross section
DESIGN S-38
Size data -
Main house 1127 sq feet first floor
Main house 431 sq feet 2nd floor incl balc.
Guest house 376 sq feet not incl balcony

# A-Frame With Its Own Guest Cottage

Here is a completely new twist to that paragon of informal living, the A-frame house.

Attached to the main structure is a separate unit with upper and lower levels of simple design. The unit is, in effect, a guest cottage that can be used for that purpose or as a rental apartment, since it includes a kitchen.

While this house is of the vacation or leisure type, it nevertheless has all the facilities of a year 'round home, including central heating, complete insulation and a full basement.

The living room at the front of the A-frame is 23′ in width and has been given a complete wall of glass. There is an impressive and bold shaft of stone from the fireplace up through the roof. On the opposite side of the chimney is a built-in barbecue, which is in addition to the regular kitchen range.

On the same floor are two bedrooms, one with a walk-in closet, one with a regular closet.

Outside the second floor of the main house is an elevated deck with a covered roof over the walkway. This deck leads to the guest cottage, a fairly small but delightful area for vacation living. Its living room has sliding glass doors which open on an outdoor deck. A prefabricated, all-in-one type of kitchen is recommended for its compact efficiency in a small space.

The lower part of the guest cottage can be used to house a boat or car or it can be utilized as a sun-shaded porch or rainy-day play area.

## Material List

**CONCRETE**
8″ — 10″ Walls, Piers, Chimney . 1763 cu. ft.
4″ Conc. base floor ........... 1050 sq. ft.
4″ Conc. Porch floor 6 x 6 mesh . 312 sq. ft.
4″ Conc. car storage ........... 410 sq. ft.

**MASONRY**
Stone veneer 4″............... 450 sq. ft.
8″ Stone wall ................ 215 sq. ft.
4″ Stone for chimney .......... 470 sq. ft.
Brick backup ..................3200 common
22 Lin. 16 x 16 flue lining
46 Lin. 8 x 13 flue lining
120 Fire brick

**CARPENTRY**
Plates - all joists — porch, carport, etc. ................ 4300 BM
Sub floor — 5/8 plyscore ........ 2400 sq. ft.
Backing-bridging .............. 450 BM
Studs, headers, etc. ........... 4400 BM
Ceiling, framing & deck ........ 980 BM
1/2 deck roof boards & catwalk . 740 sq. ft.
Roof framing ................. 2520 BM
O.S. Fin. Lbr. Cl. redwood .... 460 BM
1/2 Plyscore roof boards ........ 2500 sq. ft.
Ext. Plywood 3/8 Good 1 side .... 1260 sq. ft.
3/8 Plyscore sheathing .......... 2500 sq. ft.
1 x 12 D & M Vd siding Cl. redwood ................ 1550 BM
11 Rolls — 15 lb. Sat. felt
13/16 x 2 1/4 Cl. R. Oak floor .... 1650 BM
5/8″ Plugged Plyscore .......... 352 sq. ft.

**PLASTERBOARD**
5200 Sq. ft. — 1/2 for wall taper jt.
2500 Sq. ft. — 3/8 for ceiling

**ROOFING**
26 Sq. hand-split shakes 24″ 10″ Exp.
150 Lin. Starter & ridge
24 Roll 30 lb. Sat. felt 18″ wide
6 sq. 4 ply tar & gravel

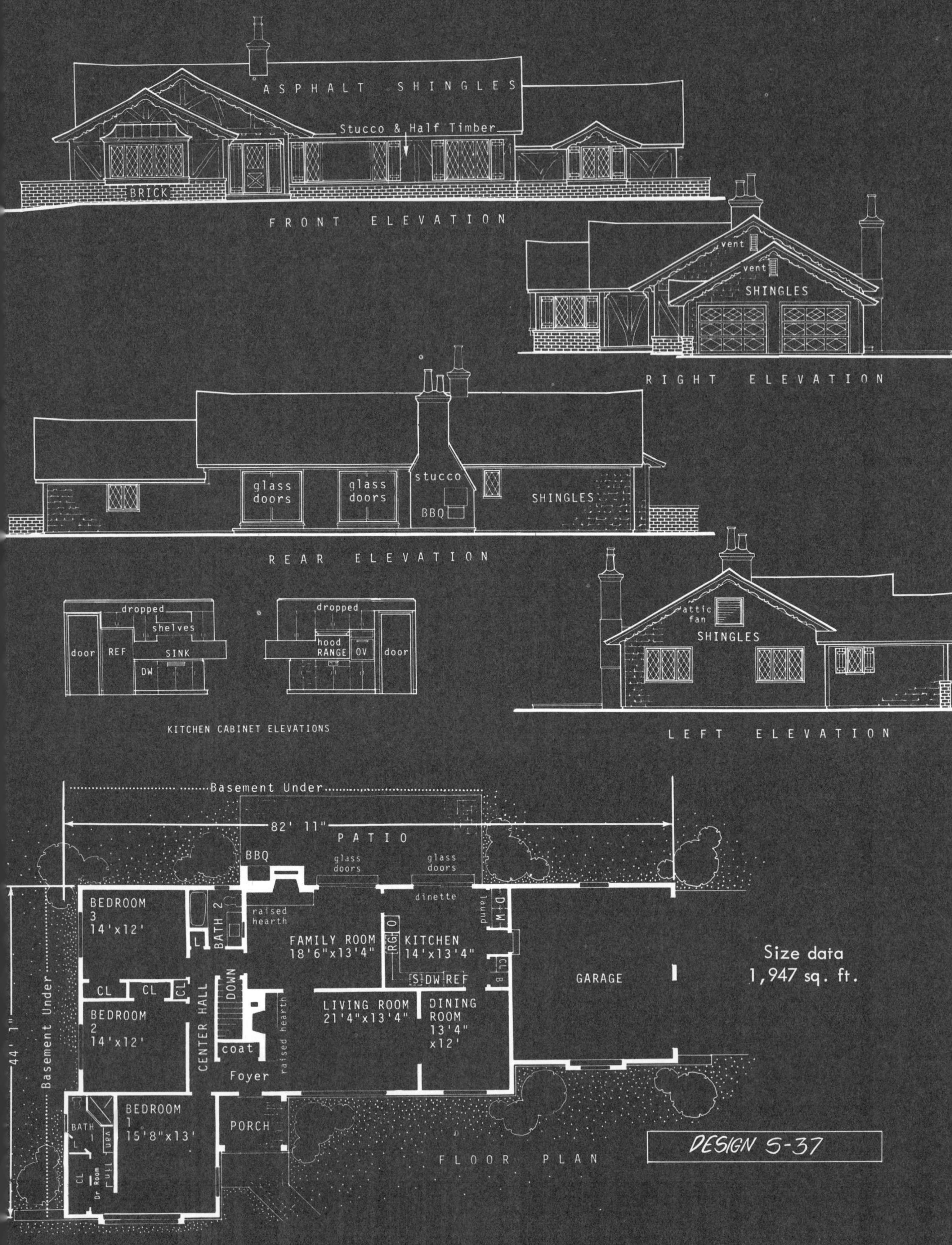


Size data
1,947 sq. ft.

DESIGN S-37

# Ranch With Swiss Chalet Styling

The traditional Swiss Chalet has a strong appeal for those who desire a substantial, comfortable dwelling with a relaxed, informal air.

This is not a true Swiss Chalet, but it combines traditional and contemporary lines, giving it a measure of individuality without interfering with its resale potentiality.

The foyer opens into an attractive living room with the dining room visable through an open arch. A wide picture window is at the front. The entire area is made to seem larger by the location of a raised-hearth fireplace on an inner wall.

At the rear of the living room-dining room combination are the family room and kitchen. The family room is especially handsome with its fireplace and sliding glass doors leading to a patio with a built-in barbecue. The patio can also be entered via sliding glass doors in the dinette area of the kitchen. Serving outdoors thus becomes easy.

Besides being accessible to the patio, the kitchen is convenient to the dining room, to the family room and to the garage.

The wing to the left of the foyer has three bedrooms. The master bedroom includes a dressing room with closets and linen shelves and a stall shower bathroom with a vanitory and corner linen closet.

## Material List

**CONCRETE WORK**

| | |
|---|---|
| Footings | 22 cu. yds. |
| Concrete Slabs | 21 cu. yds. |

**MASONRY**

| | |
|---|---|
| Concrete Blocks | |
| 8" x 8" x 16" | 516 sq. ft. |
| 12" x 8" x 16" | 1365 sq. ft. |
| 16" x 8" x 16" | 304 sq. ft. |
| Chimney Fill (Block & Brick) | 800 cu. ft. |
| Brick Facing | 458 sq. ft. |
| Flagstone Decks | 437 sq. ft. |
| Steel Beam | 184 lbs. |

**CARPENTRY**

| | |
|---|---|
| Framing Lumber | 16,040 bd. ft. |
| Hand-split shakes | 36 squares |
| Half timber and stucco | 700 sq. ft. |
| Asphalt Roof shingles | 36 squares |
| Insulation | 3362 sq. ft. |
| 5/8" Plyscore | 5500 sq. ft. |
| Resilient tile flooring | 428 sq. ft. |
| 5/8" Gypsum board | 893 sq. ft. |
| 1/2" Gypsum board | 6620 sq. ft. |
| 3/8" Exterior Plywood | 462 sq. ft. |
| 3/4" Plyscore | 2445 sq. ft. |
| 1/2" Plyscore | 1631 sq. ft. |

**EXTERIOR OPENINGS**

2 Pair Pella Sliding doors and frames
1 Front Door
10 Casement units

**MILLWORK**

24 Interior flush doors
Stairs, 14 risers
1 set of disappearing stairs

**CERAMIC TILE**

| | |
|---|---|
| Bath #1 | 145 sq. ft. |
| Bath #2 | 110 sq. ft. |

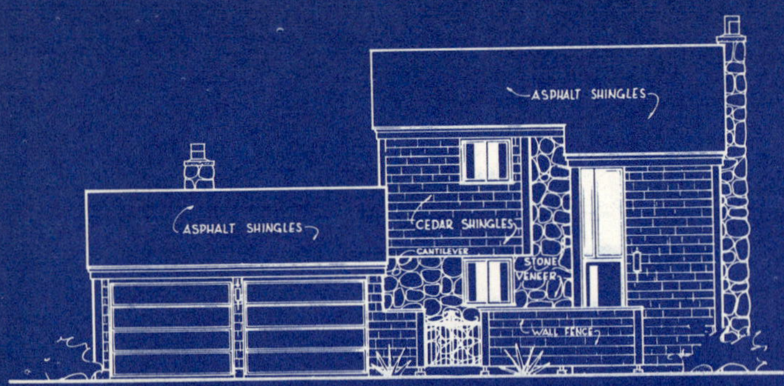


FRONT ELEVATION

LEFT SIDE ELEVATION

REAR ELEVATION

RIGHT SIDE ELEVATION

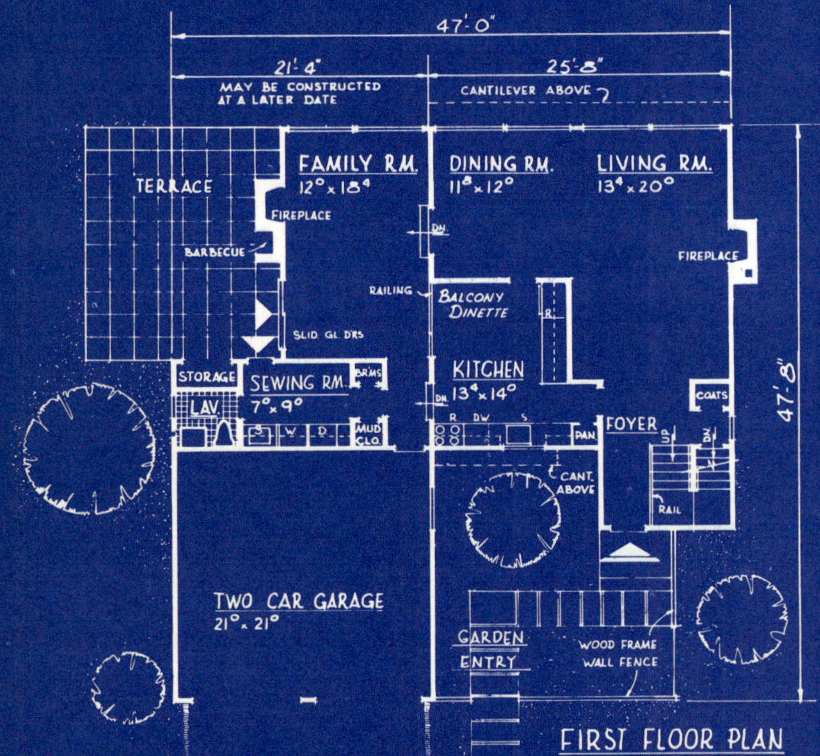


FIRST FLOOR PLAN



SECOND FLOOR PLAN

## HOUSE DATA

FIRST FL. ........ 754 SQ.FT.
FAM.& LAUN. ..... 385 SQ.FT.
SECOND FL. ...... 826 SQ.FT.
GARAGE .......... 473 SQ.FT.

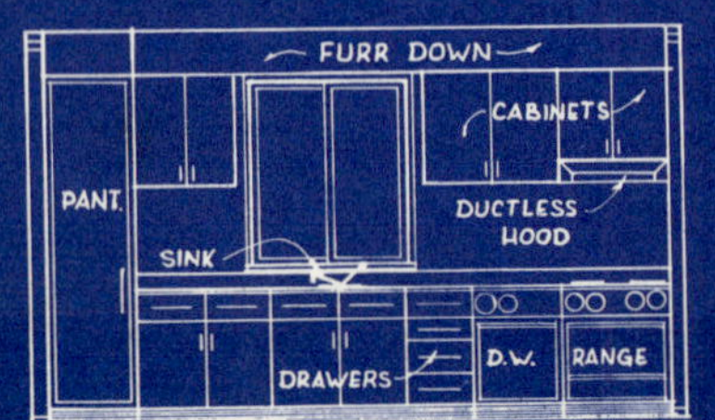

KITCHEN CABINET ELEVATIONS

DESIGN L-11

# Compact Two-Story for Modest Lot

This is a two-story house in a neat package that, despite its compact size, contains several unusual areas. Moreover, with an overall side-to-side dimension of only 47′, including a two-car garage, this house will fit on a lot of moderate size.

The position of the garage allows the use of a wall extended across the front of the garden entry. Using the same material, cedar shingles, as found on the side walls of the house, the front wall becomes an architectural design device which adds much to the attractiveness of the exterior.

Entering through grilled wrought iron gates into a garden court, an atmosphere of privacy is evident to the visitor.

The entrance foyer, with its two-storied front wall of glass, will be cheerful and spacious. There is a balcony overlooking this entrance area and an open stair which doubles back on itself, all designed with contemporary detail. Immediately on entering, one catches a long view of the rear garden because the living room is open planned and glazed at the rear. A fireplace at one side is located on the long wall, making for excellent furniture groupings.

Convenient to the front door, the kitchen overlooks the garden entry in one direction and uses a railed dinette to give the impression of being a balcony two steps above the family room floor.

## Material List

**CONCRETE WORK**

| | | |
|---|---|---|
| Concrete Walls | | 814 cu. ft. |
| Slabs | | 517 cu. ft. |
| Concrete Footings | | 197 cu. ft. |
| Misc. Concrete | | 253 cu. ft. |

**STRUCTURAL STEEL**

| | | |
|---|---|---|
| Lally Columns 3½″ diam. | | 3 pieces |
| Girder | 6″ I Beam 26′-0″ | 1 piece |

**BRICK WORK**

| | | |
|---|---|---|
| Chimney | Stone & Brick | 376 cu. ft. |
| Flue Lining | T. C. | 99 lin. ft. |
| Veneer | Stone | 127 sq. ft. |

**CARPENTRY**

| | | |
|---|---|---|
| Framing Lumber | | 6523 B.F. |
| Studs | | 4000 B.F. |
| Plates | | 1200 B.F. |
| Roof Sheathing | | 2220 sq. ft. |
| Sub Flooring | | 1614 sq. ft. |
| Side Wall Sheathing | | 2470 sq. ft. |
| Insulation Walls | | 1650 sq. ft. |
| Insulation Ceilings | | 1270 sq. ft. |
| Oak Flooring | | 1520 sq. ft. |
| Kitchen Plywood | | 168 sq. ft. |

**MILLWORK**

| | | |
|---|---|---|
| Ext. Doors & Frames Complete | | 4 pieces |
| Sliding Glass Doors | | 1 unit |
| Garage Door Complete Set | | 2 units |
| Window Units | | 28 units |
| Int. Doors & Frames Complete | | 16 pieces |
| Bi Fold Door Units | | 2 units |
| Fascia | | 262 lin. ft. |
| Louvers | | 2 pieces |
| Base | | 610 lin. ft. |
| Stairs | 12 risers | 1 set |
| Stairs | 13 risers | 1 set |

**ROOFING**

| | | |
|---|---|---|
| Shingles | 235# asphalt | 2220 sq. ft. |
| Roofing Paper | 15# felt | 2220 sq. ft. |

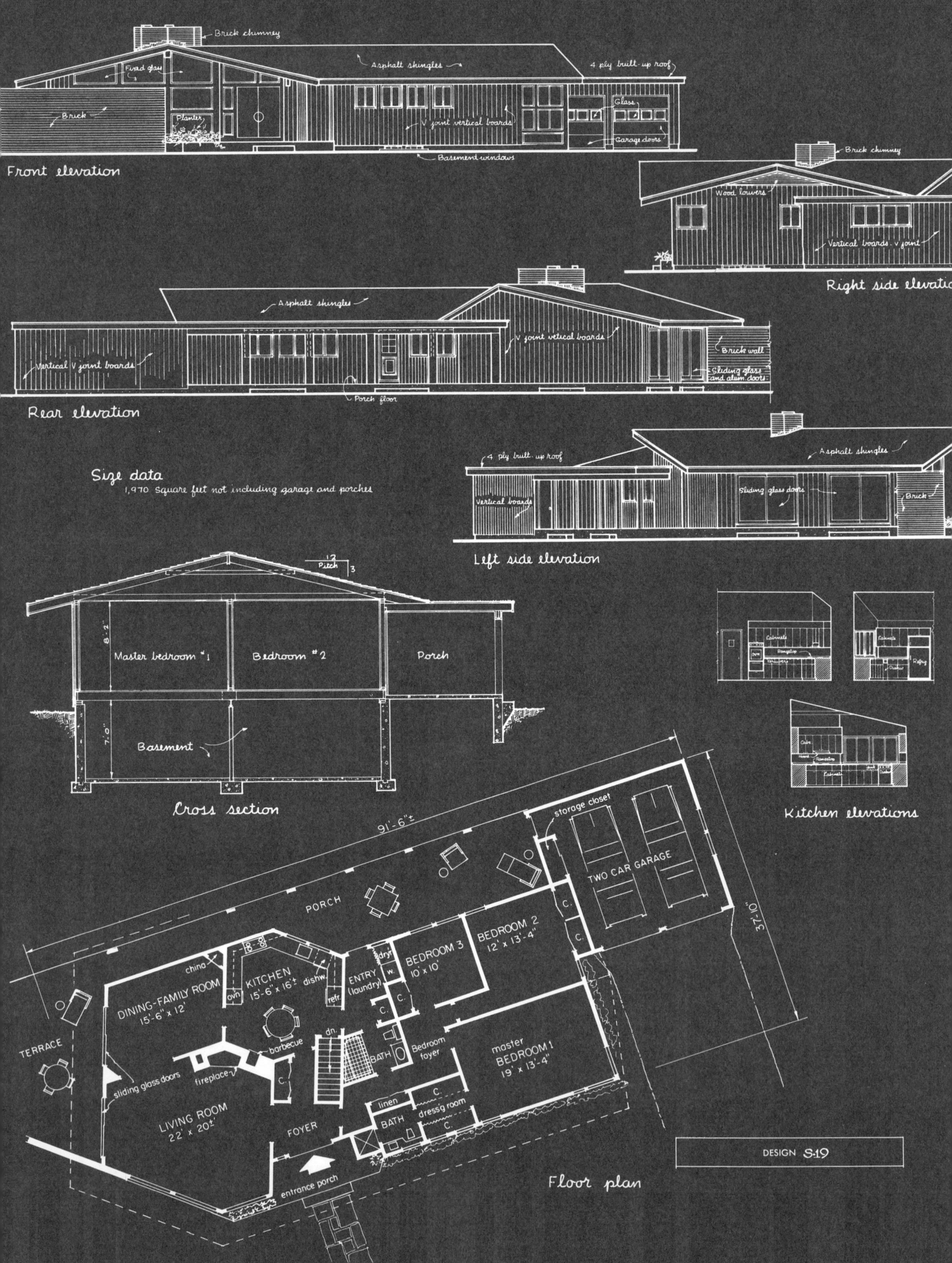

Brick chimney
Fixed glass
Asphalt shingles
4 ply built-up roof
Brick
Planter
V joint vertical boards
Glass
Garage doors
Basement windows
Front elevation
Brick chimney
Wood louvers
Vertical boards. V joint
Right side elevation
Asphalt shingles
V joint vertical boards
Vertical V joint boards
Brick wall
Sliding glass and alum doors
Porch floor
Rear elevation
Size data
1,970 Square feet not including garage and porches
4 ply built-up roof
Asphalt shingles
Vertical boards
Sliding glass doors
Brick
Left side elevation
12
Pitch
3
8'-2"
Master bedroom #1
Bedroom #2
Porch
7'-0"
Basement
Cross section
Kitchen elevations
91'-6"±
storage closet
TWO CAR GARAGE
37'-10"
PORCH
BEDROOM 2
12' x 13'-4"
BEDROOM 3
10' x 10'
dryr
w.
C.
ENTRY (laundry)
china
DINING-FAMILY ROOM
15'-6" x 12'
KITCHEN
15'-6" x 16'±
dishw.
ovn
refr.
TERRACE
barbecue
dn.
BATH
Bedroom foyer
master BEDROOM 1
19' x 13'-4"
sliding glass doors
fireplace
LIVING ROOM
22' x 20'±
linen
dress'g room
FOYER
entrance porch
DESIGN S-19
Floor plan

# Contemporary Ranch Full of Angles

The living, dining, service area of this unusual residence is designed along one axis in a hexagon shape while the sleeping area axis angles off at 20 degrees. Although the exterior does show that the house is at an angle, it is roofed in such a way as to minimize the dramatics and put on a conservative facade for more general acceptance in a neighborhood.

The combination family-dining room has an angled shape. Sliding glass doors in both the living room and family-dining room lead to an open terrace.

The kitchen is a large hexagon. It has a four-sided, U-shaped service area. The sink and appliances are set around four sides of the hexagon. Just beyond the kitchen is a bathroom with an angled wall and a large tub which follows the angle. The tub is built up and lined with mosaic tile. The room is ideally located, since it can be entered from the outside without going through any of the rooms yet can be reached easily from anywhere inside. The laundry is right inside the rear door next to the kitchen.

All three bedrooms are square-walled, and all their entrances are grouped around an interior bedroom foyer. The master bedroom connects to a seven-foot dressing room with 14′ of closet. This, in turn, connects with a private master bath with a stall shower.

Exterior materials are vertical boards, aluminum sliding windows and a panel of brick.

## Material List

**CONCRETE**

|  |  |
|---|---|
| 8″ 10″ walls piers, etc. | 2300 cu. ft. |
| 4″ Conc. floor — house garage | 2420 sq. ft. |
| 4″ Conc. floor porch 6 x 6 mesh | 880 sq. ft. |

**MASONRY**

|  |  |
|---|---|
| 4″ Brick veneer | 153 sq. ft. |
| 8″ Brick work | 72 sq. ft. |
| 4″Brick veneer chimney | 420 sq. ft. |
| Common brick back up | 5400 sq. ft. |
| 10 Lin. ft. 8 x 13 flue lining | |
| 30 Lin. ft. 13 x 13 flue lining | |

**CARPENTRY**

|  |  |
|---|---|
| Floor framing plates etc. | 4000 BM |
| Sub floor ⅝ plyscore | 2050 sq. ft. |
| Studs - headers etc. - house & garage | 5100 BM |
| Ceiling joist | 850 BM |
| Rafters, etc. | 4050 BM |
| 16″ Beams (Sel. Fir) | 850 BM |
| ½″ plyscore roof boards | 4100 sq. ft. |
| O.S. Fin. Lumber Cl. redwood | 400 BM |
| 1 x 8 D&M Vd Vert siding | 1950 BM |
| ¾ x 1⅛ Cove | 430 lin. ft. |
| Sheathing - ⅜ plyscore | 2050 sq. ft. |
| ⅜ Exterior-plywood good 1 side | 1420 sq. ft. |
| 10 Rolls 15 lb. Sat. Felt | |
| 13/16 x 2¼ Cl. R Oak Floor | 1700 BM |

**ROOFING**

36 sq. 300 lb. self sealing asphalt shingle
140 Indiv. shingle for ridge
8 rolls 15 lb. sat. felt
10 sq. - 4 ply tar & gravel roof 400 lb. gravel

**PLASTERBOARD**

½ for walls 3700 sq. ft. tapered
⅜ for ceiling 2150 sq. ft. tapered

FRONT ELEVATION

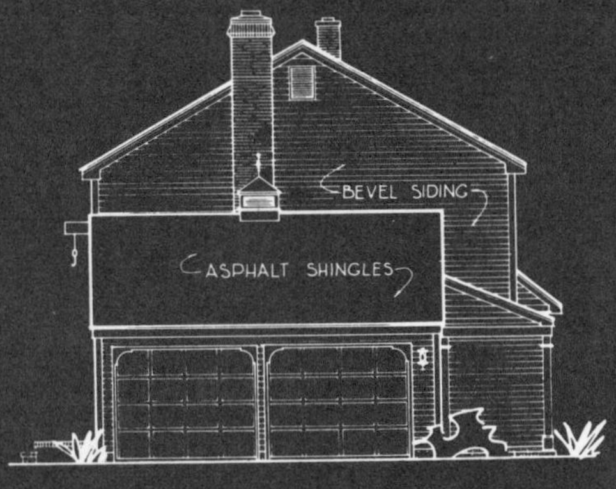


RIGHT SIDE ELEVATION

REAR ELEVATION

LEFT SIDE ELEVATION

## HOUSE DATA

| | |
|---|---|
| FIRST FLOOR | 870 SQ. FT. |
| FAMILY ROOM | 283 SQ. FT. |
| SECOND FLOOR | 831 SQ. FT. |
| GARAGE | 474 SQ. FT. |

BED RM. #1
$13^4$ x $15^0$
CL.
DRESS. RM
CL.
LIN
DN
CL.
HALL
BED RM. #2
$11^0$ x $11^8$
BED RM. #3
$10^0$ x $14^0$
CL.
27'-8"
30'-0"

SECOND FLOOR PLAN

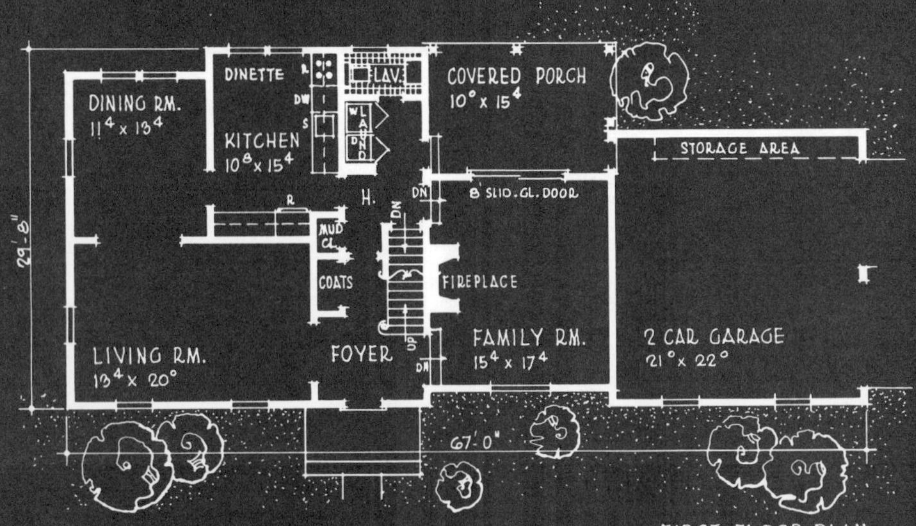


FIRST FLOOR PLAN

DESIGN S-68

# Two-Story Boasts Livability and Naturalness

Authentically New England in character, this two-story house has simplicity of form.

The floor plan attempts to provide the kind of livability needed by the contemporary American family. All of the rooms on both floors are generous in size. The entrance foyer is an example of the theme of spaciousness which is followed throughout.

The living room, dining room and kitchen are normally related one to another, but the family room is accessible from the front foyer, kitchen area and rear covered porch. It has an advantage over the family room which can only be approached from the kitchen.

Many women prefer to have the laundry on the first floor level instead of in the basement. In this plan, it is placed neatly adjacent to the plumbing of the kitchen and lavatory.

The kitchen has generous space for a dinette with a large glass area overlooking the rear garden.

Upstairs are three large bedrooms. The master bedroom has a private dressing alcove, split bath and two huge closets. The hall bathroom has a double vanity.

This kind of side hall layout makes it a simple matter, should it be desired, to provide a fourth bedroom over the family room raising the ridge of this section of the roof. The door to this fourth bedroom will be in the upstairs hallway at the head of the stair and convenient to the hall bathroom.

## Material List

**CONCRETE WORK**

|  | |  |
|---|---|---|
| Concrete Walls | | 757 cu. ft. |
| Foundation Footings | | 187 cu. ft. |
| Slabs | | 560 cu. ft. |
| Misc. Concrete | | 178 cu. ft. |

**STRUCTURAL STEEL**

|  | |  |
|---|---|---|
| Lally Columns | 3½" diam. | 4 pieces |

**BRICK WORK**

|  | |  |
|---|---|---|
| Chimney | Brick | 193 cu. ft. |
| Flue Lining | T. C. | 26 lin. ft. |

**CARPENTRY**

|  | |  |
|---|---|---|
| Framing Lumber | | 8400 B.F. |
| Studs | | 4160 B.F. |
| Plates | | 1260 B.F. |
| Roof Sheathing | | 2560 sq. ft. |
| Sub Flooring | | 1660 sq. ft. |
| Side Wall Sheathing | | 2600 sq. ft. |
| Insulation Walls | | 1750 sq. ft. |
| Insulation Ceilings | | 1051 sq. ft. |
| Wood Flooring | | 1343 sq. ft. |
| Kitchen Plywood | | 160 sq. ft. |

**MILLWORK**

|  | |  |
|---|---|---|
| Ext. Doors & Frames Complete | | 3 |
| Garage Door Complete Set | | 2 |
| Sliding Glass Door Unit | | 1 |
| Int. Doors & Frames Complete | | 9 |
| Bi Fold Doors | | 8 |
| Sliding Doors | | 4 |
| Window Units | | 19 |
| Fascia | | 280 lin. ft. |
| Shutters | | 26 pieces |
| Base | | 640 lin. ft. |
| Stairs | 12 risers | 1 set |
| Stairs | 14 risers | 1 set |

**ROOFING**

|  | |  |
|---|---|---|
| Shingles | 235# asphalt | 2560 sq. ft. |
| Roofing Paper | 15# felt | 2560 sq. ft. |

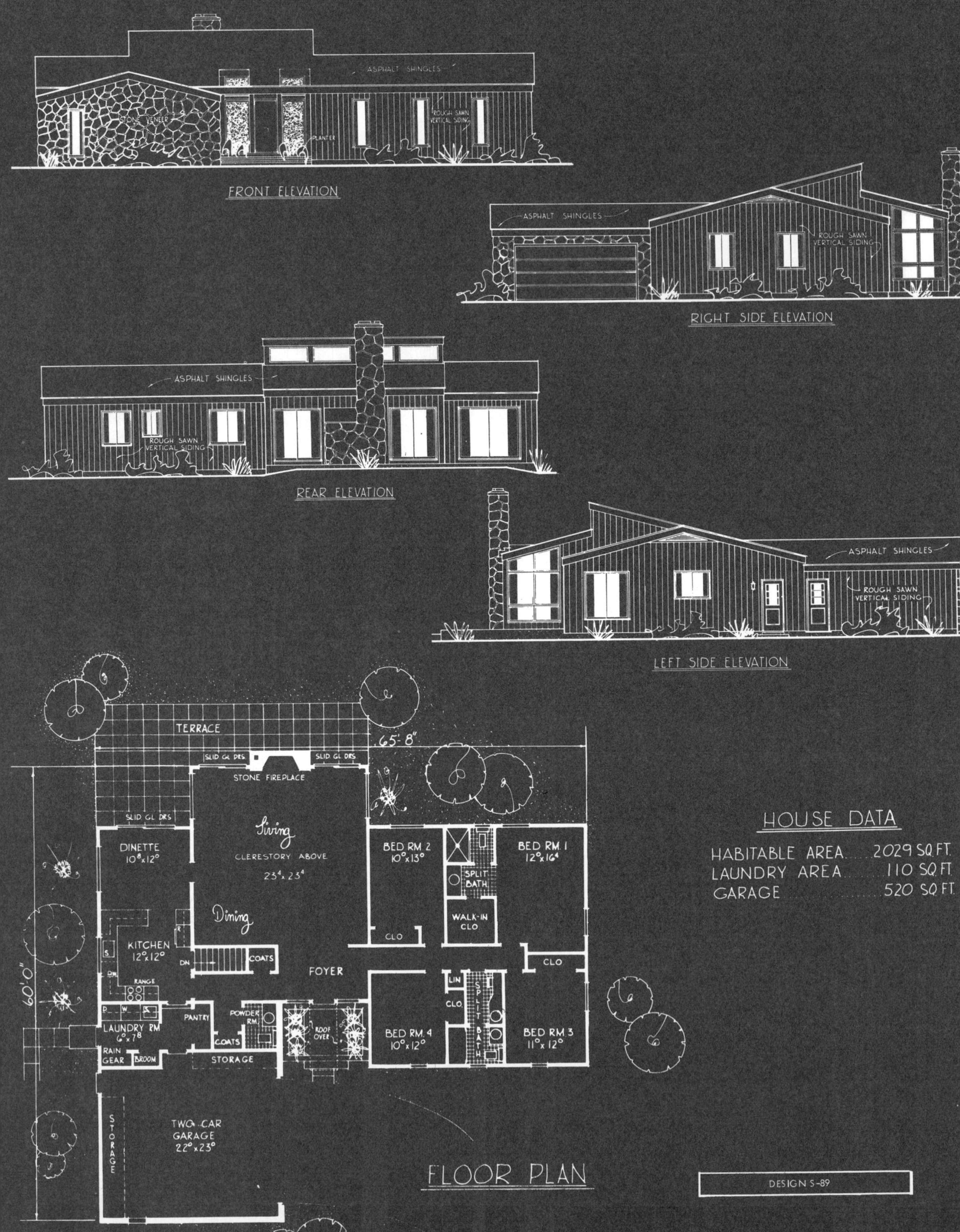
ASPHALT SHINGLES
STONE VENEER
ROUGH SAWN VERTICAL SIDING
PLANTER
FRONT ELEVATION
ASPHALT SHINGLES
ROUGH SAWN VERTICAL SIDING
RIGHT SIDE ELEVATION
ASPHALT SHINGLES
ROUGH SAWN VERTICAL SIDING
REAR ELEVATION
ASPHALT SHINGLES
ROUGH SAWN VERTICAL SIDING
LEFT SIDE ELEVATION
TERRACE
65'-8"
SLID GL DRS
STONE FIREPLACE
SLID GL DRS
SLID GL DRS
DINETTE 10⁸x12⁰
Living
CLERESTORY ABOVE
23⁴x23⁴
Dining
KITCHEN 12⁰x12⁰
S
DW
R
DN
COATS
FOYER
RANGE
D W S
LAUNDRY RM 6⁰x7⁸
PANTRY
POWDER RM
COATS
RAIN GEAR
BROOM
STORAGE
ROOF OVER
BED RM. 2 10⁰x13⁰
CLO
SPLIT BATH
WALK-IN CLO
BED RM. 1 12⁰x16⁴
CLO
LIN
CLO
SPLIT BATH
BED RM. 4 10⁰x12⁰
BED RM. 3 11⁰x12⁰
60'-0"
STORAGE
TWO-CAR GARAGE 22⁰x23⁰
FLOOR PLAN
HOUSE DATA
HABITABLE AREA 2029 SQ.FT.
LAUNDRY AREA 110 SQ.FT.
GARAGE 520 SQ.FT.
DESIGN S-89

# Family Living Room Surrounded by House

Boldness is the keynote of this house plan, a four-bedroom structure for a large family that likes informal living on a grand scale.

Any room 23′4″ in length can be considered large. Consider then the size of the huge room that is the hub of this design. It's 23′4″ square. Add to that the fact that it has a high clere-story ceiling, a stone fireplace and sliding glass doors leading to the rear terrace and garden — and you get some idea of its dramatic effect.

It's a kind of living room, family room and dining area all rolled into one. Although the plan can be called "open," it is so designed to conceal from view the kitchen, for there are solid walls between the main room and the kitchen. The latter is also something special, since it is 22′8″ by 12′ in combination with a dinette, which has sliding glass doors leading to the rear.

In the bedroom wing are four bedrooms of generous size. The architect uses a "split bath" plan for the hall bath inasmuch as it will be used by the occupants of three of the bedrooms. The owners' bathroom is also in a split configuration to give dual use to this facility.

The boldness of the front entrance, with a see-through roof, gives a hint of what happens beyond, for certainly a living room as large as this one should be approached through an appropriate entrance.

## Material List

|  | |  |
|---|---|---|
| **CONCRETE WORK** | | |
| Concrete Walls | | 1284 cu. ft. |
| Slabs | | 811 cu. ft. |
| Foundation Footings | | 415 cu. ft. |
| Misc. Concrete | | 335 cu. ft. |
| **STRUCTURAL STEEL** | | |
| Lally Columns | 3½″ Diam. | 12 pieces |
| Girder | 7″ I Beam | 105.50 lin. ft. |
| Girder | 12 WF Beam | 24 lin. ft. |
| **BRICK WORK** | | |
| Chimney | Brick & Stone | 215 cu. ft. |
| Veneer | | 364 sq. ft. |
| **CARPENTRY** | | |
| Framing Lumber | | 9109 B.F. |
| Studs | | 3733 B.F. |
| Plates | | 1200 B.F. |
| Roof Sheathing | | 2520 sq. ft. |
| Sub Flooring | | 2007 sq. ft. |
| Side Wall Sheathing | | 2400 sq. ft. |
| Wall Insulation | | 1312 sq. ft. |
| Ceiling Insulation | | 2118 sq. ft. |
| Wood Flooring | | 1886 sq. ft. |
| Kitchen Plywood | | 144 sq. ft. |
| **MILLWORK** | | |
| Ext. Doors & Frames Complete | | 3 |
| Sliding Glass door units | | 3 |
| Interior Doors & Frames | | 19 |
| Bi Fold Doors | | 12 |
| Window Units | | 23 |
| Fascia | | 342 lin. ft. |
| Base | | 650 lin. ft. |
| Stairs | 12 risers | 1 set |
| **ROOFING** | | |
| Shingles | 235# Asphalt | 2520 sq. ft. |
| Roofing Paper | 15# Felt | 2520 sq. ft. |
| **SHEET METAL** | | |
| Saddle & Counter Flashing 16 oz. | | Copper |

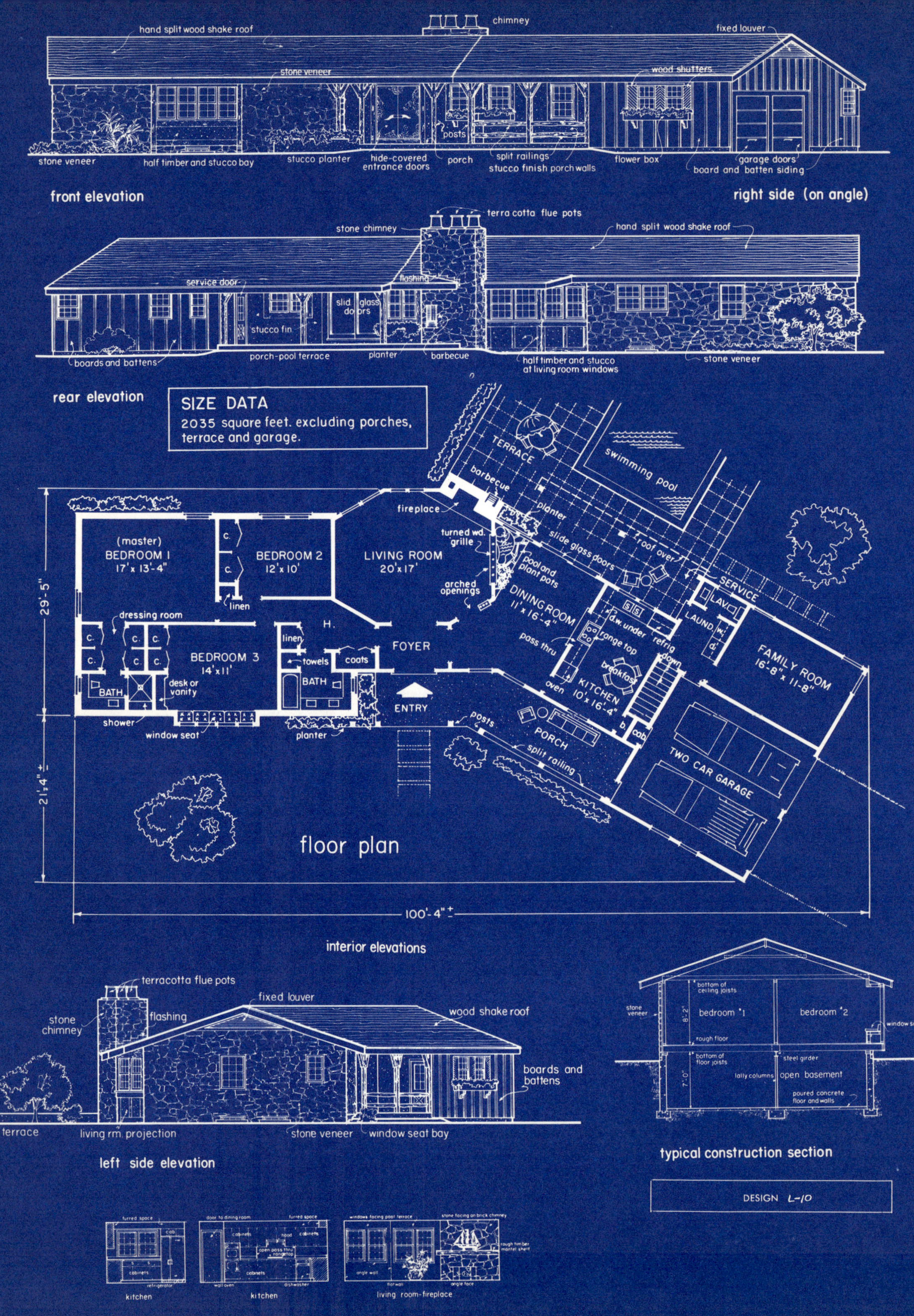

chimney
hand split wood shake roof
fixed louver
stone veneer
wood shutters
posts
stone veneer
half timber and stucco bay
stucco planter
hide-covered entrance doors
porch
split railings
stucco finish porch walls
flower box
garage doors
board and batten siding
front elevation
right side (on angle)
terra cotta flue pots
stone chimney
hand split wood shake roof
service door
flashing
slid glass doors
stucco fin
boards and battens
porch-pool terrace
planter
barbecue
half timber and stucco at living room windows
stone veneer
rear elevation
SIZE DATA
2035 square feet. excluding porches, terrace and garage.
TERRACE
swimming pool
barbecue
fireplace
planter
(master) BEDROOM 1 17' x 13'-4"
BEDROOM 2 12' x 10'
LIVING ROOM 20' x 17'
turned wd. grille
arched openings
slide glass doors
roof over
pool and plant pots
SERVICE
LAV.
LAUND.
DINING ROOM 11' x 16'-4"
d.w. under
range top
refrig
pass thru
breakfast
down
KITCHEN 10' x 16'-4"
oven
FAMILY ROOM 16'-8" x 11'-8"
linen
dressing room
H.
FOYER
BEDROOM 3 14' x 11'
linen
towels
coats
desk or vanity
BATH
BATH
ENTRY
shower
window seat
planter
posts
PORCH
split railing
TWO CAR GARAGE
29'-5"
21'-4"±
floor plan
100'-4"±
interior elevations
terracotta flue pots
fixed louver
stone chimney
flashing
wood shake roof
boards and battens
pool terrace
living rm. projection
stone veneer
window seat bay
left side elevation
bottom of ceiling joists
bedroom #1
bedroom #2
stone veneer
window seat
rough floor
bottom of floor joists
steel girder
lally columns
open basement
poured concrete floor and walls
typical construction section
DESIGN L-10
kitchen
kitchen
living room-fireplace
interior elevations

# Western Ranch for the Discriminating

An angled plan was used to accentuate the rambling aspect of this low-slung and luxurious Western ranch house.

To maintain authenticity, typically Western exterior materials have been used, such as the rough cut stone, rough hewn timber posts and brackets, split rails, stucco and the often used boards and battens. The rough irregular wood shingles top off the structure perfectly. Two refinements are also added for the decor — clay chimney pots and the horsehide-covered door. Long-horned steer horns are used here for door pulls.

The front porch is a huge 32′ long and complements the front entrance; double posts frame the opening. Inside, the large foyer introduces a visitor to an exceptional interior. A large octagonal living room lies to the rear with a two-sided window wall allowing a full view of the rear lawn. The adjoining wall is the stone fireplace containing a built-in outdoor barbecue. A fourth wall consists of a timber-framed opening with large Spanish-style turned poles providing decoration to both living and dining rooms. By looking through, one can view the pool terrace in the rear.

A triangular, indoor-outdoor planter-pool is on the dining side of the framed opening. This shapes the room as well as providing a very decorative arrangement. The kitchen is L-shaped with a large breakfast area.

## Material List

|  |  |
|---|---|
| **CONCRETE** | |
| 8″ & 10″ Walls Piers Chimney | 1900 cu. ft. |
| Porch Floor Smooth 6 x 6 Mesh | 270 sq. ft. |
| Porch Floor Scored 6 x 6 Mesh | 200 sq. ft. |
| Basement & Garage Floor | 1900 sq. ft. |
| Water Proofing | 2000 sq. ft. |
| **MASONRY** | |
| Stone Veneer Walls | 920 sq. ft. |
| Stone Veneer Chimney | 300 sq. ft. |
| Brick Backup | 2800 Com. Brick |
| **CARPENTRY** | |
| Plates & Floor Joists | 4260 BM |
| 5/8 Plyscore Sub Floor | 2120 sq. ft. |
| All Studs — Plates — Headers — Porches | 4750 BM |
| Backing — Bridging, etc. | 780 BM |
| Ceiling Joists | 2150 BM |
| Porch Beams 6 x 6 & 6 x 8 Sel. Fir. | 540 BM |
| Roof Framing | 4160 BM |
| 1/2 Plyscore Roof Boards | 3700 sq. ft. |
| Outside Finishing Lumber | 420 BM |
| 3/8 Exterior Plywood | 910 sq. ft. |
| 3/8 Exterior Plyscore Sheathing | 2400 sq. ft. |
| 3/8 Exterior Plywood Siding Good 1 Side | 670 sq. ft. |
| 3/4 Closet shelving good 2 sides | 240 sq. ft. |
| 360 Lin. 5/8 x 1 1/2 Battens | |
| 11 Rolls 15 lb. Saturated Felt | |
| 13/16 x 2 1/4 Cl. Red Oak Floor | 1760 BM |
| 5/8 Plugged Plyscore | 500 sq. ft. |
| Basement Stairs | 160 BM |
| **ROOFING** | |
| 41 1/2 sq. hand-split wood shakes | |
| 19 Rolls 30 lb. 18″ sat. felt | |
| **PLASTERBOARD** | |
| 1/2 For Walls Taper Joint | 5600 sq. ft. |
| 3/8 For Ceiling Taper Joint | 2100 sq. ft. |
| 5/8 For Entire Garage Taper Joint | 1030 sq. ft. |

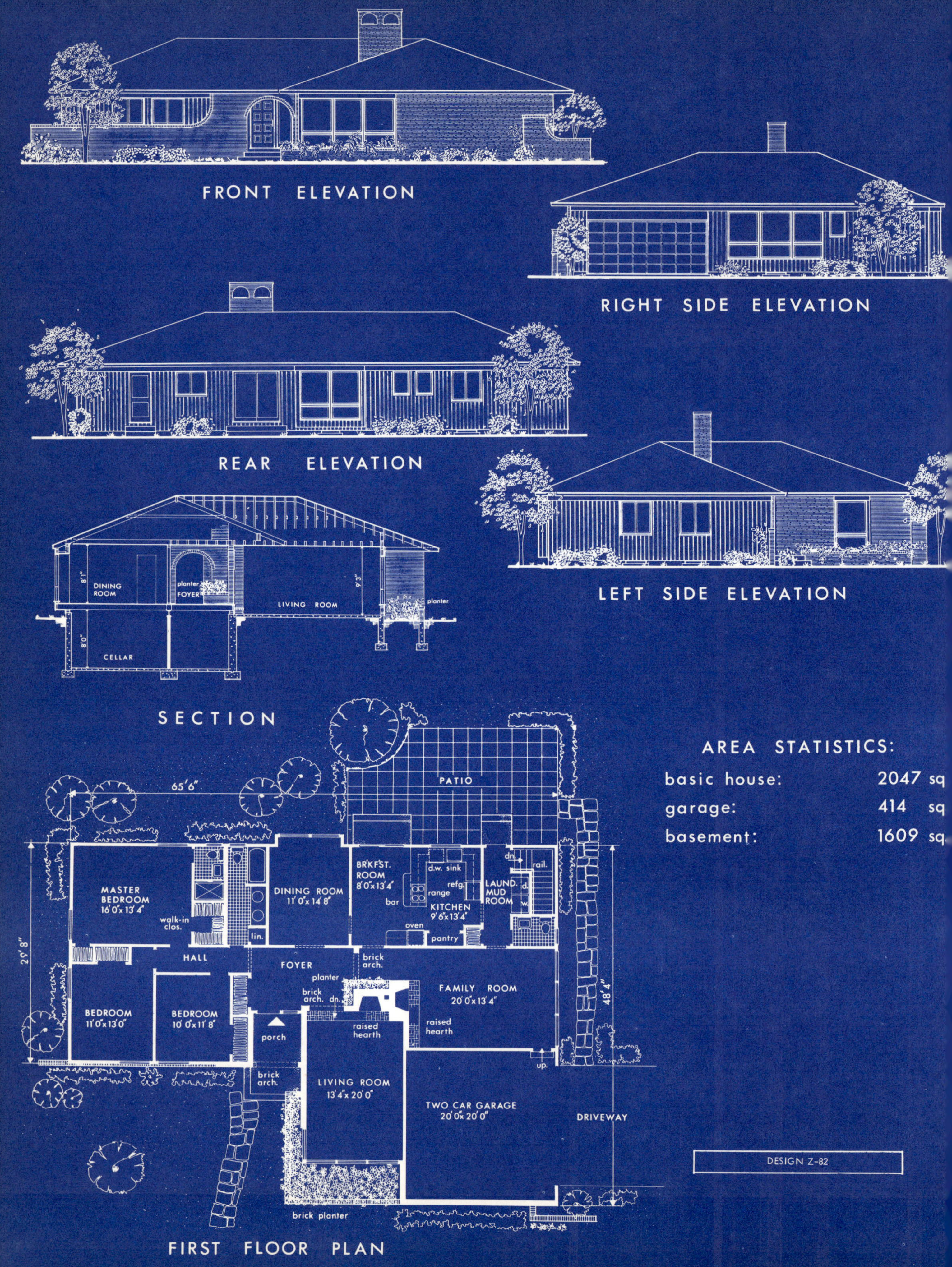
FRONT ELEVATION
RIGHT SIDE ELEVATION
REAR ELEVATION
LEFT SIDE ELEVATION
DINING ROOM
planter
FOYER
LIVING ROOM
planter
CELLAR
SECTION
AREA STATISTICS:
basic house: 2047 sq
garage: 414 sq
basement: 1609 sq
PATIO
65'6"
MASTER BEDROOM 16'0"x13'4"
walk-in clos.
lin.
DINING ROOM 11'0"x14'8"
BRK'FST. ROOM 8'0"x13'4"
d.w. sink
refg.
range
bar
KITCHEN 9'6"x13'4"
oven
pantry
LAUND. MUD ROOM
dn.
rail.
HALL
FOYER
brick arch.
planter
brick arch.
dn.
FAMILY ROOM 20'0"x13'4"
29'8"
48'4"
BEDROOM 11'0"x13'0"
BEDROOM 10'0"x11'8"
porch
raised hearth
raised hearth
up.
brick arch.
LIVING ROOM 13'4"x20'0"
TWO CAR GARAGE 20'0"x20'0"
DRIVEWAY
DESIGN Z-82
brick planter
FIRST FLOOR PLAN

# Brick Arches Dominate Stylized Ranch

Although contemporary in styling, this three-bedroom ranch utilizes a popular feature of Mediterranean design to give it the kind of individuality found only in custom homes.

Its unusual facade has a theme of curved brick arches extended to the inside as well, thus providing a continuity of design.

The exterior combines a sweeping, low hip roof with brick veneer, large front windows and vertical siding to provide what architects and builders call "street appeal."

A brick arch serves as a distinctive entry to a covered portico, which in turn leads to the main foyer. The foyer has a glamour of its own and is a true center hall, affording access to all parts of the house.

The living room is two steps down from the foyer. These brick steps are an extension of the fireplace hearth. Across from the living room, the dining room overlooks the rear garden.

One end of the foyer leads to the kitchen and the family room. Adjoining is a separate breakfast space with sliding glass doors leading to a rear patio.

The family room, with its own brick fireplace and raised hearth, is informal and cheery.

The wing at the left side of the house features three bedrooms, two full baths and plenty of closet space.

## Material List

**CONCRETE WORK**

|  |  |  |
|---|---|---|
| Foundations, Footings, Slabs, etc. | | 142 cu. yds. |

**STEEL**

|  |  |  |
|---|---|---|
| Girder | 8B10# | 65 lin. ft. |
| Lally Cols. | 3½" dia. | 7 pieces |
| Reinforcing Mesh. | W.W.M. | 309 sq. ft. |

**MASONRY**

|  |  |
|---|---|
| Brick Veneer Walls | 686 sq. ft. |
| Brick Fireplace & Chimney | 639 sq. ft. |
| Brick Planter Walls | 78 sq. ft. |
| Interior Brick Walls | 22 sq. ft. |

**FRAMING LUMBER**

|  |  |
|---|---|
| Total sills, joists, rafters, studs, plates, etc. | 15,303 B.F.M. |

**SHEATHING, INSULATION**

|  |  |
|---|---|
| Sub Flooring | 1796 sq. ft. |
| Wall Sheathing | 1610 sq. ft. |
| Roof Sheathing | 3588 sq. ft. |
| Wall Insulation | 1610 sq. ft. |
| Ceiling Insulation | 2057 sq. ft. |

**FINISH, INTERIOR (other than masonry)**

|  |  |
|---|---|
| Vinyl Tile | 591 sq. ft. |
| Oak Flooring | 1326 sq. ft. |
| Ceramic Tile, floors | 86 sq. ft. |
| Ceramic Tile, walls | 312 sq. ft. |
| Gypsum Board - House | 6335 sq. ft. |
| Gypsum Board - Garage | 985 sq. ft. |

**DOOR SCHEDULE**

|  |  |
|---|---|
| Exterior Glazed Sliding | 1 |
| Exterior Hardwood Paneled | 1 |
| Exterior Hardwood Glazed | 1 |
| Flush Fireproof | 1 |
| Interior Hardwood Flush | 10 |
| Interior Hardwood Bi-fold | 1 |
| Interior Hardwood Sliding | 4 |
| Wood Overhead Garage | 1 |

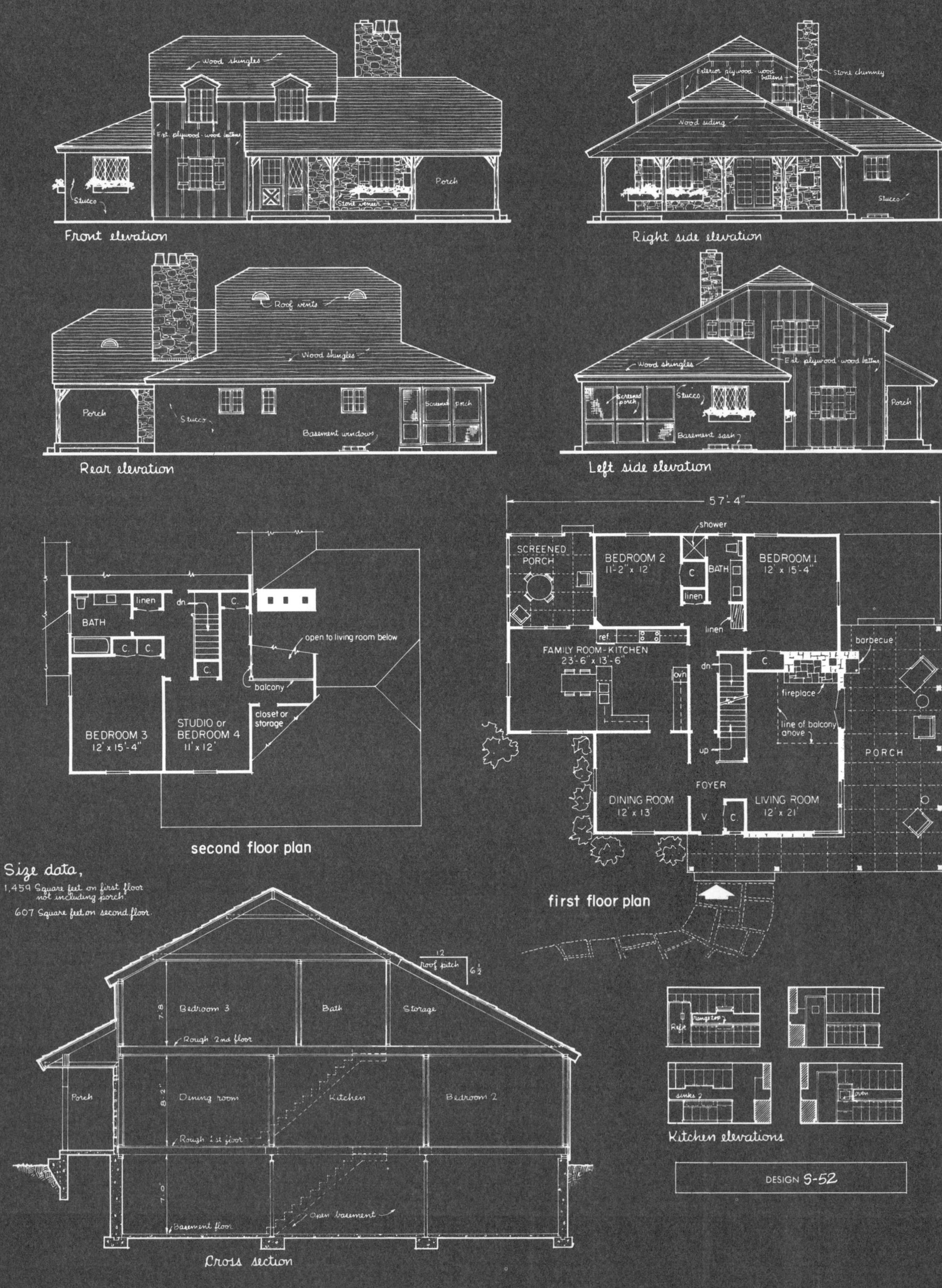

Wood shingles
Ext. plywood-wood battens
Porch
Stucco
Stone veneer
Front elevation
Exterior plywood-wood battens
Stone chimney
Wood siding
Stucco
Right side elevation
Roof vents
Wood shingles
Porch
Stucco
Screened porch
Basement windows
Rear elevation
Wood shingles
Ext. plywood-wood battens
Screened porch
Stucco
Basement sash
Porch
Left side elevation
57'-4"
SCREENED PORCH
BEDROOM 2
11'-2" x 12'
shower
C
BATH
linen
BEDROOM 1
12' x 15'-4"
linen
ref.
FAMILY ROOM-KITCHEN
23'-6" x 13'-6"
ovn
dn.
C.
barbecue
fireplace
line of balcony above
up
PORCH
FOYER
DINING ROOM
12' x 13'
V.
C
LIVING ROOM
12' x 21'
first floor plan
linen
dn.
C.
BATH
C.
C.
C.
open to living room below
balcony
closet or storage
BEDROOM 3
12' x 15'-4"
STUDIO or BEDROOM 4
11' x 12'
second floor plan
Size data,
1,459 Square feet on first floor not including porch.
607 Square feet on second floor.
12
roof pitch
6½
Bedroom 3
Bath
Storage
7'-8"
Rough 2nd floor
Porch
Dining room
Kitchen
Bedroom 2
8'-2"
Rough 1st floor
7'-0"
Open basement
Basement floor
Cross section
Refr.
Range top
sinks
oven
Kitchen elevations
DESIGN S-52

# Balconied Living Room in Woodsy Traditional

This house has a woodsy style reminiscent of many Early American homes.

It not only uses boards and battens on all sides, but it has a wood-shingled roof. Such shingles, of course, should be of the fire-resistant type. In areas where wood roofs are not acceptable, you may select harmonizing asphalt shingles from the wide variety of available colors.

There are four bedrooms, but since two of these are on the first floor, a smaller family need not finish the second floor unless and until it were required. Both of the first-floor bedrooms are at the rear of the house with a bathroom conveniently between them. Typical of the old floor plan styling, this layout has the living and dining rooms on opposite sides of the foyer with an open-sided stair separating the two rooms.

A dramatic touch has been given to the 21-foot living room with an open-balconied ceiling. The balcony is centered on a stone fireplace wall.

To the rear of the dining room is an old-fashioned combined kitchen and family room, stretching to 23′6″.

The dramatic atmosphere of the living room is matched on the exterior by wrapping a large side porch around the front. By building the family room and rear porch on the other side of the house as a one-story structure, any boxiness of the basic design is eliminated.

## Material List

**CONCRETE**

| | |
|---|---|
| Concrete 8″-10″-12″ walls, etc. | 1812 cu. ft. |
| Basement floor | 1372 sq. ft. |
| Porch floor 6 x 6 mesh scored | 690 sq. ft. |
| Waterproofing | 1300 sq. ft. |

**STEEL**

9 - 3½ Lally columns 6′10″
80 Lin. 6″ I 12½ lb.
20 Lin. ft. 4 x 3 - 5/16 angle iron

**MASONRY**

210 sq. ft. - 4″ Stone veneer
480 sq. ft. - 4″ Stone chimney
4200 Common brick backup
46 Lin. 13 x 13 flue lining
18 Lin. 8 x 13 flue lining

**CARPENTRY**

| | |
|---|---|
| Plates, joists, both floors | 4500 BM |
| ⅝ Plyscore sub floor | 2850 sq. ft. |
| Bridging - backing | 600 BM |
| All studs - headers, etc. | 4100 BM |
| Ceiling joists 2 x 6 | 2100 BM |
| Roof framing | 3540 BM |
| ½ Plyscore roof boards | 2370 sq. ft. |
| ⅜ Plyscore sheathing | 2200 sq. ft. |
| ⅜ Ext. hard board siding & ceilings | 1100 sq. ft. |
| 350 Lin. 1 x 2 battens | |
| 11 Rolls - 15 lb. sat. felt | |
| 13/16 x 2¼ Red oak floor | 2000 BM |
| ⅝ Plugged plyscore | 360 sq. ft. |

**ROOFING**

26 sq. 24″ Wood shingles
2″ Min. thickness
126 Lin. hips & ridges
12 Rolls 30 lb. felt

**PLASTERBOARD**

| | |
|---|---|
| ½ for walls | 5700 sq. ft. |
| ⅜ for ceiling | 2000 sq. ft. |

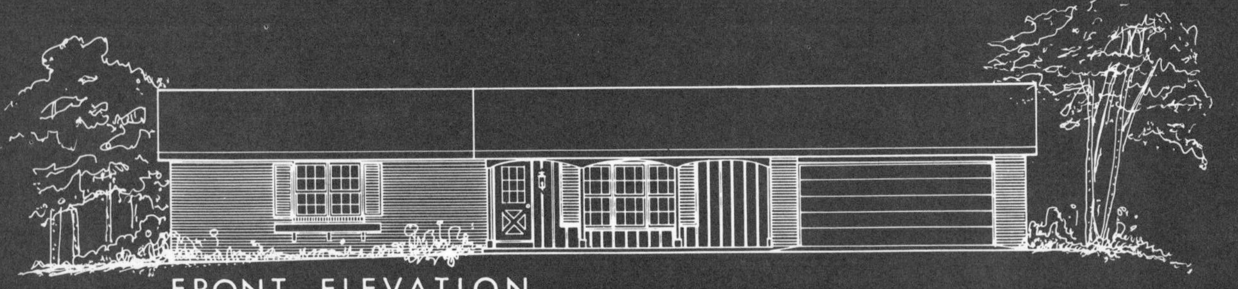

FRONT ELEVATION

RIGHT SIDE ELEVATIO

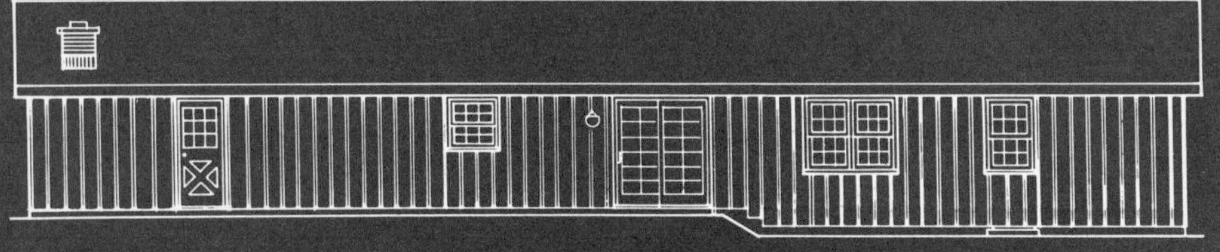

REAR ELEVATION

LEFT SIDE ELEVATION

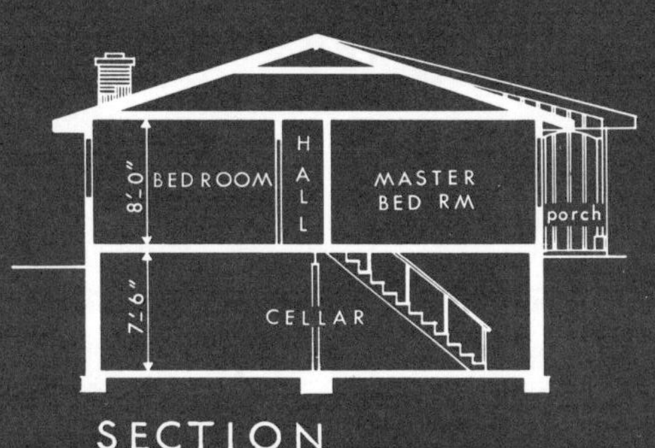


SECTION

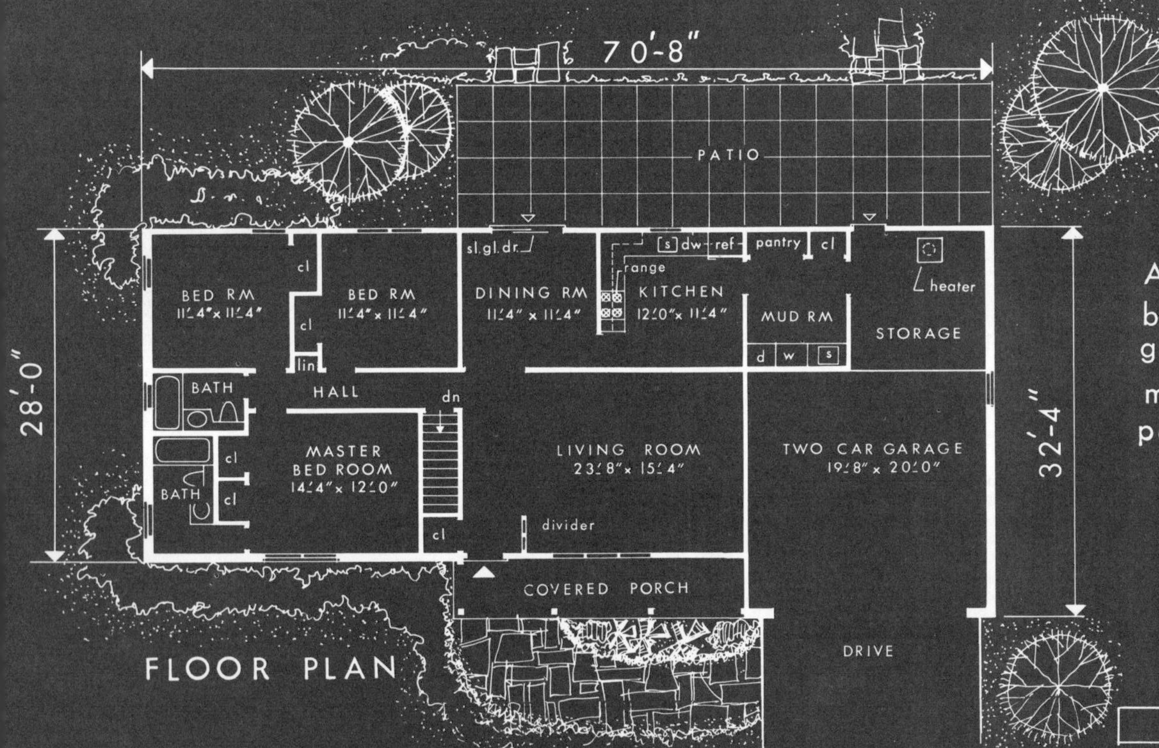


FLOOR PLAN

## AREA STATISTICS

basic house ..... 1,383 sq

garage storage &

mud rm ........... 640

partial cellar... 710 "

DESIGN L-47

# Seems Large But Has Modest Living Area

The modest area of this one-story house has been effectively combined with an exterior that gives the appearance of bigness.

Economy was considered in the design. With only 1383 square feet of livable area, the house nevertheless stretches nearly 71′ across the front. The covered entrance porch, with its arched motif, and the vertical siding and large sheltered window emphasize the entrance.

As you enter the house from the covered porch, there is a vestibule defined by a closet on the left and a divider on the right which can also be used as a planter. The spacious living room merges with the entry and adjoins the dining room.

The square dining room has quick access to a rear patio through sliding glass doors. In the adjoining kitchen is an L-shaped arrangement of counters and appliances.

To the left of the living room is the hall leading to the three bedrooms. The master bedroom has a wall of closets and direct access to its own private bathroom.

Behind the two-car garage is a storage room. It contains space for heater equipment. There is an entry directly from the garage to the mud room.

The simplicity of construction supports the economical cost. A bearing partition almost down the middle of the 28-foot-deep house enables small-sized lumber to be used.

## Material List

**CONCRETE WORK**
Foundations, footings, slabs, etc. ... 80 cu. yds.

**MASONRY**
Exterior Walls 4″ Brick Veneer ... 685 sq. ft.

**FRAMING LUMBER**
Total Sills, Joists, Rafters, Studs, Plates, etc. ... 9064 B.F.M.

**SHEATHING, INSULATION**
Sub Flooring ... 700 sq. ft.
Wall Sheathing ... 1673 sq. ft.
Roof Sheathing ... 2736 sq. ft.
Wall Insulation ... 1600 sq. ft.
Ceiling Insulation ... 2000 sq. ft.

**FINISHES, INTERIOR**
Finish Flooring ... 2023 sq. ft.
Ceramic Tile Floors ... 59 sq. ft.
Ceramic Tile Walls ... 248 sq. ft.
Gypsum Board — house walls ... 3550 sq. ft
Gypsum Board — house ceilings ... 1300 sq. ft.
Gypsum Board — Garage ... 1325 sq. ft.

**FINISHES, EXTERIOR (other than masonry)**
Vertical Siding ... 988 sq. ft.
Asphalt Shingle Roofing ... 2736 sq. ft.
Plywood Eave & Porch Soffits ... 558 sq. ft.

**WINDOW SCHEDULE**
Wood Double Hung ... 13 units
Basement ... 3 units

**DOOR SCHEDULE**
Ext. Hardwood, paneled & glazed ... 2 units
Aluminum Sliding ... 1 unit
Wood Overhead garage ... 1 unit
Int. Hardwood, flush, staingrade ... 6 units
Int. Hardwood, louvered, bi-folding ... 7 units

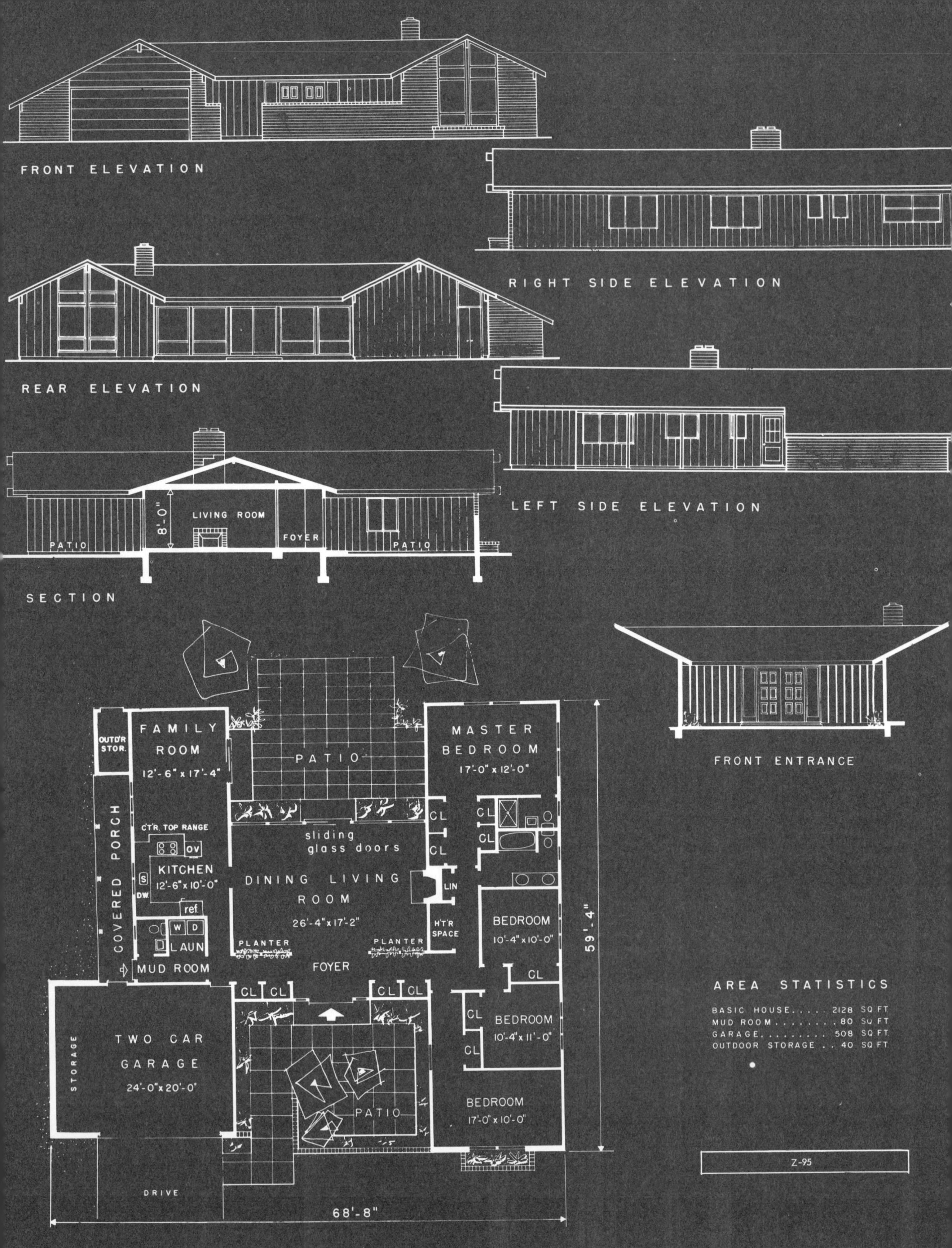
FRONT ELEVATION
RIGHT SIDE ELEVATION
REAR ELEVATION
LEFT SIDE ELEVATION
LIVING ROOM
8'-0"
PATIO
FOYER
PATIO
SECTION
FRONT ENTRANCE
FAMILY ROOM
12'-6" x 17'-4"
OUTD'R STOR.
PATIO
MASTER BEDROOM
17'-0" x 12'-0"
CL
CL
CL
CL
CTR. TOP RANGE
OV
KITCHEN
12'-6" x 10'-0"
S
DW
ref.
W D
LAUN
COVERED PORCH
sliding glass doors
DINING LIVING ROOM
26'-4" x 17'-2"
LIN
H'T'R SPACE
PLANTER
PLANTER
BEDROOM
10'-4" x 10'-0"
CL
MUD ROOM
FOYER
CL CL
CL CL
59'-4"
CL
CL
BEDROOM
10'-4" x 11'-0"
STORAGE
TWO CAR GARAGE
24'-0" x 20'-0"
PATIO
BEDROOM
17'-0" x 10'-0"
DRIVE
68'-8"
FLOOR PLAN
AREA STATISTICS
BASIC HOUSE..... 2128 SQ FT
MUD ROOM........ 80 SQ FT
GARAGE......... 508 SQ FT
OUTDOOR STORAGE .. 40 SQ FT
Z-95

# Privacy Theme of Contemporary Ranch

An H-shaped design gives this contemporary ranch substantial amounts of privacy on the outside and a roomy, well-planned design within. The view from the street incorporates a long, low silhouette composed of large masses of brick, including a privacy wall, and several smart frame gable roofs.

The H-shape of the interior layout provides for an excellent zoning arrangement: daily activity and service facilities occupy the left arm of the H, private sleeping quarters occupy the right arm of the H, with the entertaining area composing the horizontal bar of the H. The private exterior spaces created by this plan are a striking front courtyard and a sheltered rear patio "wrapped" in the wings of the house.

The spacious reception foyer has four closets. Directly ahead is an enormous living-dining room.

The efficient, U-shaped kitchen has a built-in oven and dishwasher, plenty of cabinet space and a breakfast bar facing the adjacent family room. The latter has dual exposures, one of which leads to the rear patio. A continuous covered porch runs along the side of the house and can be a covered sitting or snack area. A good-sized outdoor storage room is adjacent to the porch.

The entire right wing of the house is a sleep zone incorporating four bedrooms, two full baths and plenty of closet space.

## Material List

**CONCRETE WORK**

| | |
|---|---|
| Foundations, Footings, Slabs, etc. | 100 cu. yds. |

**MASONRY**

| | |
|---|---|
| Brick Fireplace, Barbecue, Chimneys | 14 cu. yds. |

**FRAMING LUMBER**

| | |
|---|---|
| Total Sills, Joists, Rafters, Studs, Plates | 5000 B.F.M. |

**SHEATHING, INSULATION**

| | |
|---|---|
| Wall Sheathing | 4000 sq. ft. |
| Roof Sheathing | 3800 sq. ft. |
| Wall Insulation | 2900 sq. ft. |

**FINISHES, EXTERIOR**

| | |
|---|---|
| Horizontal Siding | 32 sq. ft. |
| Vertical Siding | 2200 sq. ft. |
| Asphalt Shingle Roofing | 4000 sq. ft. |
| Plywood Eave & Soffit | 1000 sq. ft. |

**FINISHES, INTERIOR**

| | |
|---|---|
| Vinyl Tile | 600 sq. ft. |
| Ceramic Tile Floor | 90 sq. ft. |
| Ceramic Tile Walls | 197 sq. ft. |
| Gypsum Board House Walls | 3316 sq. ft. |
| Gypsum Board House Ceiling | 2240 sq. ft. |
| Gypsum Board Garage | 1140 sq. ft. |

**DOOR SCHEDULE**

| | |
|---|---|
| Exterior Glazed Sliding | 2 |
| Exterior Glazed Hardwood | 2 |
| Exterior Hardwood Flush | |
| Interior Fireproof Flush | 1 |
| Wood Overhead Garage | 1 |
| Interior Hardwood Flush | 19 |
| Interior Hardboard Bi-Folding | 7 pair |

**WINDOW SCHEDULE**

| | |
|---|---|
| Wood Casement | 16 |
| Awning | 8 |
| Fixed | 4 |

FRONT ELEVATION

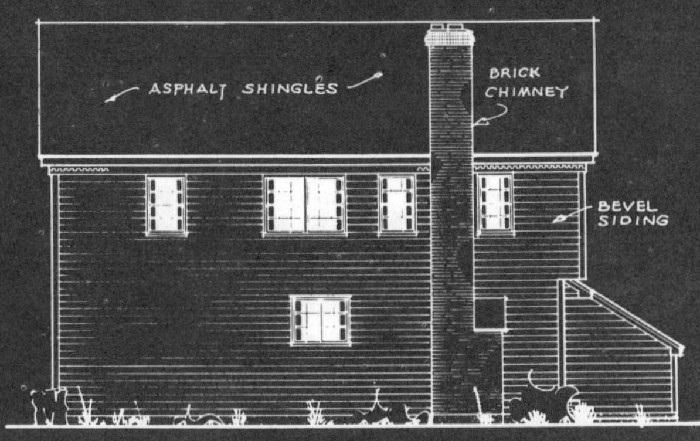


RIGHT SIDE ELEVATION

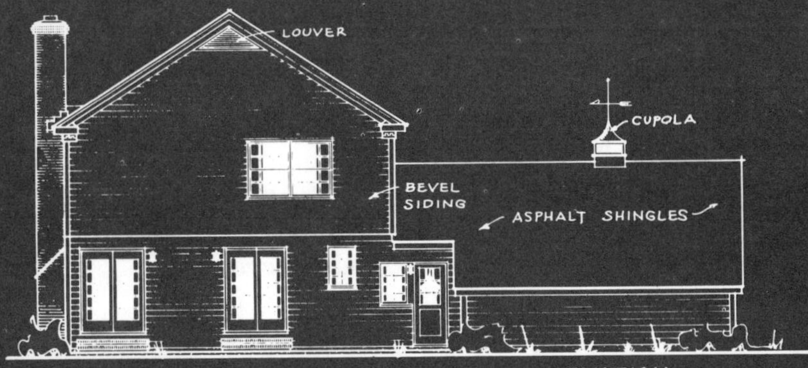


REAR ELEVATION

LEFT SIDE ELEVATION

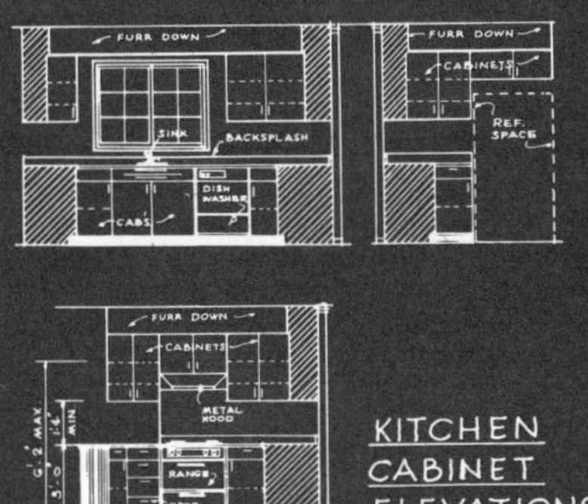


KITCHEN CABINET ELEVATIONS

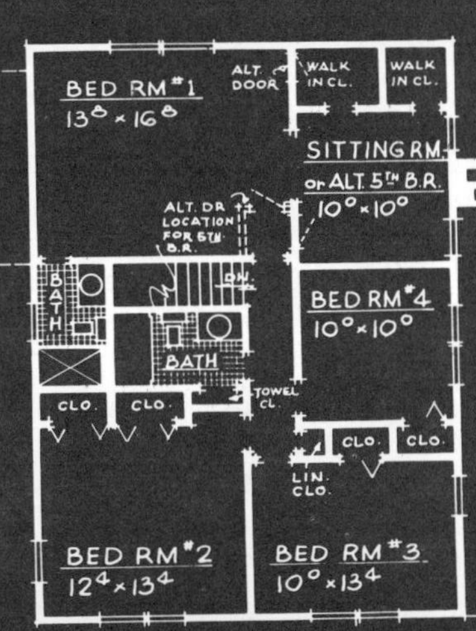


SECOND FLOOR PLAN

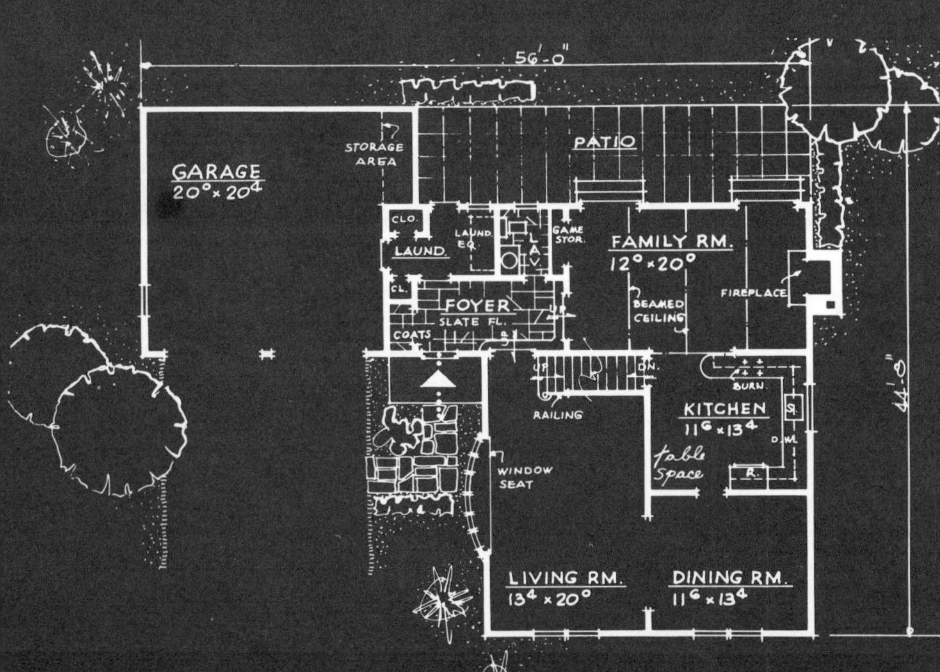


FIRST FLOOR PLAN

## HOUSE DATA

FIRST FLOOR .....1123 SQ. FT.
SECOND FLOOR...1052 SQ. FT.
GARAGE ...........451 SQ. FT.

Z-97

# A Two-Story with Nantucket Flavor

This modern-as-tomorrow house has exterior styling reminiscent of Colonial residences on the island of Nantucket.

Recognizing the preference for traditional architectural lines, the architect has placed within such a framework the popular family room, the large slate-floored foyer, the laundry room and the two-car garage. The modest dimensions of 56′ by 44′8″ include these features as well as a patio that stretches along the rear of the house.

As one approaches the entrance door, an immediate feature is brought into focus — the large bow window of the living room.

The entrance foyer, spacious and charming, gains additional allure because it is slightly below the living and family rooms.

At the rear of the house and adjacent to the patio is the family room, with two pairs of French doors, beamed ceiling and fireplace. This room can be decorated in Colonial style and doubtless would become the focal point of the house because of its location.

The kitchen has all work centers in the efficient U-shape. The dining room is separate but partially open to the living room, which has a window seat in the bow window.

Upstairs, there are four bedrooms and a sitting room. The latter can be a fifth bedroom, but if used as part of the owner's bedroom, becomes part of a master suite with two walk-in closets and private bath.

## Material List

|  | |  |
|---|---|---|
| **CONCRETE WORK** | | |
| Concrete Walls | | 784 cu. ft. |
| Found. Foot. | | 196 cu. ft. |
| Slabs | | 578 cu. ft. |
| **STRUCTURAL STEEL** | | |
| Lally Columns | | 6 each |
| Girder | 7″ I | 45 lin. ft. |
| Girder | 7″ B | 10 lin. ft. |
| **BRICK WORK** | | |
| Brick | Chimney | 181 cu. ft. |
| Flue Lining | T.C. | 57 lin. ft. |
| **CARPENTRY** | | |
| Framing Lumber | | 7890 B.F. |
| Studs | | 4000 B.F. |
| Plates | | 1400 B.F. |
| Roof Sheathing | | 2274 sq. ft. |
| Side Wall Sheathing | | 3114 sq. ft. |
| Wall Insulation | | 1822 sq. ft. |
| Ceiling Insulation | | 1215 sq. ft. |
| Wood Flooring | | 1757 sq. ft. |
| Resilient Flooring | | 76 sq. ft. |
| Kitchen Flooring | | 154 sq. ft. |
| **MILLWORK** | | |
| Exterior Doors & Frames Complete | | 7 pieces |
| Garage Door Complete Set | | 2 sets |
| Interior Doors & Frames Compl. | | 16 pieces |
| Bi-Fold Doors | | 8 sets |
| Fascia | | 270 lin. ft. |
| Base | | 750 lin. ft. |
| Windows | | 20 units |
| **KITCHEN CABINETS** | | |
| Base Cabinets | | 10′4″ long |
| Wall Cabinets | | 16′2″ long |
| **ROOFING** | | |
| Shingles | 235# asphalt | 2274 sq. ft. |
| Roofing Paper | 15# felt | 2274 sq. ft. |

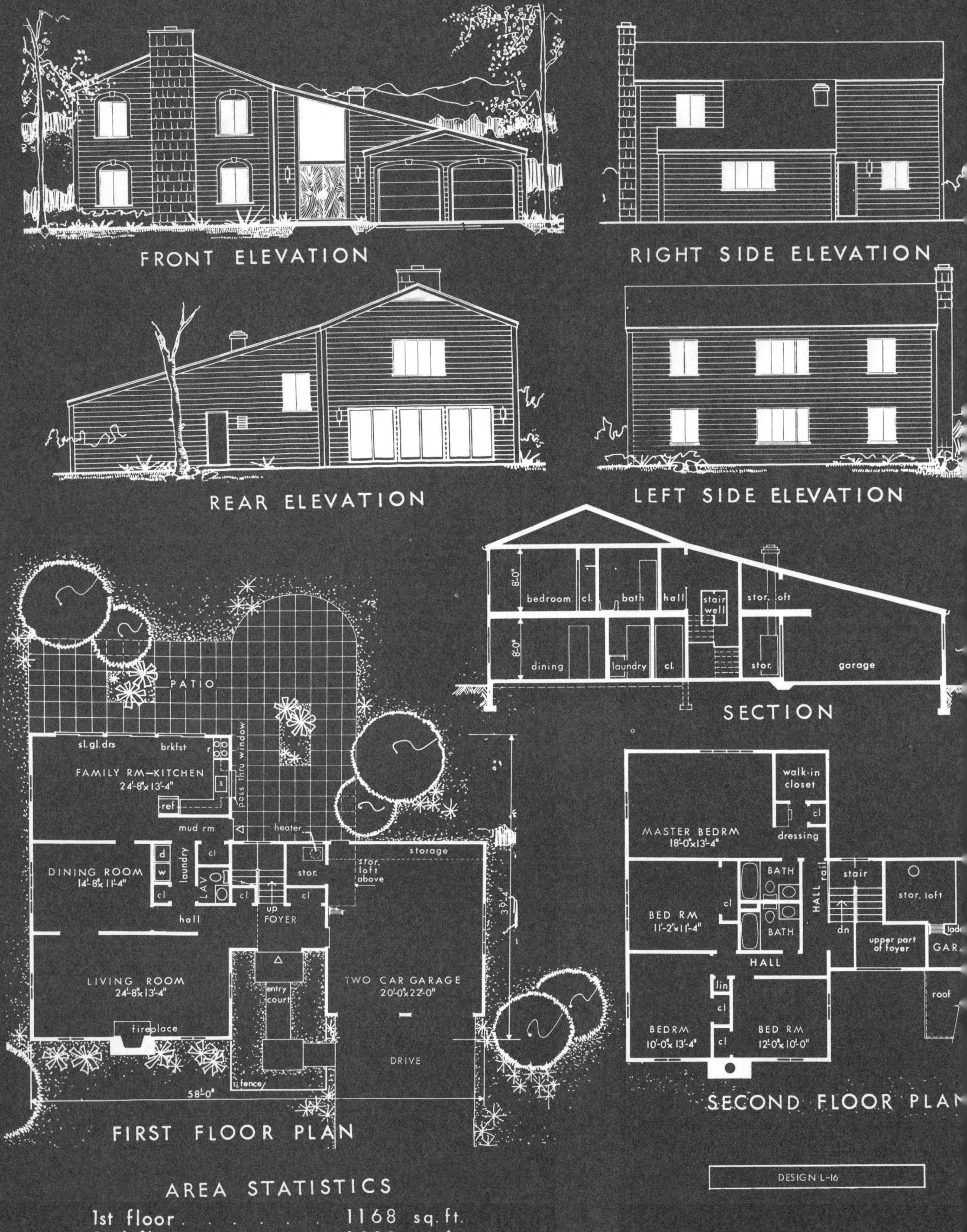


DESIGN L-16

## AREA STATISTICS

1st floor . . . . . 1168 sq. ft.
2nd floor . . . . . 1010 sq. ft.
garage . . . . . . . 427 sq. ft.

# Two-Story Farmhouse Blends Old and New

The nostalgia provoked by a real old-fashioned Colonial — with its wings, additions and picket fence enclosure — bespeaks all the attributes one seeks in a home: warmth, coziness and a sense of belonging.

Sheer imitation of the past, however, is not the true desire of the family of today. While it wants the characteristics embodied in the old styles, it also wants them in tune with current modes of living. This creation blends the best of the past with the best of the present. Its format is a two-story farmhouse Colonial in shape, form and basic feeling. But in materials, windows and the interior layout, it is in step with today.

A charming three-sided yard serves as a forecourt to the house. Inside, the contemporary qualities of the home burst forth. The foyer is a dramatic cathedral-ceilinged space soaring up to a second-floor balcony.

The entire front of the first floor comprises a huge living room — 24′8″ by 13′4″ — with three exposures and a central formal fireplace.

The entire rear of the first floor is a light, cheery and spacious family room-kitchen, equal in size to the living room. The rear wall is made up of sliding glass doors.

On the second floor is a lavish master suite complete with a dressing area, walk-in closets and private bath, and a separate children's suite comprising three bedrooms, plenty of closets and another full bath.

## Material List

**CONCRETE WORK**

| | |
|---|---|
| Foundations, Footings, Slabs, etc. | 41 cu. yds. |

**FRAMING LUMBER**

| | |
|---|---|
| Total Sills, Rafters, Studs, Plates, etc. | 10,337 B.F.M. |

**DRYWALL**

| | |
|---|---|
| ½″ Gypsum Board | 9170 sq. ft. |

**SHEATHING, INSULATION, SIDING**

| | |
|---|---|
| Sub Flooring | 1095 sq. ft. |
| Wall Sheathing | 2800 sq. ft. |
| Insulation | 3525 sq. ft. |
| Roof Sheathing | 1750 sq. ft. |
| Bevel Siding | 2550 sq. ft. |

**FLOORING**

| | |
|---|---|
| Oak | 820 sq. ft. |
| Vinyl Tile | 500 sq. ft. |
| Carpet | 56 sq. yds. |

**EXTERIOR FINISH**

Bevel Wood Siding and Wood Shingles

**WINDOWS, DOORS**

16 Wood Casement units
4 Exterior Solid Core Doors
3 Aluminum Sliding units
3 Interior Sliding units
2 Interior bi-fold units
14 Interior hollow core doors

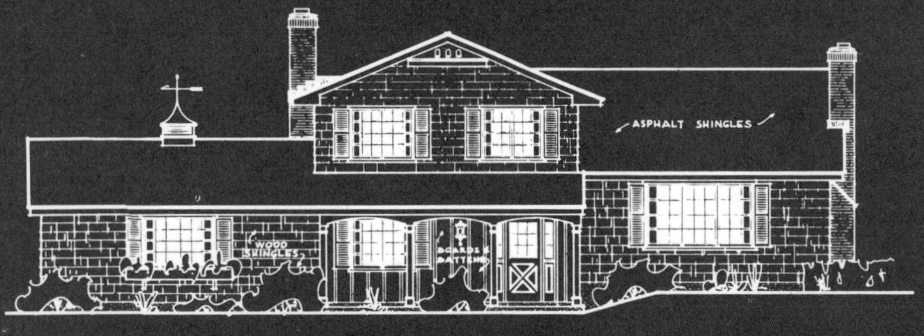

ASPHALT SHINGLES
WOOD SHINGLES
BOARDS & BATTENS

~FRONT ELEVATION~

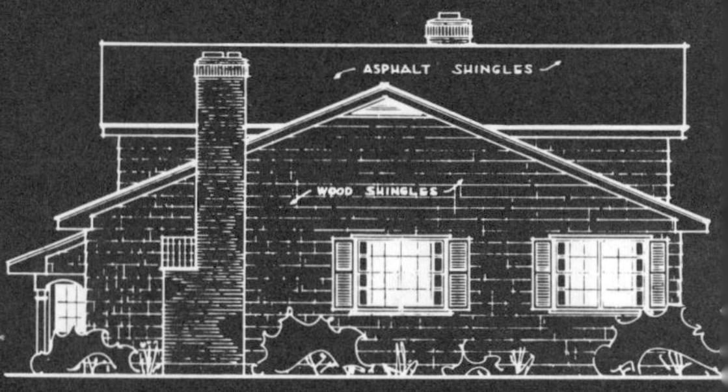

ASPHALT SHINGLES
WOOD SHINGLES

~RIGHT SIDE ELEVATION~

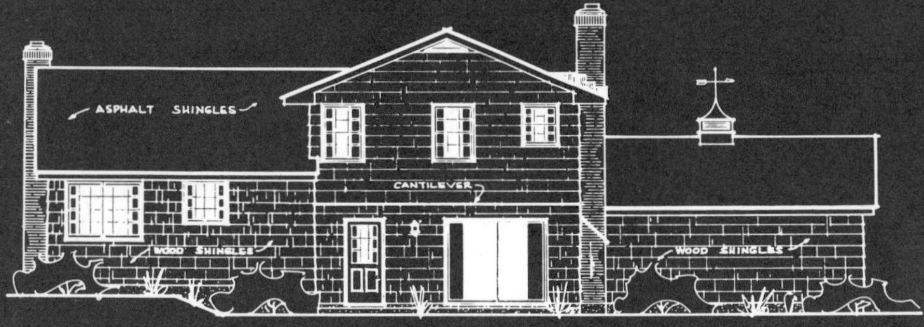

ASPHALT SHINGLES
CANTILEVER
WOOD SHINGLES
WOOD SHINGLES

~REAR ELEVATION~

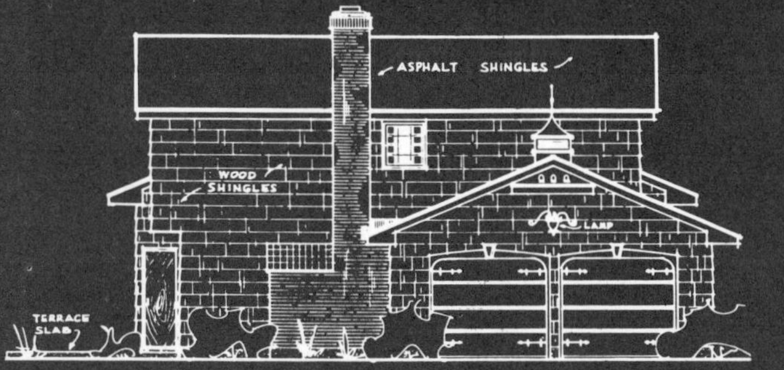

ASPHALT SHINGLES
WOOD SHINGLES
LAMP
TERRACE SLAB

~LEFT SIDE ELEVATION~

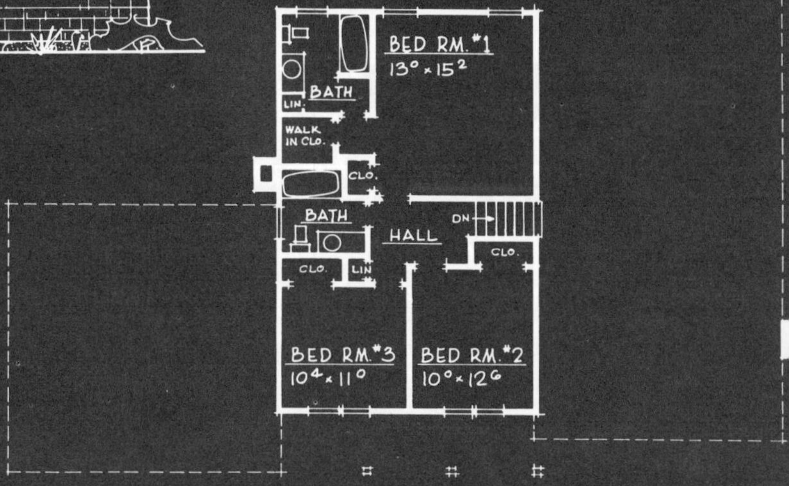

BED RM. #1
13⁰ × 15²
BATH
LIN.
WALK IN CLO.
CLO.
BATH
HALL
DN.
CLO.
CLO.
LIN
BED RM. #3
10⁴ × 11⁰
BED RM. #2
10⁰ × 12⁶

~UPPER LEVEL PLAN~

64'-0"
PATIO
OUTDOOR PACKAGE RECEIVER
SLIDING GL. DRS.
R. D.W. S.
DINETTE
8⁰ × 11⁴
KITCH.
11⁰ × 11⁴
UP
FAMILY RM.
12⁰ × 20⁰
LAUND.
D. W. S
PANTRY
REF.
DESK
LOG STORAGE
FIREPLACE
LAV.
UPPER FOYER
BRICK WALL
DN.
UP
UP
DINING RM.
11⁴ × 14⁸
COATS
W.I. RAILING
40'-3"
TWO CAR GARAGE
22⁰ × 22⁰
DEN or 4TH B.R.
10⁰ × 11⁴
CLO
LOWER FOYER
LIVING RM.
13⁴ × 20⁴
ALTERNATE GARAGE DOOR LOCATION
COVERED PORTICO

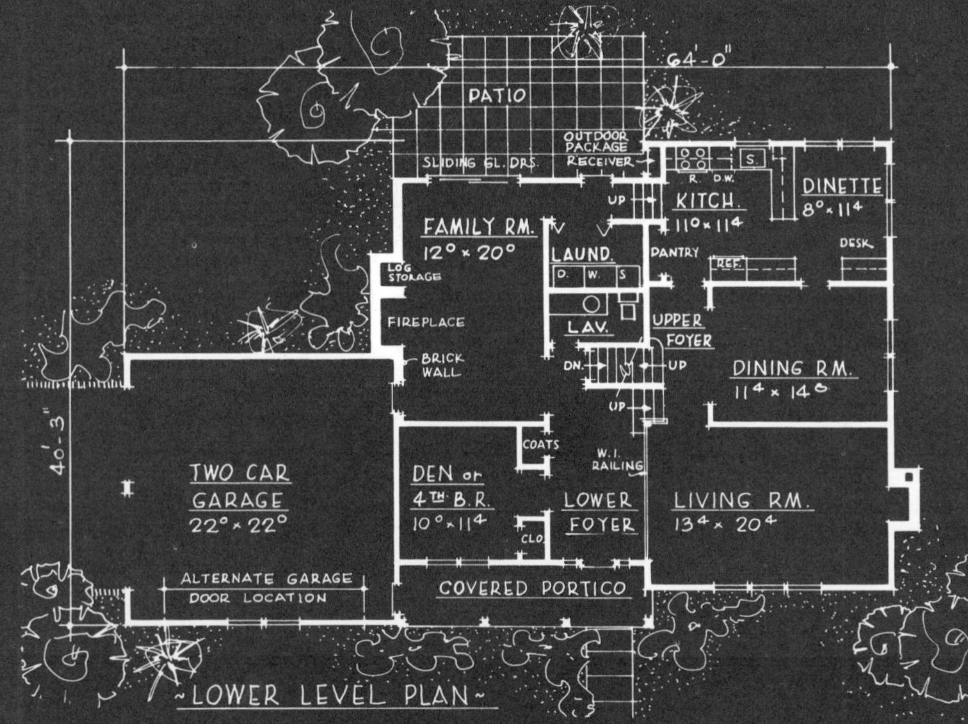

~LOWER LEVEL PLAN~

## ~HOUSE DATA~

| | |
|---|---|
| LOWER LEVEL . . . | 1450 SQ. FT. |
| UPPER LEVEL . . . | 730 SQ. FT. |
| GARAGE . . . . . . . . | 506 SQ. FT. |

L-14

# Split-Level With Distinctive Floor Plan

Technically speaking, this design qualifies equally well as either a split-level or a two-story house. The concept here is to reduce the number of steps between the areas most frequently used, hence only four short steps connect the family room level with the kitchen level, whereas nine risers lead to the bedrooms above. A split-level it is but with the characteristics of a two-story.

The foyer flows visually into the living room because it is separated only by an attractive railing and the living room level is only 32 inches higher.

The family room, with its large fireplace, is within easy reach of the front door, rear patio, kitchen and lavatory.

There is an advantage to the downstairs den or bedroom, for here a member of the family can relax or take an afternoon nap.

In the living room, there is a fireplace with a chimney large enough to handle the furnace flue.

The dining room is separate off the upper foyer, with direct access to the dinette. There are four large windows in the dinette with room for a writing or planning desk. A large pantry in the kitchen creates generous storage space.

On the upper level there are three bedrooms and two bathrooms, one private for the owners. The owners' bedroom is very large with an unusual amount of wall space.

## Material List

**CONCRETE WORK**

| | |
|---|---|
| Concrete Walls | 720 cu. ft. |
| Slabs | 722 cu. ft. |
| Foundation Footings | 247 cu. ft. |
| Misc. Concrete | 233 cu. ft. |

**STRUCTURAL STEEL**

| | | |
|---|---|---|
| Lally Columns | 3½" diam. | 6 pcs. |
| Girder | 6" I Beam | 21 lin. ft. |

**BRICK WORK**

| | | |
|---|---|---|
| Chimney | Brick & Block | 265 cu. ft. |

**CARPENTRY**

| | |
|---|---|
| Framing Lumber | 7498 B.F. |
| Studs | 4267 B.F. |
| Plates | 1280 B.F. |
| Roof Sheathing | 2868 sq. ft. |
| Sub Flooring | 1504 sq. ft. |
| Side Wall Sheathing | 2950 sq. ft. |
| Wall Insulation | 2187 sq. ft. |
| Ceiling Insulation | 1404 sq. ft. |
| Wood Flooring | 1273 sq. ft. |
| Resilient Flooring | 678 sq. ft. |
| Kitchen Underlayment | 125 sq. ft. |

**MILLWORK**

| | |
|---|---|
| Exterior Doors & Frame Compl. | 4 pcs. |
| Garage Door Complete Set | 2 units |
| Sliding Glass Doors | 1 unit |
| Int. Doors & Frames Compl. | 19 pcs. |
| Bi Fold Doors | 3 units |
| Rake | 350 lin. ft. |
| Base | 750 lin. ft. |
| Windows | 18 units |
| Stairs ....... 8 risers | 1 set |
| Stairs ....... 4 risers | 1 set |
| Stairs ....... 9 risers | 1 set |

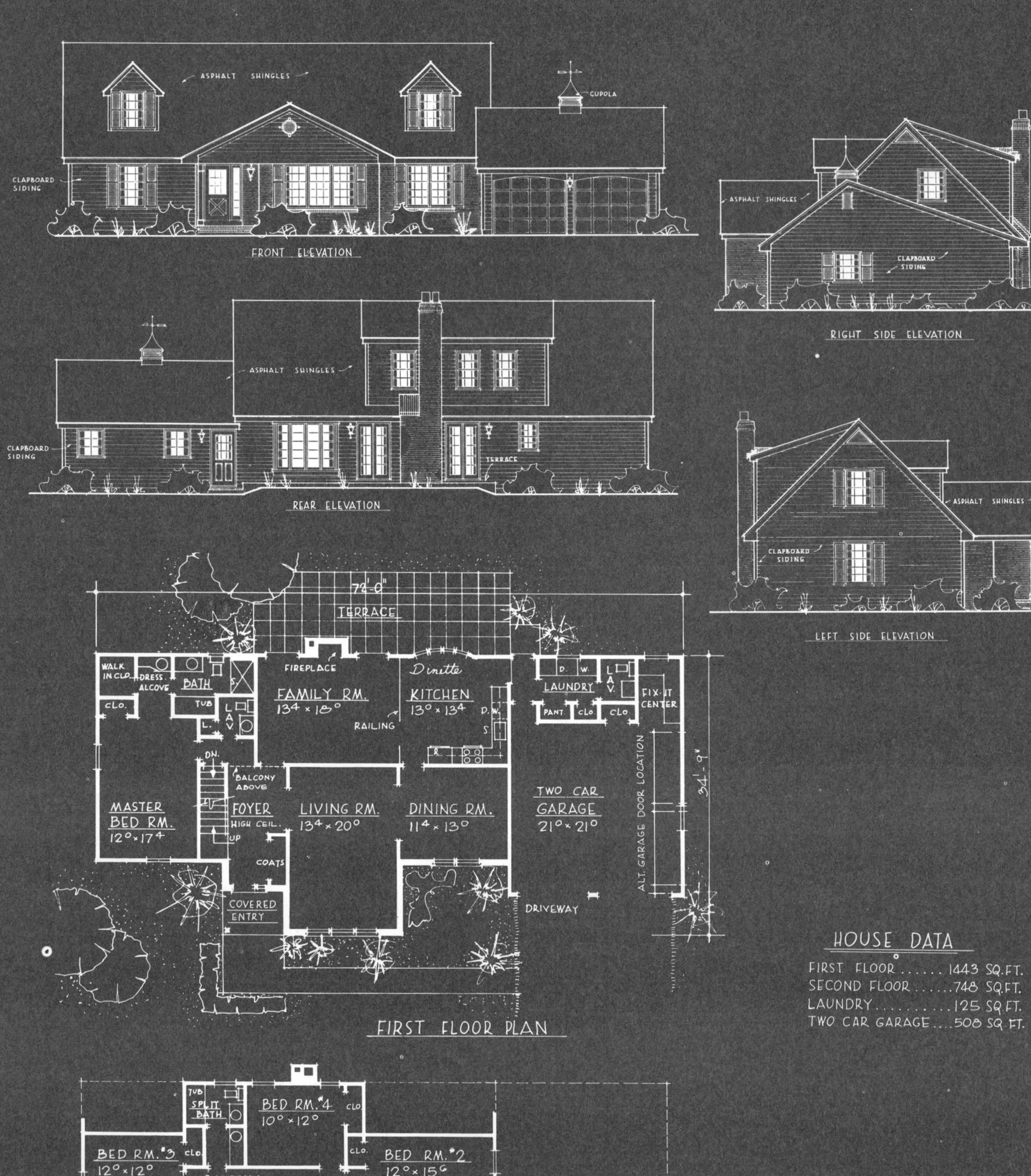


TUB
SPLIT BATH
BED RM. #4
10⁰ × 12⁰
CLO.
BED RM. #3
12⁰ × 12⁰
CLO.
CLO.
BED RM. #2
12⁰ × 15⁶
BALCONY
RAILING
LINEN
CLO.
UPPER FOYER
STORAGE AREA

SECOND FLOOR PLAN

DESIGN L-21

# Master Bedroom Gets Choice Location

While the uncluttered exterior design of this house is traditional, the floor plan is contemporary in concept.

Not seen too often is the location of the master bedroom in a wing of its own on the first floor and completely separate from the three other bedrooms on the upper level. There are several reasons for this location. It gives complete privacy to the owners, eliminates stair climbing and provides good balance between both floor areas, particularly in a one and one-half story layout.

Another unusual feature of the plan is the balcony upper hall facing the two-storied entrance foyer, creating an immediate impression of space.

The foyer is dramatic because of its ceiling height and large size.

The living room, although of modest size, will furnish well because traffic crosses only one end. The dining room is in a "dead end" arrangement, desirable for a room which is used less frequently than most others.

The fireplace in the family room is the focal point as one enters from the foyer. In this plan, the family room is combined with the kitchen, making an area 13′4″ x 31′, a huge space for the gathering of the "clan." Two pairs of sliding glass doors connect with the rear terrace.

Upstairs three large bedrooms are arranged to give an element of privacy for each.

## Material List

**CONCRETE WORK**

| | | |
|---|---|---|
| Concrete Walls | | 952 cu. ft. |
| Slabs | | 770 cu. ft. |
| Foundation Footings | | 291 cu. ft. |
| Misc. Concrete | | 327 cu. ft. |

**STRUCTURAL STEEL**

| | | |
|---|---|---|
| Lally Columns | | 9 pieces |
| Girder | 6″ & 7″ I Beam | 79 lin. ft. |

**BRICK WORK**

| | | |
|---|---|---|
| Chimney | Brick | 120 cu. ft. |
| Flue Lining | T. C. | 44 lin. ft. |
| Veneer | Brick | 372 sq. ft. |

**CARPENTRY**

| | |
|---|---|
| Framing Lumber | 8943 B.F. |
| Studs | 4000 B.F. |
| Plates | 1200 B.F. |
| Roof Sheathing | 3400 sq. ft. |
| Sub Flooring | 2500 sq. ft. |
| Side Wall Sheathing | 2500 sq. ft. |
| Insulation Walls | 2000 sq. ft. |
| Insulation Ceilings | 1600 sq. ft. |
| Wood Flooring | 2100 sq. ft. |
| Kitchen Plywood | 174 sq. ft. |

**MILLWORK**

| | | |
|---|---|---|
| Ext. Doors & Frames Complete | | 6 pieces |
| Interior Doors & Frames Compl. | | 15 units |
| Bi Fold Door Units | | 12 units |
| Windows | | 22 units |
| Base | | 730 lin. ft. |
| Stairs | 12 risers | 1 set |
| Stairs | 13 risers | 1 set |

**KITCHEN CABINETS**

| | |
|---|---|
| Base Cabinets | 8.75 lin. ft. |
| Wall Cabinets | 17 lin. ft. |

**ROOFING**

| | | |
|---|---|---|
| Shingles | 235# asphalt | 3400 sq. ft. |
| Roofing Paper | 15# Felt | 3400 sq. ft. |

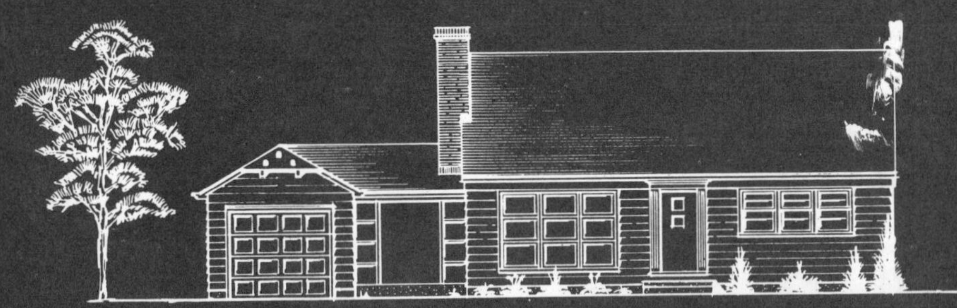

FRONT ELEVATION

RIGHT SIDE ELEVATION

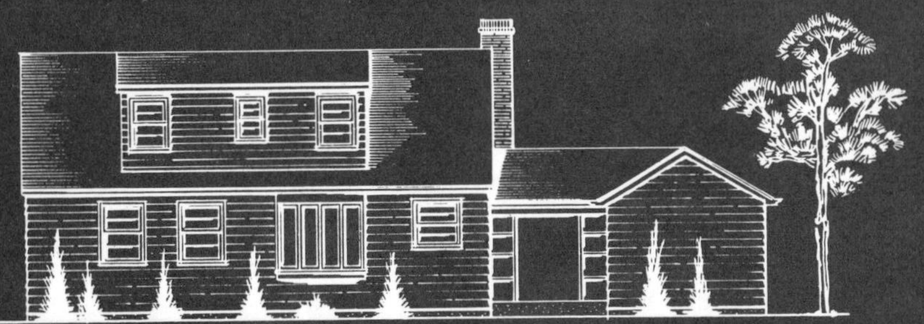

REAR ELEVATION

LEFT SIDE ELEVATION

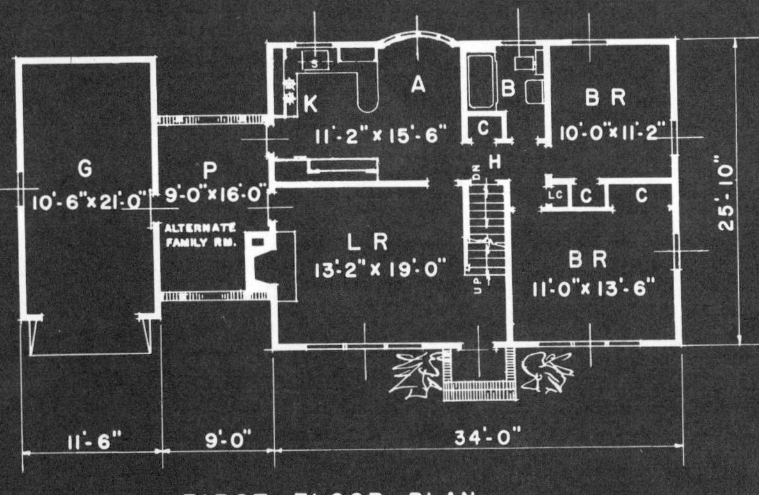


FIRST FLOOR PLAN

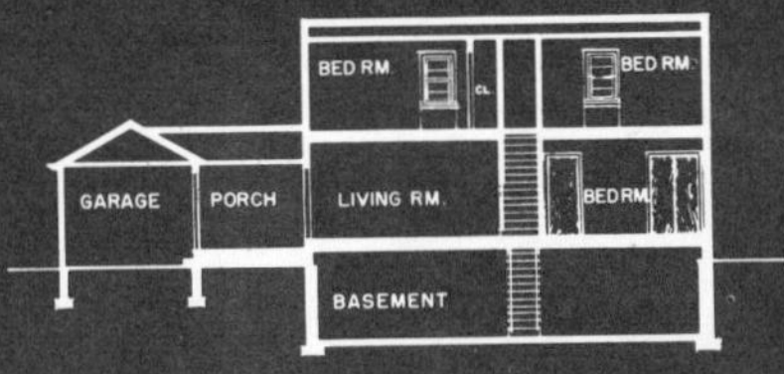


LONGITUDINAL SECTION

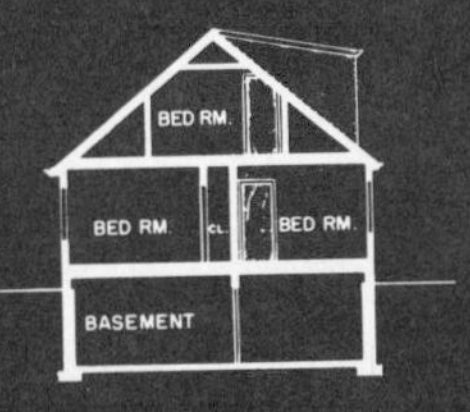


CROSS SECTION

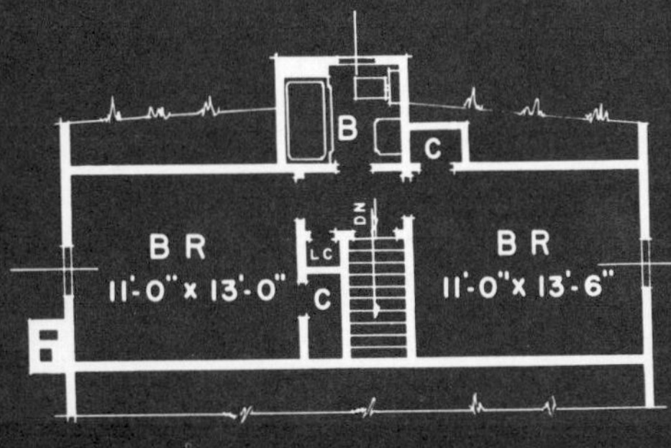


SECOND FLOOR PLAN

AREA STATISTICS:

SQ.FT. FIRST FLOOR....... 877 SQ.FT.
SQ.FT. SECOND FLOOR..... 450 SQ.FT.
SQ.FT. BASEMENT........ 877 SQ.FT.
SQ.FT. GARAGE........ 253 SQ.FT.

DESIGN L-9

# Small House Also Handles Large Family

Basically a modest house for a couple or a small family, this wood-shingled traditional can be transformed into a sizable home for a large family by utilizing its easily expandable features.

The main plan has a living room, kitchen-dinette, two bedrooms and a bath, totaling 877 square feet of living area within an economy-geared rectangular shape.

Two bedrooms are at the right side of the house. The bathroom, with built-in tub, is well-placed between the living and sleeping areas.

If it is desired to have a den or family room on this floor, the open porch between the house and the one-car garage can be enclosed. Being near the kitchen, it is convenient for serving refreshments or light snacks. It might also be used as a home office for a professional man, as a study for a school child or as an extra bedroom. Or, to carry the possibilities even further, it can be left as a kind of breezeway or eliminated altogether.

Plans are included for two extra bedrooms and a second bath should the owners decide to finish the upstairs area.

Outside, Design L-9 has a cottage-like charm, with a full-height glass-paneled window setting off the front door. Machine-split wood shingles are used on the garage as well as the house, with shutters on the front bedroom window.

## Material List

**CONCRETE WORK**

Footings, floors, etc. .......... 40 cu. yds.

**MASONRY**

8" Block ..................... 1700 blocks
Pancake ...................... 91 blocks
Brick ........................ 190 sq. ft.

**FRAMING LUMBER**

Sills, joists, rafters, studs, plates, etc. ................ 8100 BFM

**EXTERIOR SHEATHING**

½" Plyscore wall sheathing (or qyplap) ................. 2000 sq. ft.
½" Plyscore roof sheathing ..... 1750 sq. ft.

**SUB FLOORING**

⅝" Plyscore .................. 1400 sq. ft.

**INSULATION**

Rockwool semi-thick ........... 1500 sq. ft.
Rockwool full-thick ............ 1200 sq. ft.

**DOOR SCHEDULE**

(1) 3-0 x 6-8 x 1¾" wp ent. doors
(3) 2-6 x 6-8 SD
(16) Flush interior veneered doors
(1) 8-0 x 7-0 overhead

**WINDOW SCHEDULE**

(4) 1-8 x 2-8 Basement sash
(1) 2-0 x 3-0 DH
(2) 3-0 x 3-2 DH
(2) 3-6 x 3-10 DH
(4) 3-0 x 4-2 DH
(2) 3-0 x 3-10 DH
(3) 2-8 x 3-6 DH
(1) 3-4 x 4-2 DH
(1) 9-4 x 6-4 Awning

**EXTERIOR FINISHES**

Wood siding .................. 2200 sq. ft.
Asphalt shingles .............. 17 squares

**INTERIOR FINISHES**

Ceramic Tile floor ............ 32 sq. ft.
Ceramic Tile walls ............ 120 sq. ft.
Gypsum wall & ceiling boards .... 6000 sq. ft.
Oak Flooring .................. 1300 sq. ft.

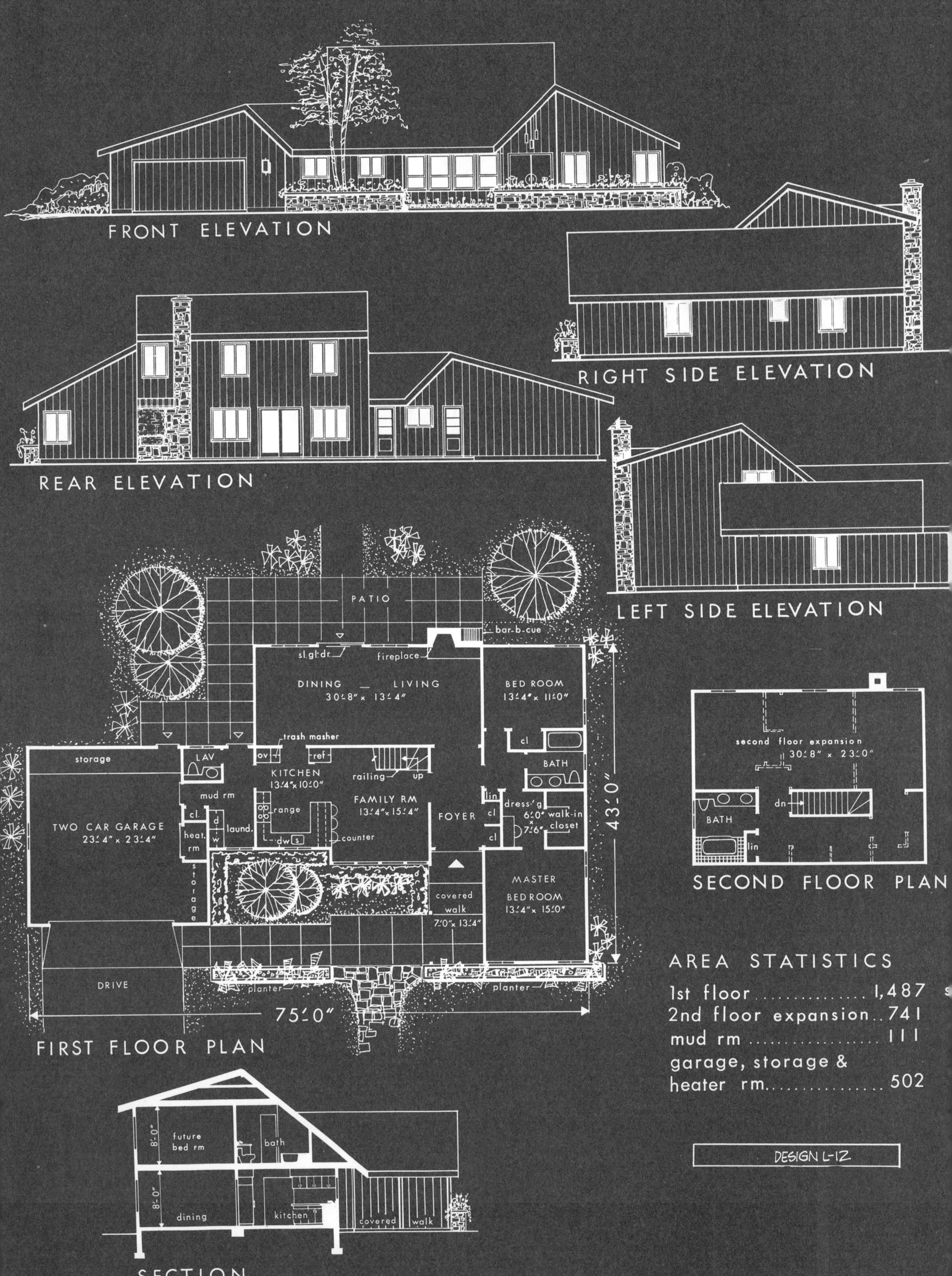


## AREA STATISTICS

1st floor ............. 1,487 s
2nd floor expansion .. 741
mud rm ................ 111
garage, storage & heater rm............. 502

DESIGN L-12

# Informal Living In Simple Contemporary

This two-bedroom home can be tailored to the needs of almost any family because of an expandable second floor area that can be made into two or three extra bedrooms and bath. The first floor is a complete house in itself.

The entrance court, with its covered walk leading to the front door, says welcome in a gracious manner. On entering the house, one views a fireplace in a rear corner of the living room. The foyer has two closets and leads to the main rooms. A turn to the left brings one into the family room which is off the efficient U-shaped kitchen. The two rooms, although separated by counter space and hung cabinets, flow together into one large space.

The living-dining rooms at the rear, together, are almost 31′ in length and 13′4″ in depth. They feature an abundance of windows as well as the fireplace. A large patio with barbecue is reached from the living room through sliding glass doors. The patio connects with the service entrance.

To the right of the foyer is the bedroom wing, which consists of the master bedroom, second bedroom and a central bath with two basins.

The exterior portrays a rustic effect with its vertical rough-sawn redwood siding and grooves between boards. The small stone walls which define the courtyard space contribute to this rustic appearance.

## Material List

**CONCRETE WORK**

| | |
|---|---|
| Foundations, footings, slabs, etc. | 64 cu. yds. |

**STEEL**

| | |
|---|---|
| Reinforcing mesh | 1488 sq. ft. |

**MASONRY**

| | |
|---|---|
| Brick Backup | 210 sq. ft. |
| Stone Chimney | 256 sq. ft. |
| Stone Planter Walls | 261 sq. ft. |

**FRAMING LUMBER**

| | |
|---|---|
| Total Sills, Joists, Rafters, Studs, Plates, etc. | 10,501 B.F.M. |

**SHEATHING, INSULATION**

| | |
|---|---|
| Sub Flooring | 741 sq. ft. |
| Roof Sheathing | 2092 sq. ft. |
| Wall Insulation | 2256 sq. ft. |
| Ceiling Insulation | 1430 sq. ft. |

**FINISHES, INTERIOR**

| | |
|---|---|
| Vinyl Tile | 504 sq. ft. |
| Oak Flooring | 835 sq. ft. |
| Ceramic Tile Floors | 96 sq. ft. |
| Ceramic Tile Walls | 275 sq. ft. |
| Gypsum Board — house walls | 4160 sq. ft. |
| Gypsum Board — house ceilings | 2339 sq. ft. |
| Gypsum Board — garage | 1089 sq. ft. |

**FINISHES, EXTERIOR (other than masonry)**

| | |
|---|---|
| Vertical Siding (Texture 1-11 Reverse Board & Batten) | 2256 sq. ft. |
| Asphalt Shingle Roofing | 2092 sq. ft. |
| Plywood Eave & Porch Soffits | 450 sq. ft. |

**DOOR SCHEDULE**

| | |
|---|---|
| Ext. Hardwood, glazed | 2 units |
| Ext. Hardwood, paneled | 2 units |
| Aluminum Sliding | 1 unit |
| Wood Overhead garage | 1 unit |
| Int. Hardwood, flush, staingrade | 10 units |
| Int. Hardwood, louvered, bi-folding | 5 units |
| Int. Fireproof door | 1 unit |

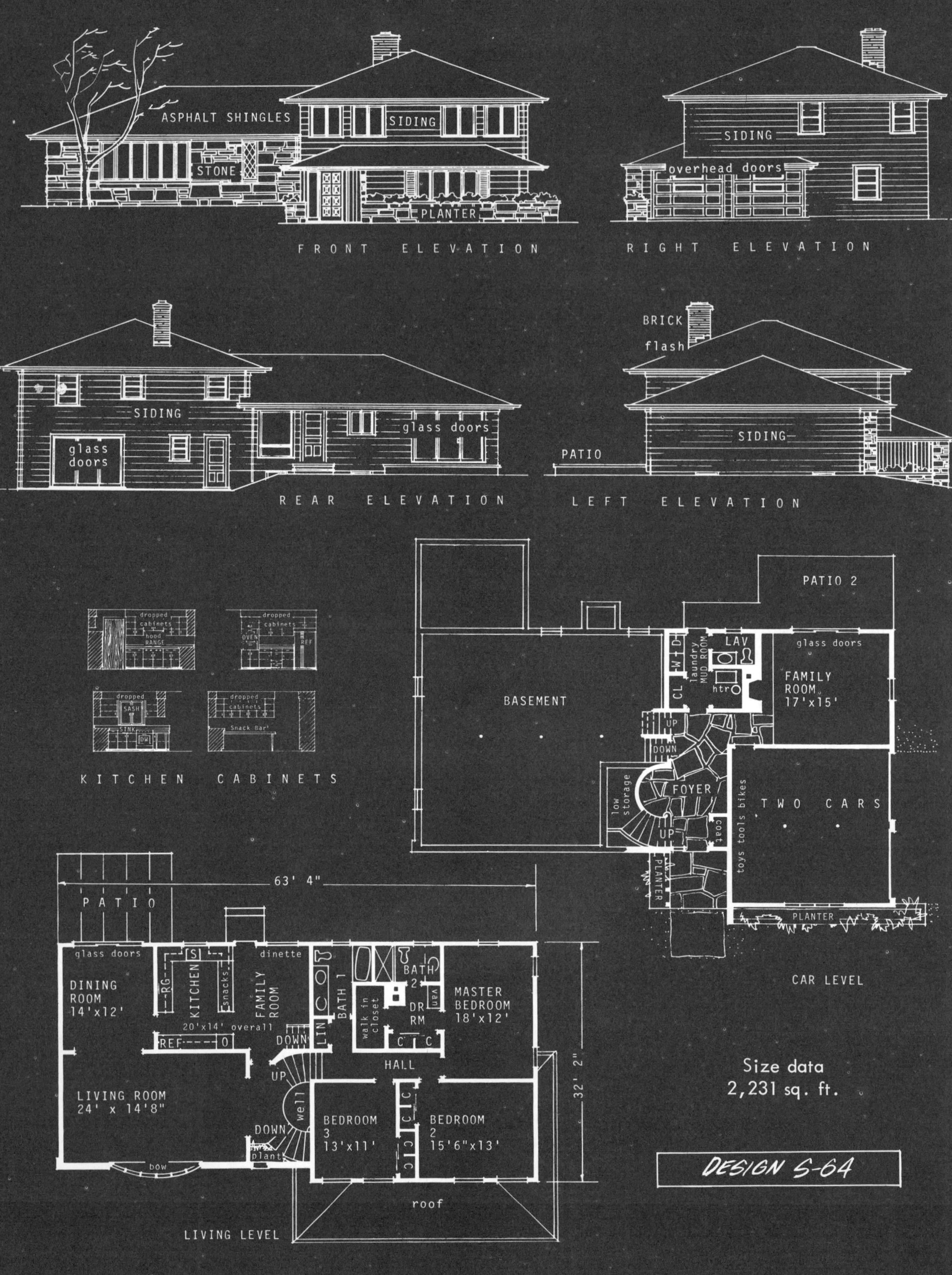

ASPHALT SHINGLES
SIDING
STONE
PLANTER
FRONT ELEVATION
SIDING
overhead doors
RIGHT ELEVATION
SIDING
glass doors
glass doors
REAR ELEVATION
BRICK
flash
SIDING
PATIO
LEFT ELEVATION
dropped cabinets
hood RANGE
dropped cabinets
OVEN
REF
dropped
SASH
SINK
DW
dropped cabinets
Snack Bar
KITCHEN CABINETS
PATIO 2
BASEMENT
laundry MUD ROOM
W D
CL
LAV
htr
glass doors
FAMILY ROOM 17'x15'
UP
DOWN
low storage
FOYER
UP
coat
TWO CARS
toys tools bikes
PLANTER
PLANTER
CAR LEVEL
63' 4"
PATIO
glass doors
dinette
DINING ROOM 14'x12'
RG
S
KITCHEN
snacks
FAMILY ROOM
20'x14' overall
REF
O
DOWN
LIN
BATH 1
walk in closet
BATH 2
van
DR RM
C C
MASTER BEDROOM 18'x12'
HALL
UP
LIVING ROOM 24' x 14'8"
well
DOWN
plant
bow
BEDROOM 3 13'x11'
C C
C C
BEDROOM 2 15'6"x13'
32' 2"
roof
LIVING LEVEL
Size data
2,231 sq. ft.
DESIGN S-64

# Suggestion of Oriental in Split-Level

There's an air of serenity about this split-level.

The reason for its distinguished yet informal appearance is not immediately apparent — there's a bit of the Orient in the exterior structure. The feeling is generated by the hip roof lines, which suggest a pagoda and an impression of tranquility and space in the best Oriental tradition.

The foyer level, which has been labeled the car level on the floor plans, is actually the family-room level, too, since the flagstone entry leads to these as well as to the upper floor. There's a stairway down to the basement level under the living room.

The downstairs family room is almost square and well laid out. The sweep of the raised hearth makes the room seem larger, and the sliding glass door wall at the back visually includes the patio.

Up the stairway, past an alcove planter with its own diamond-paned window is the spacious living room. The dining room opens behind it for better than a 28-foot sweep from the bow window in the living room to the glass doors at the back and the living level patio beyond.

The sleeping wing is half a flight up the curved stair just outside the kitchen. Two front bedrooms share a four-closet wall. The master bedroom, with its own dressing room, vanity and bathroom, is large and has ample closet space.

## Material List

**CONCRETE WORK**

| | |
|---|---|
| Footings | 21 cu. ft. |
| Floors (basement & platforms) | 30 cu. ft. |

**MASONRY**

| | |
|---|---|
| 4 x 8 x 18 CB | 230 lin. ft. |
| 8 x 8 x 18 CB | 255 sq. ft. |
| 10 x 8 x 18 CB | 760 sq. ft. |
| 12 x 8 x 18 CB | 310 sq. ft. |
| Chimney brick | 65 sq. ft. |
| Chimney fill | 215 cu. ft. |
| Flagstone | 60 sq. ft. |
| Stone veneer | 435 sq. ft. |

**CARPENTRY**

| | |
|---|---|
| 4" Lally columns | 6 pieces |
| Framing lumber | 8320 B.F. |
| Studs & Plates | 4238 B.F. |
| Roof sheathing 1 x 6 | 3780 sq. ft. |
| plywood | 3120 sq. ft. |
| Fascia #1 pine 1 x 6 | 300 lin. ft. |
| Soffit 3/8" x 2' plywood | 300 lin. ft. |
| Shingle mould | 270 lin. ft. |
| Sheathing 1 x 6 | 3420 sq. ft. |
| plywood | 2825 sq. ft. |
| Siding | 3045 sq. ft. |
| Roofing 210# asphalt | 29 squares |

| | |
|---|---|
| Insulation | 5230 sq. ft. |
| Finish wood flooring | 2030 sq. ft. |

**DRY WALL**

| | |
|---|---|
| Ceilings 3/8" | 3265 sq. ft. |
| Walls 1/2" | 7870 sq. ft. |

**MILLWORK**

| | |
|---|---|
| Base | 815 lin. ft. |
| Shelving 12" pine | 85 lin. ft. |
| Windows | 20 pieces |
| Exterior doors | 7 pieces |
| Interior doors | 19 pieces |

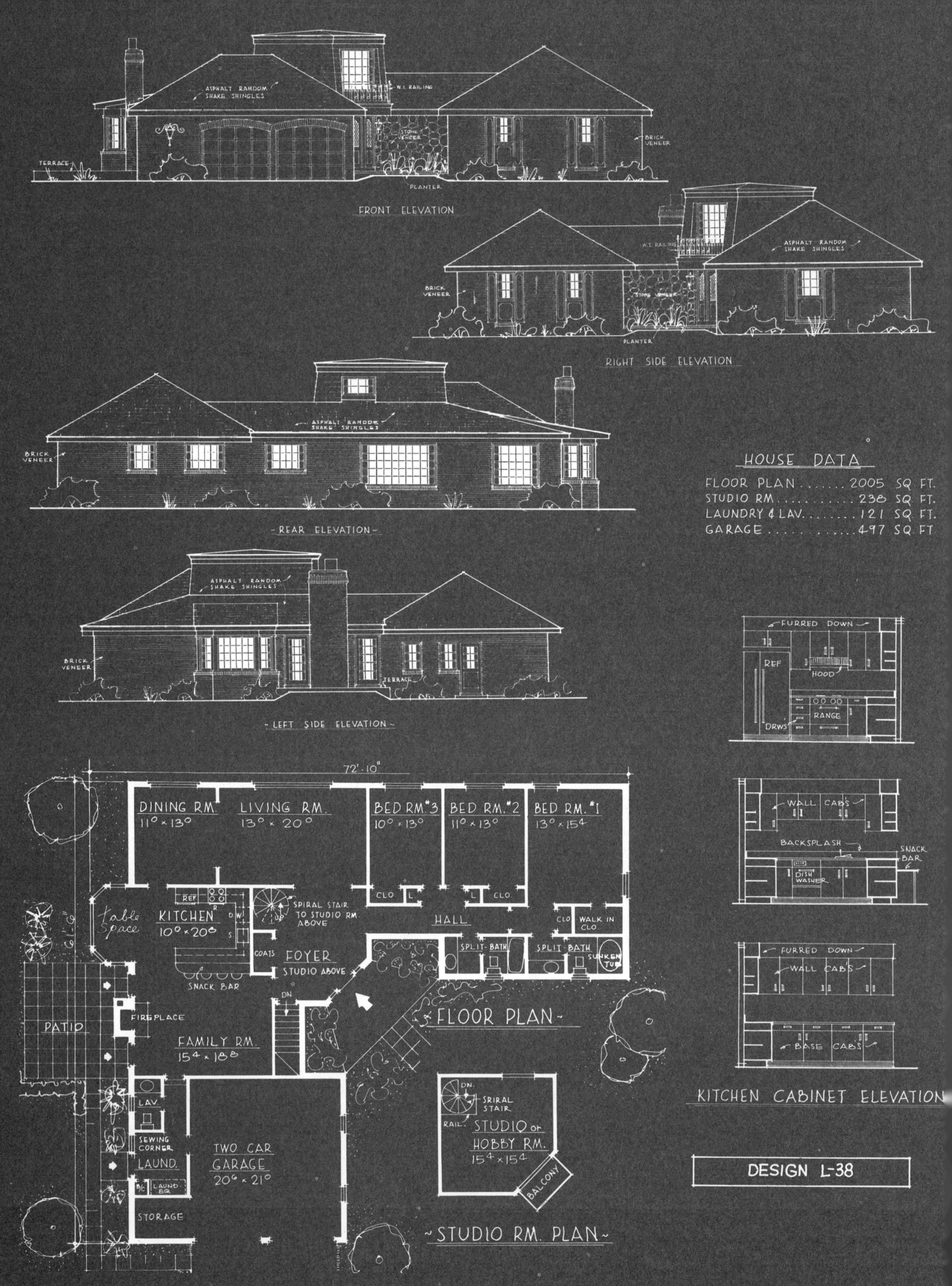
ASPHALT RANDOM SHAKE SHINGLES
W.I. RAILING
STONE VENEER
BRICK VENEER
TERRACE
PLANTER
FRONT ELEVATION
RIGHT SIDE ELEVATION
REAR ELEVATION
LEFT SIDE ELEVATION
HOUSE DATA
FLOOR PLAN ....... 2005 SQ. FT.
STUDIO RM ........... 238 SQ. FT.
LAUNDRY & LAV. ....... 121 SQ. FT.
GARAGE ............. 497 SQ. FT.
FURRED DOWN
REF
HOOD
RANGE
DRWS
WALL CABS
BACKSPLASH
SNACK BAR
DISH WASHER
BASE CABS
KITCHEN CABINET ELEVATION
72'-10"
DINING RM. 11⁰ × 13⁰
LIVING RM. 13⁰ × 20⁰
BED RM. #3 10⁰ × 13⁰
BED RM. #2 11⁰ × 13⁰
BED RM. #1 13⁰ × 15⁴
CLO
HALL
WALK IN CLO.
SPLIT-BATH
SUNKEN TUB
Table Space
KITCHEN 10⁰ × 20⁸
SPIRAL STAIR TO STUDIO RM. ABOVE
COATS
FOYER
STUDIO ABOVE
SNACK BAR
DN
FIREPLACE
PATIO
FAMILY RM. 15⁴ × 18⁸
FLOOR PLAN
LAV.
SEWING CORNER
LAUND.
LAUND. EQ.
STORAGE
TWO CAR GARAGE 20⁶ × 21⁰
DN.
SPIRAL STAIR
RAIL
STUDIO or HOBBY RM. 15⁴ × 15⁴
BALCONY
STUDIO RM. PLAN
DESIGN L-38

# Two-Story Tower in One-Story House

Although basically a one-story, this house has a two-story stair hall tower in which a studio overlooks the front and the rear. This upstairs room, ideal for a hobbyist, is reached by an attractive open spiral stair that lends a distinctive air to the oversized front foyer. At the same time, the tower gives the exterior of the house the flavor of a French Provincial, emphasized by the roof lines.

The sizable family room and the big kitchen are separated only by a snack bar. And the living room and dining room, at the rear, have combined dimensions of 31′ by 13′, an ideal setup for entertaining guests. A beautiful bay window in the kitchen provides space for a table without interfering with the U-shaped food preparation area.

Between the family room and adjacent garage are a wash-up toilet, laundry and sewing corner. The two-car garage has a garden tool storage corner.

In the bedroom wing, all principle rooms have windows toward the rear garden. The owners' bedroom is large with generous space for twin beds and several pieces of furniture. It has two closets, one a walk-in type. The private bath is split to provide dual use. The bathing facility is a sunken tub of generous proportions.

The hall bath is also divided. It is private yet close enough to the front foyer to serve as a guest powder room.

## Material List

**CONCRETE WORK**

| | | |
|---|---|---|
| Concrete Walls | | 1453 cu. ft. |
| Concrete Footings | | 451 cu. ft. |
| Slabs | | 890 cu. ft. |
| Misc. Concrete | | 81 cu. ft |

**STRUCTURAL STEEL**

| | | |
|---|---|---|
| Lally Columns | 3½″ Diam. | 12 pieces |
| Girder | 6″ I Beam | 97 lin. ft. |

**BRICK WORK**

| | | |
|---|---|---|
| Walls | 4″ Brick & Stone | 1976 sq. ft. |
| Chimney | Brick | 150 cu. ft. |
| Flue Lining | T. C. | 40 lin. ft. |

**CARPENTRY**

| | |
|---|---|
| Framing Lumber | 11,925 B.F. |
| Studs | 4321 B.F. |
| Plates | 1290 B.F. |
| Roof Sheathing | 4230 sq. ft. |
| Sub Flooring | 2099 sq. ft. |
| Side Wall Sheathing | 2670 sq. ft. |
| Insulation Walls | 2026 sq. ft. |
| Insulation Ceilings | 2220 sq. ft. |
| Wood Flooring | 1753 sq. ft. |
| Kitchen Underlayment | 207 sq. ft. |
| Resilient Tile | 100 sq. ft. |

**MILLWORK**

| | |
|---|---|
| Exterior Doors & Frames Compl. | 4 units |
| Garage Door Complete Set | 2 units |
| Interior Doors & Frames Compl. | 16 pieces |
| Bi Fold Door Units | 7 units |
| Windows | 23 units |
| Fascia | 372 lin. ft. |

**ROOFING**

| | | |
|---|---|---|
| Shingles | 345# Asphalt | 3780 sq. ft. |
| Roofing Paper | 15# Felt | 3780 sq. ft. |
| Built up | 4 Ply | 450 sq. ft. |

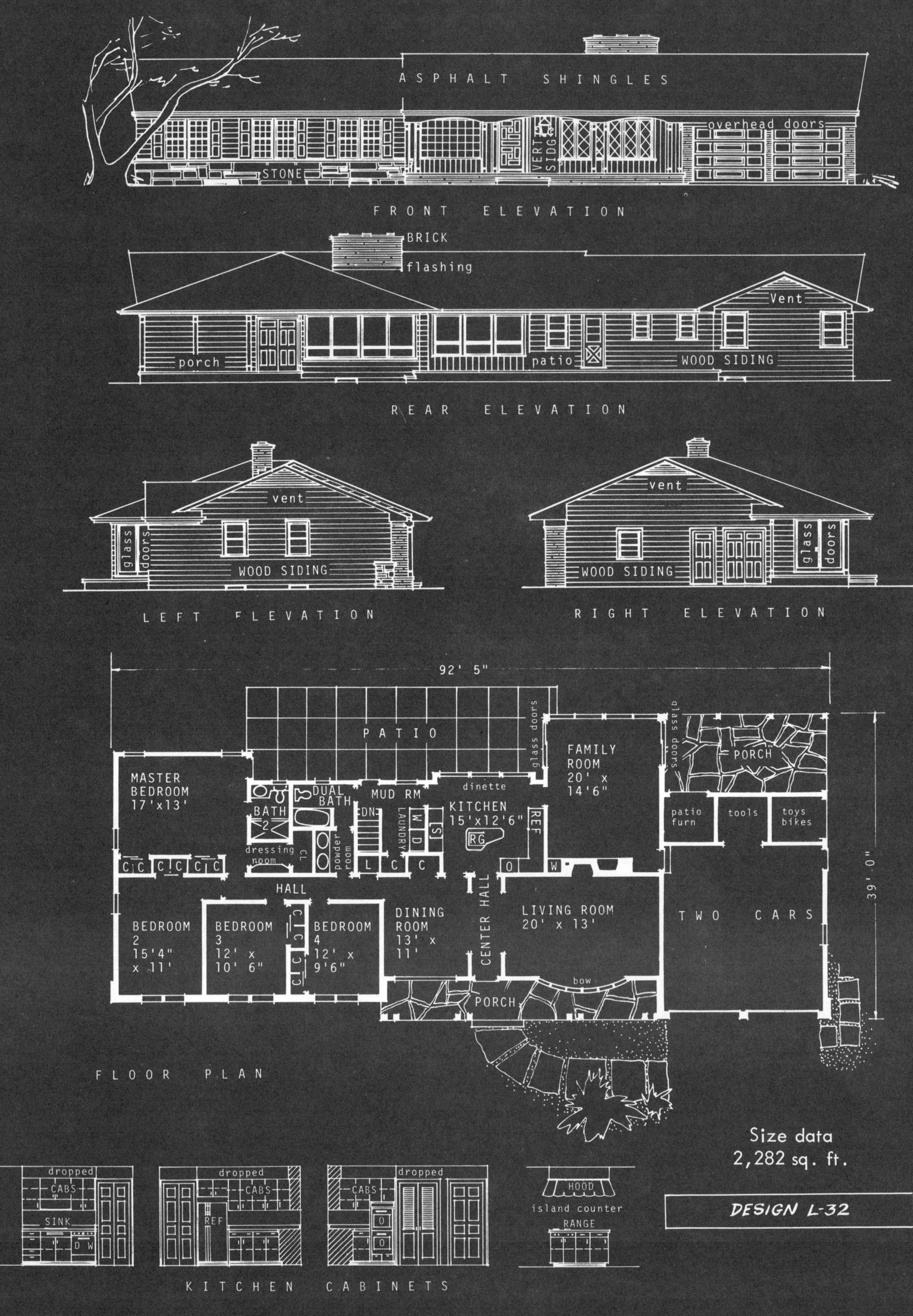


Size data
2,282 sq. ft.

DESIGN L-32

# Spacious Living Around Center Hall

Certain details in a house become linked with the best traditions of home design.

The center hall is one of these. Usually it is a part of a traditional two-story house, but it appears in this expansive ranch plan.

With double columns, the porch provides a handsome introduction to the house within. The patio at the back is an extra living area in moderate or warm climates or in summer anywhere. Access from the master bedroom and family room make this area a multipurpose outdoor room. A flagstone porch on the other side of the family room allows two generations to plan activities outside without getting in each other's way.

The kitchen and mud room form a convenient and wide zone. The counter arrangement in the kitchen is a stepsaver with a sink counter on one wall, L-shaped counters for refrigerator and oven, and an island counter with the range on it, all surrounded by good work tops. The back end of the room has a dinette area with a patio view.

The master bedroom is at the rear, bright and airy, amply supplied with closets, a dressing room and bath. The door to the patio suggests a private corner, screened or latticed if desired.

Wide doorways for both the living and dining rooms serve to make each room seem larger even though separate functions are preserved by the center hall.

## Material List

| CONCRETE WORK | |
|---|---|
| Footings | 21 cu. yds. |
| Floors (basement & platforms) | 42 cu. yds. |
| **MASONRY** | |
| 4 x 8 x 18 CB | 300 lin. ft. |
| 8 x 8 x 18 CB | 405 sq. ft. |
| 10 x 8 x 18 CB | 1480 sq. ft. |
| 12 x 8 x 18 CB | 370 sq. ft. |
| Brick veneer - chimney fill | 165 sq. ft. |
| | 540 cu. ft. |
| Flagstone - stone veneer | 425 sq. ft. |
| | 165 sq. ft. |
| **CARPENTRY** | |
| 4" Lally columns | 8 pieces |
| Framing lumber | 12,125 B.F. |
| Stud & plates | 4436 B.F. |
| Roof sheathing 1 x 6 | 4745 sq. ft. |
| Plywood | 3915 sq. ft. |
| Fascia #1 pine 1 x 6 | 200 lin. ft. |
| Soffit ⅜" x 2' plywood | 200 lin. ft. |
| Shingle mould | 200 lin. ft. |
| Wall sheathing 1 x 6 | 3900 sq. ft. |
| Plywood | 3225 sq. ft. |
| Siding | 3275 sq. ft. |
| Roofing 210# asphalt | 40 squares |
| Insulation - walls | 2025 sq. ft. |
| ceiling | 2565 sq. ft. |
| Finish wood flooring | 2600 sq. ft. |
| **DRY WALL** | |
| Ceilings ⅜" | 6095 sq. ft. |
| Walls ½" | 8010 sq. ft. |
| **MILLWORK** | |
| Base | 760 lin. ft. |
| Shelving 12" pine | 55 lin. ft. |
| Windows | 23 pieces |
| Exterior doors | 10 pieces |
| Interior doors | 24 pieces |

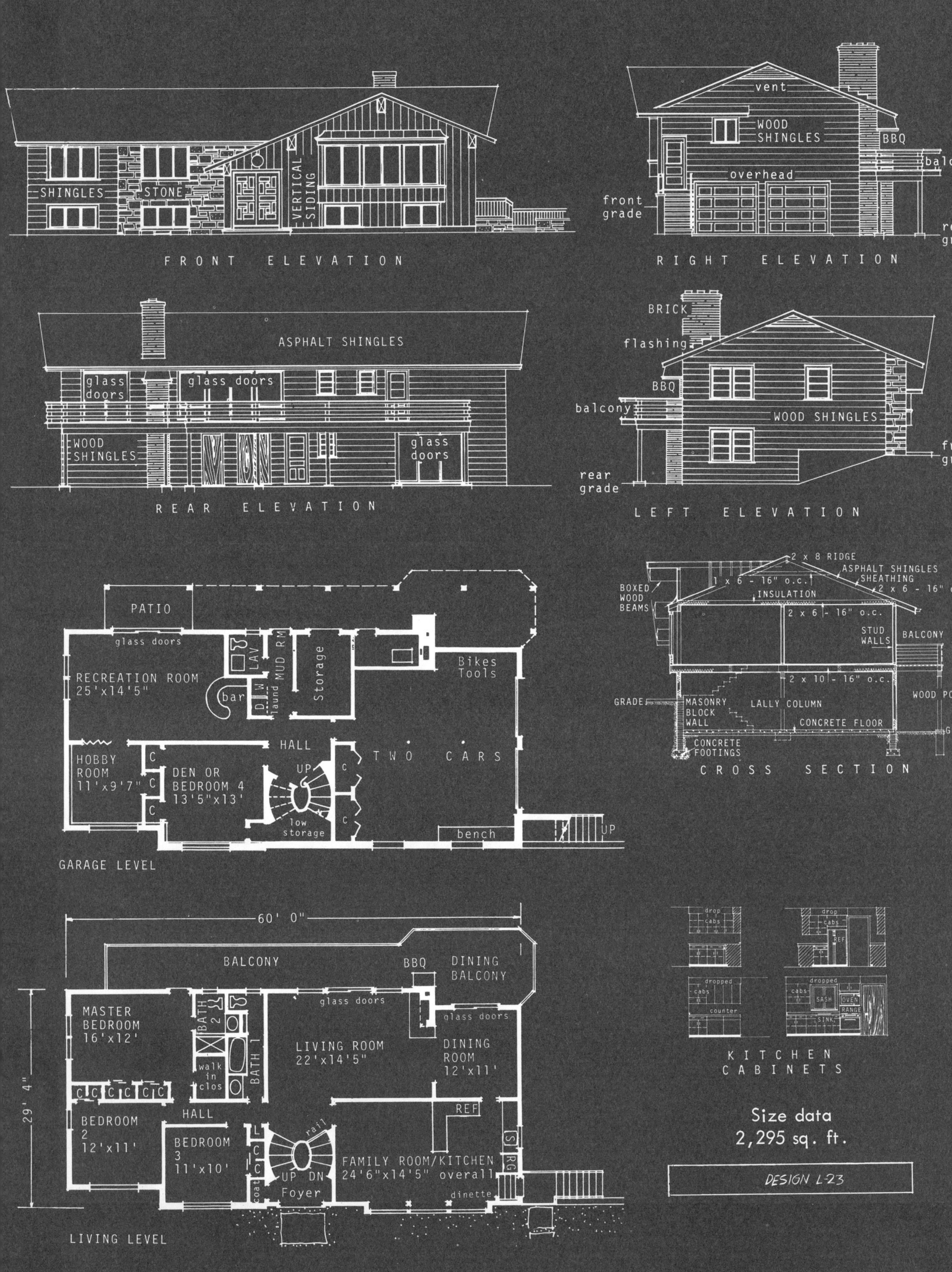


Size data
2,295 sq. ft.

DESIGN L-23

# High Ranch Gets Distinctive Exterior

Not much doubt about this house having a distinctive appearance.

Although a conventional high ranch in the sense of having a livable lower level, it has a contemporary look with the flavor of a chalet.

Design L-23 is not too big for a moderate property and, since the lower level can be finished at the owners' convenience, there can be an economy factor in building now with a plan for future use.

On the entry level, decorator double doors open to a mid-level foyer, where a curving staircase leads up half a flight to the living level or down to the garage level.

Three bedrooms, sharing closet walls, and two bathrooms are in the area to the left of the upper landing. The master bedroom has a door out to the back balcony that stretches the width of the house overlooking the rear garden. Sliding glass doors in a window wall in the living room afford a balcony view and good weather access. At the dining section, the balcony widens to form a delightful outdoor dining area, with privacy assured.

The family room and kitchen are at the front, sharing an expanse almost 25′ across, highlighted by the big bay window.

Three big rooms are shown on the lower level. The purchaser can elect to add one or all of them as his needs dictate.

## Material List

| CONCRETE WORK | |
|---|---|
| Footings | 10 cu. ft. |
| Floors (basement & platforms) | 21 cu. ft. |

| MASONRY | |
|---|---|
| 4 x 8 x 18 CB | 180 lin. ft. |
| 8 x 8 x 18 CB | 50 sq. ft. |
| 10 x 8 x 18 CB | 615 sq. ft. |
| 12 x 8 x 18 CB | 90 sq. ft. |
| Brick veneer - Chimney fill | 235 sq. ft. |
| | 375 cu. ft. |
| Flagstone - Stone veneer | 40 sq. ft. |
| | 205 sq. ft. |

| CARPENTRY | |
|---|---|
| 4″ Lally columns | 2 pieces |
| Framing lumber | 8500 B.F. |
| Stud & Plates | 4652 B.F. |
| Roof Sheathing | 1760 sq. ft. |
| Plywood | 1450 s.f. |
| Fascia #1 pine 1 x 6 | 115 lin. ft. |
| Soffit ⅜″ x 3′ plywood | 115 lin. ft. |
| Shingle Mould | 115 lin. ft. |
| Sheathing 1 x 6 - wall | 3835 sq. ft. |
| Plywood | 3165 s.f. |
| Siding | 3475 sq. ft. |
| Roofing 210# asphalt | 15 squares |

| | |
|---|---|
| Insulation | 5675 sq. ft. |
| Finish Wood Flooring | 2080 sq. ft. |

| DRY WALL | |
|---|---|
| Ceilings ⅜″ | 3575 sq. ft. |
| Walls ½″ | 9100 sq. ft. |

| MILLWORK | |
|---|---|
| Base | 940 lin. ft. |
| Shelving 12″ pine | 50 lin. ft. |
| Windows | 16 pieces |
| Exterior Doors | 11 pieces |
| Interior Doors | 24 pieces |

FRONT ELEVATION

RIGHT SIDE ELEVATION

REAR ELEVATION

LEFT SIDE ELEVATION

## HOUSE DATA

| | |
|---|---|
| FIRST FLOOR | 1653 SQ. FT. |
| LAUND. AREA | 105 SQ. FT. |
| GARAGE | 482 SQ. FT. |
| BED RM. LEVEL | 656 SQ. FT. |

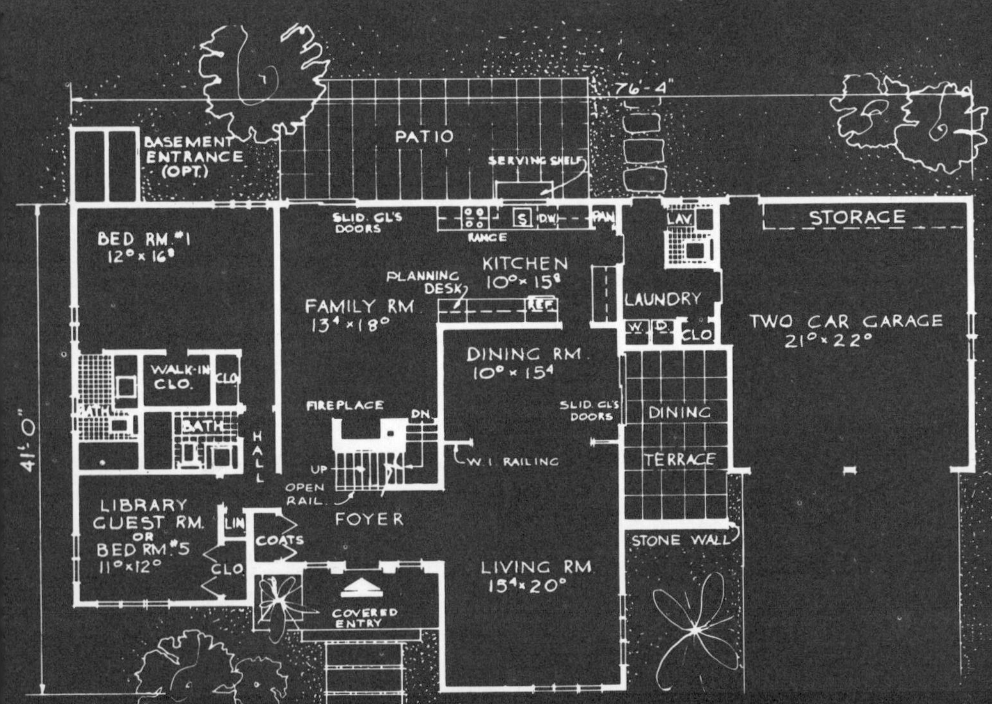


FIRST FLOOR PLAN

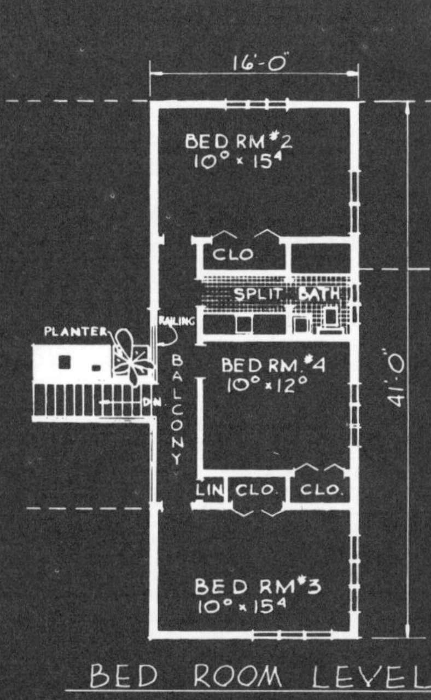


BED ROOM LEVEL

DESIGN S-26

# Two-Story With Contemporary Exterior

There are several unusual design ideas here.

The most apparent departure from normal two-story architecture is the roof line, giving the contemporary plan an individual character. Other features add dramatic interest, among them a two-storied entrance foyer with decorative translucent panels framed by vertical stone piers.

In the rear there's an intimate dining terrace, open to the sky but enclosed on four sides in atrium fashion. Sliding glass doors from the dining room make this terrace an informal extension of the dining room.

Another interesting idea is the location of the owners' bedroom on the first floor and completely away from the other bedrooms.

The family room has been screened from the foyer, using the stair and fireplace as a buffer, yet is convenient to the patio.

Upstairs is an open-balconied hallway overlooking the foyer. Here, three bedrooms are provided with a split bathroom.

There is a generous use of stone and glass in the well-balanced front elevation. The low, long garage wing is proportioned to complement the owners' bedroom wing and two-storied center section. The windows in the structure are placed asymmetrically to provide balance with the larger decorative panels of the foyer. A massive stone chimney, somewhat down the roof, adds interest to the intriguing exterior.

## Material List

|  | |  |
|---|---|---|
| **CONCRETE WORK** | | |
| Concrete Walls | | 1163 cu. ft. |
| Concrete Footings | | 280 cu. ft. |
| Slabs | | 886 cu. ft. |
| Misc. Concrete | | 255 cu. ft. |
| **STRUCTURAL STEEL** | | |
| Lally Columns | | 8 pieces |
| Girder | 6" & 7" I Beam | 62 lin. ft. |
| **BRICKWORK** | | |
| Chimney | Brick & Stone | 260 cu. ft. |
| Flue Lining | T. C. | 48 lin. ft. |
| Walls | 4" Stone | 78 sq. ft. |
| Walls | 8" Stone | 63 sq. ft. |
| **CARPENTRY** | | |
| Framing Lumber | | 11,499 B.F. |
| Studs | | 4267 B.F. |
| Plates | | 1400 B.F. |
| Roof Sheathing | | 3652 sq. ft. |
| Sub Flooring | | 2313 sq. ft. |
| Side Wall Sheathing | | 2900 sq. ft. |
| Insulation Walls | | 2900 sq. ft. |
| Insulation Ceilings | | 1765 sq. ft. |
| Wood Floor | | 2042 sq. ft. |
| Kitchen Plywood | | 157 sq. ft. |
| **MILLWORK** | | |
| Exterior Doors & Frames Compl. | | 3 pieces |
| Sliding Glass Doors | | 2 units |
| Garage Door Complete Set | | 2 units |
| Int. Doors & Frames Complete | | 16 pieces |
| Sliding & Pocket Doors | | 3 pieces |
| Bi Fold Doors | | 5 units |
| Window Units | | 23 units |
| Fascia | | 360 lin. ft. |
| Base | | 700 lin. ft. |
| Stairs (12 risers) | | 1 set |
| Stairs (13 risers) | | 1 set |

FRONT ELEVATION

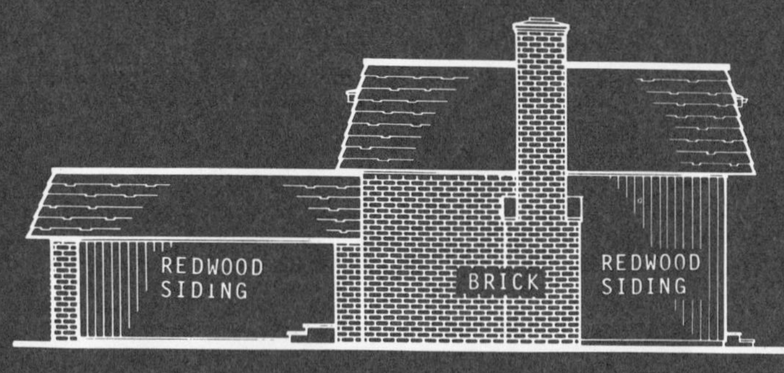


RIGHT ELEVATION

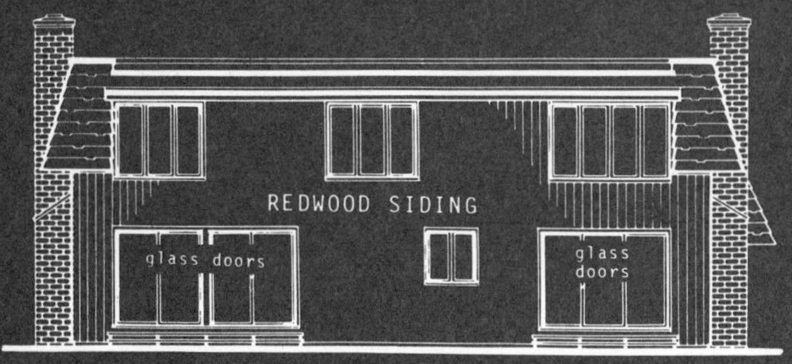


REAR ELEVATION

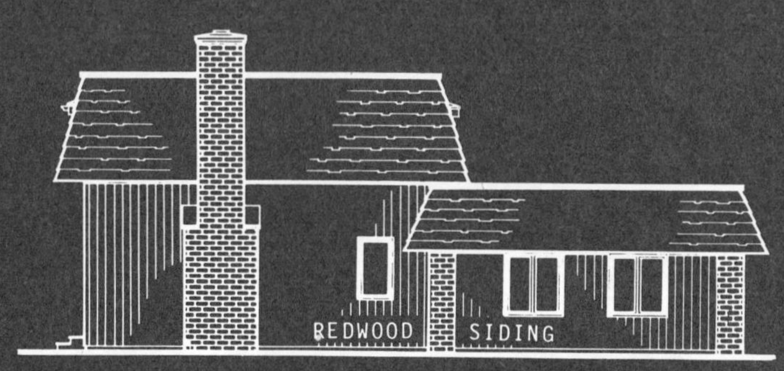


LEFT ELEVATION

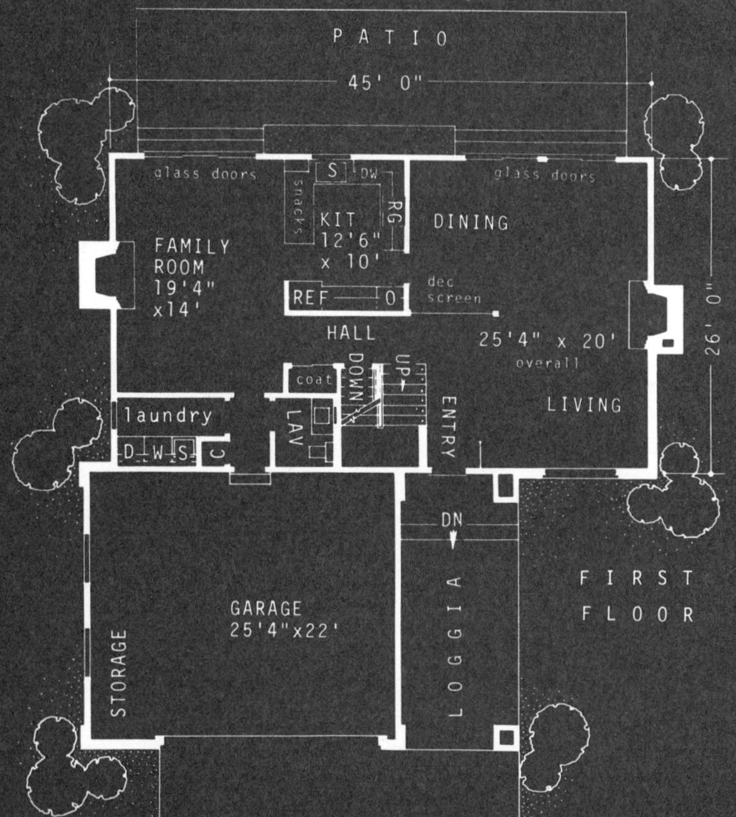


MASTER BEDROOM 16'6" x15'8"
CL
CL
BEDROOM 2 12'5" x10'6"
CL
CL
LIN
BEDROOM 3 12'5" x10'6"
balcony hall
cab
van
cab
DR RM
CL
BATH 1
DOWN
DOWN
BATH 2
CL
BEDROOM 4 12'7" x10'6"
OVERHANG 2'
ROOF DECK
SECOND FLOOR

Size data
2,340 sq. ft.

**DESIGN L-27**

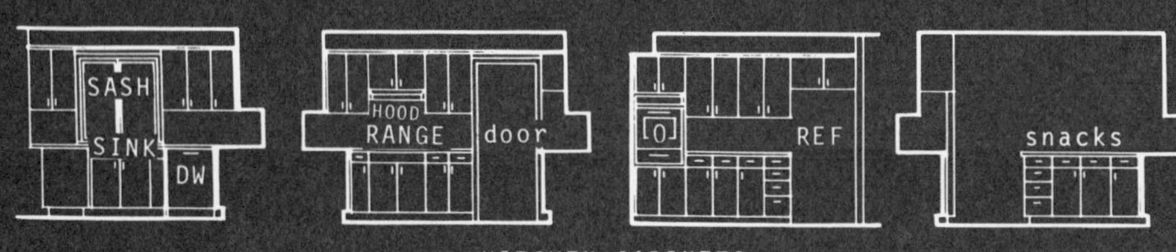


KITCHEN CABINETS

# Mansard Roof Provides Distinction

The mansard roof line has a long history in architecture and over the years has been adapted from its original formal French origins into provincial homes as well.

Today, the high roof slope is found as a contemporary keynote in many buildings and adapts particularly well to residential styling, adding distinction to small or large homes. The two-foot overhang, mansard all around, gives the feeling of a much larger home, formal or informal, depending on siting and landscaping.

The low mansard roof over the garage is an important addition to the look of the house at the front, carrying the line horizontally to the entry portico and minimizing the two-story height.

The loggia, a covered platform to the front door and vestibule, is a handsome effect with purely practical application. The living and dining areas are combined, 25′4″ by 20′ overall. Steps outside a window wall lead down to the rear patio, extending the living-dining area outdoors in a gracious manner.

The kitchen is in the heart of everything and open on the left to the family room across a snack-and-work counter. The family room has its own fireplace.

Upstairs, the hall has the air of an open balcony overlooking the full stairwell. The master bedroom suite is designed for privacy, quiet and a garden view.

## Material List

**CONCRETE WORK**

| | |
|---|---|
| Foundation footings | 431 cu. ft. |
| Slabs | 621 cu. ft. |

**STRUCTURAL STEEL**

| | |
|---|---|
| Columns | 11 pieces |
| Girders | 183 lin. ft. |
| Fireplace Lintel | 10 lin. ft. |

**BRICKWORK**

| | |
|---|---|
| Chimney | 450 cu. ft. |
| Flue lining | 57 lin. ft. |
| Veneer | 470 S.F. |
| Foundation walls | 2225 S.F. |
| Veneer | 111 S.F. |
| Loggia floor | 168 S.F. |

**CARPENTRY**

| | |
|---|---|
| Framing lumber | 12,406 B.F. |
| Studs | 4160 B.F. |
| Plates | 1334 B.F. |
| Roof sheathing | 3400 S.F. |
| Sub-flooring | 2340 S.F. |
| Side wall sheathing | 1500 S.F. |
| Wall insulation | 1850 S.F. |
| Ceiling insulation | 1170 S.F. |
| Wood flooring | 2000 S.F. |
| Kitchen flooring | 144 S.F. |

**MILLWORK**

| | |
|---|---|
| Ext. doors & frames complete | 5 pieces |
| Garage door complete set | 1 unit |
| Int. doors & frames complete | 15 pieces |
| Sliding doors | 14 pieces |
| Windows | 17 units |
| Shutters (door) | 2 pieces |
| Base | 720 lin. ft. |

**ROOFING**

| | | |
|---|---|---|
| Shingles | Red Cedar | 500 S.F. |
| Built-up Roofing | 5 ply | 3400 S.F. |
| Roofing Paper | 15# felt | 3900 S.F. |

FRONT ELEVATION

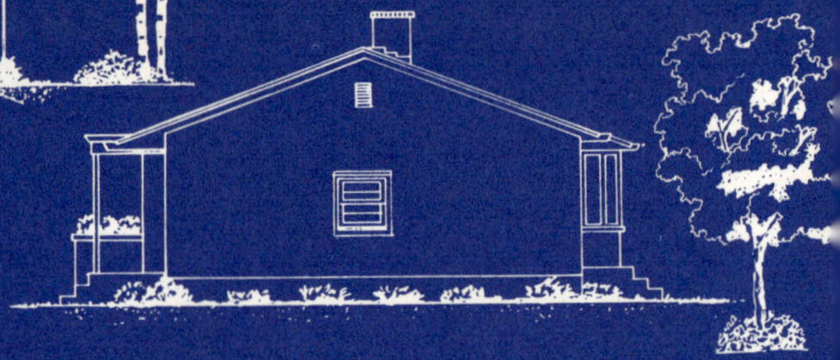

RIGHT SIDE ELEVATION

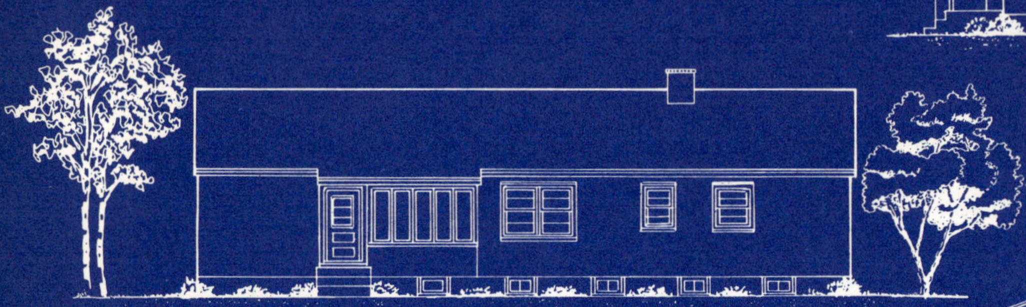

REAR ELEVATION

LEFT SIDE ELEVATION

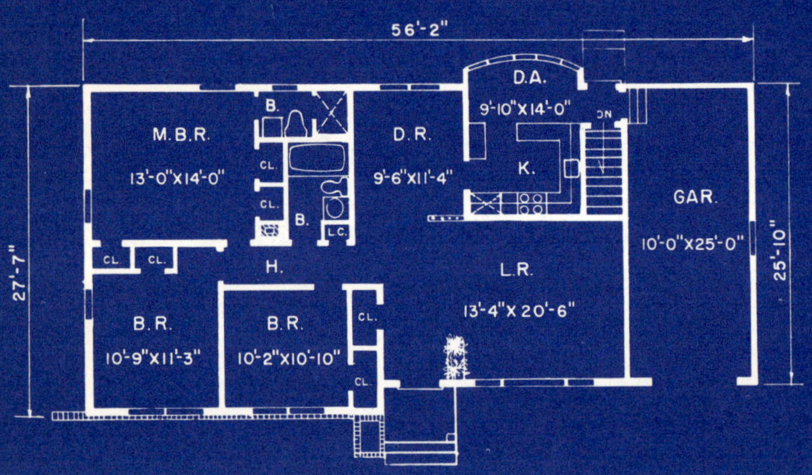


FIRST FLOOR PLAN

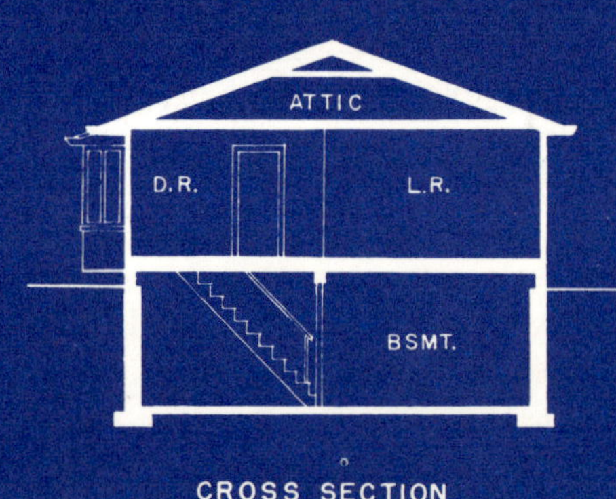


CROSS SECTION

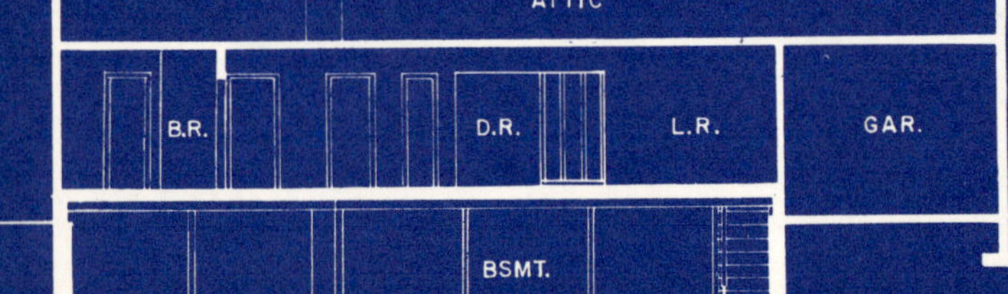


LONGITUDINAL SECTION

## AREA STATISTICS:

| | | |
|---|---|---|
| FIRST FLOOR..... | 1225 | SQ.FT. |
| BASEMENT....... | 1225 | SQ.FT. |
| GARAGE......... | 270 | SQ.FT. |

DESIGN S-73

# Three Bedrooms In Economy Ranch

Simple in design, with value a prime ingredient, this ranch gives a first impression of a pleasant traditional. A closer look reveals its straight, clean lines and modern style.

At the front of the house brick veneer has been effectively combined with V-joint vertical siding. These bricks are on either side of the living room picture window and at windowsill height along the left front. The roof, extending along the house and the one-car garage, makes this three-bedroom ranch seem longer than its 56′2″.

A sheltered entry provides ample protection from the weather. To the left of the front door is a guest closet. Straight ahead and a little to the right is the living room, a respectable 20′-6″ long with lots of wall space. A turned-wood divider is at the entrance to the dining room at the rear. The kitchen, which can be entered from the dining room or a rear entrance, features a wide-view, circular bay-windowed dinette. There is a practical L-shaped counter arrangement with refrigerator, eye-level oven range, dishwasher and sink.

Three bedrooms are at the left side of the house. The master bedroom, facing the rear, has two exposures, good wall space and a separate bathroom with a stall shower. A second full bathroom, with luminous ceiling and mechanically ventilated exhaust, is convenient to the two other bedrooms as well as to the rest of the house.

## Material List

**CONCRETE WORK**
Footings, floors, etc. .......... 23 cu. yds.

**MASONRY**
8″ Block .................... 1000 Blocks
Pancake ..................... 150 Blocks
Brick Veneer ................ 250 sq. ft.

**FRAMING LUMBER**
Sills, joists, rafters, studs, plates, etc. ................ 7600 BFM

**EXTERIOR SHEATHING**
½″ plyscore wall sheathing (or gyplap) ................ 1600 sq. ft.
½″ plyscore roof sheathing ..... 1800 sq. ft.

**SUB FLOORING**
⅝″ plyscore ................. 1250 sq. ft.

**DOOR SCHEDULE**
(1) 3-0 x 6-8 x 1¾ wp entrance doors
(1) 2-8 x 6-8 SD
(13) flush interior veneered doors
(1) 9-0 x 7-0 overhead
(1) 2-8 x 6-8 S.C.F.P.

**WINDOW SCHEDULE**
(8) 1-2 x 2-9 Basement sash
(8) 3-0 x 3-2 DH
(2) 5-0 x 5-2 Fixed
(1) 2-4 x 3-2 DH
(2) 2-4 x 5-2 DH
(5) 1-6 x 5-2 Csmt.

**EXTERIOR FINISHES**
Wood shingles ............... 1600 sq. ft.
Asphalt shingles ............. 18 squares

**INTERIOR FINISHES**
Ceramic Tile floor ............ 50 sq. ft.
Ceramic Tile walls ............ 240 sq. ft.
Gypsum wall and ceiling boards . 5000 sq. ft.
Oak flooring ................. 1100 sq. ft.
Vinyl tile ................... 120 sq. ft.

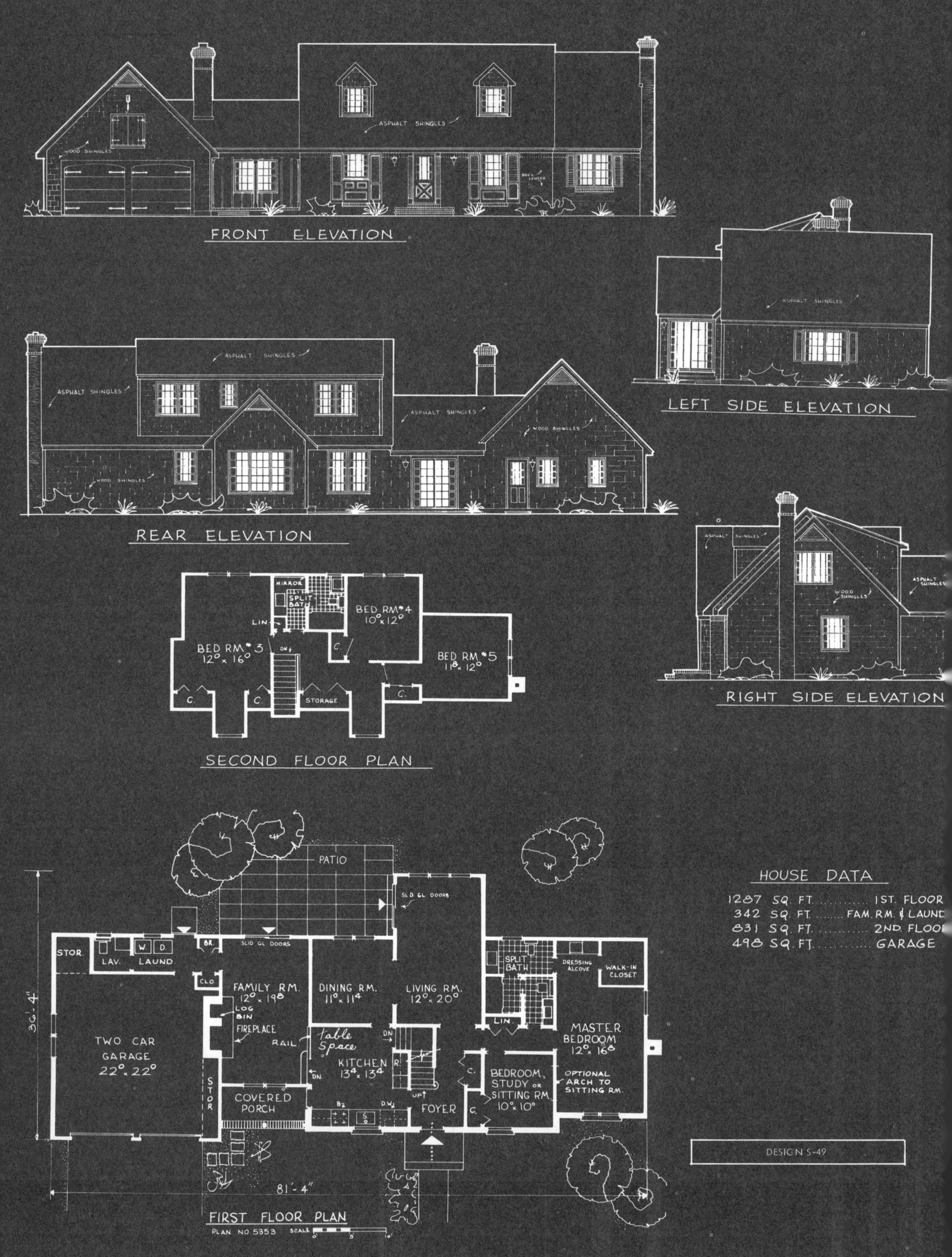

FRONT ELEVATION
ASPHALT SHINGLES
WOOD SHINGLES
LEFT SIDE ELEVATION
REAR ELEVATION
RIGHT SIDE ELEVATION
SECOND FLOOR PLAN
MIRROR
SPLIT BATH
LIN.
BED RM #3 12⁰ x 16⁰
BED RM #4 10⁰ x 12⁰
BED RM #5 11⁸ x 12⁰
STORAGE
PATIO
SLD GL DOORS
SLID GL DOORS
STOR.
LAV.
LAUND.
W. D.
CLO.
FAMILY RM. 12⁰ x 19⁸
LOG BIN
FIREPLACE
RAIL
DINING RM. 11⁰ x 11⁴
LIVING RM. 12⁰ x 20⁰
Table Space
KITCHEN 13⁴ x 13⁴
SPLIT BATH
DRESSING ALCOVE
WALK-IN CLOSET
MASTER BEDROOM 12⁰ x 16⁸
BEDROOM, STUDY OR SITTING RM. 10⁰ x 10⁰
OPTIONAL ARCH TO SITTING RM.
TWO CAR GARAGE 22⁰ x 22⁰
COVERED PORCH
FOYER
36'-4"
81'-4"
FIRST FLOOR PLAN
PLAN NO. 5353
HOUSE DATA
1287 SQ. FT. .......... 1ST. FLOOR
342 SQ. FT. ....... FAM. RM. & LAUND
831 SQ. FT. .......... 2ND. FLOOR
498 SQ. FT. .......... GARAGE
DESIGN S-49

# Emphasize Relaxation in Tasteful Colonial

A touch of elegance is added to the natural simplicity of this design by combining brick, hand-split shingles, a black roof and white trim in just the right proportions. The brick color provides enough contrast to add interest in a manner reserved for a house whose owners have discriminating taste. The two principal living spaces, the living room and the family room, have immediate access to a patio, a popular gathering place.

The family room is open to the kitchen, with a railing along one side of the kitchen eating space. A separate entrance at the rear leads to a laundry room with lavatory.

There seems to be an increasing demand for the master bedroom to be located on the first floor. This has been done here. There is a second room which can be used as a study, a sitting room with an opening to the owners' bedroom or as a guest room. This flexibility is possible because there are three more bedrooms on the second floor.

Upstairs, in addition to the three bedrooms, there is another split bathroom with four fixtures arranged for dual use.

While the garage doors can be moved to the side of the house, some of the charm of the elevation as shown is in the barn loft door in the front gable of the garage. There is a full cellar under the main living portion of the house, providing many additional cubic feet of storage or possible indoor recreational use.

## Material List

**CONCRETE WORK**

| | |
|---|---|
| Concrete Walls | 953 cu. ft. |
| Foundation Footings | 262 cu. ft. |
| Slabs | 800 cu. ft. |
| Misc. | 257 cu. ft. |

**STRUCTURAL STEEL**

| | | |
|---|---|---|
| Lally Columns | 3½" diam. | 7 ea. |
| Girder | 7" I | 76 lin. ft. |

**BRICK WORK**

| | | |
|---|---|---|
| Chimney & Fireplace Brick | | 319 cu. ft. |
| Flue Lining | T. C. | 48 lin. ft. |
| Walls | 4" Brick | 350 sq. ft. |

**CARPENTRY**

| | |
|---|---|
| Framing Lumber | 10,291 B.F. |
| Studs | 4000 B.F. |
| Plates | 1200 B.F. |
| Roof Sheathing | 3242 sq. ft. |
| Sub Floor | 2158 sq. ft. |
| Side Wall Sheathing | 2559 sq. ft. |
| Wall Insulation | 1960 sq. ft. |
| Ceiling Insulation | 1627 sq. ft. |
| Wood Flooring | 1833 sq. ft. |
| Resilient Flooring | 276 sq. ft. |

**MILLWORK**

| | |
|---|---|
| Exterior Doors & Frames Compl. | 2 pieces |
| Garage Door Complete Set | 2 pieces |
| Int. Doors & Frames Compl. | 19 pieces |
| Stairs — 12 risers | 1 set |
| Stairs — 13 risers | 1 set |
| Fascia | 350 lin. ft. |
| Louvers | 3 pieces |

**KITCHEN CABINETS**

| | |
|---|---|
| Base Cabinets | 15'-6" long |
| Wall Cabinets | 14'-6" long |

**ROOFING**

| | | |
|---|---|---|
| Shingles | 235# asphalt | 3242 sq. ft. |
| Roofing Paper | 15# felt | 3242 sq. ft. |

Front elevation

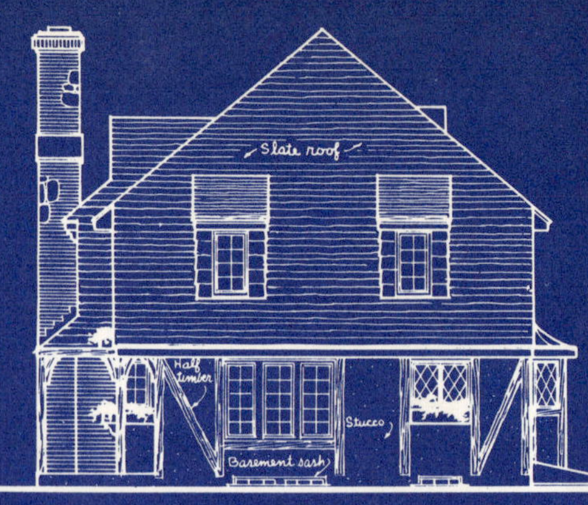

Right side elevation

Rear elevation

Left side elevation

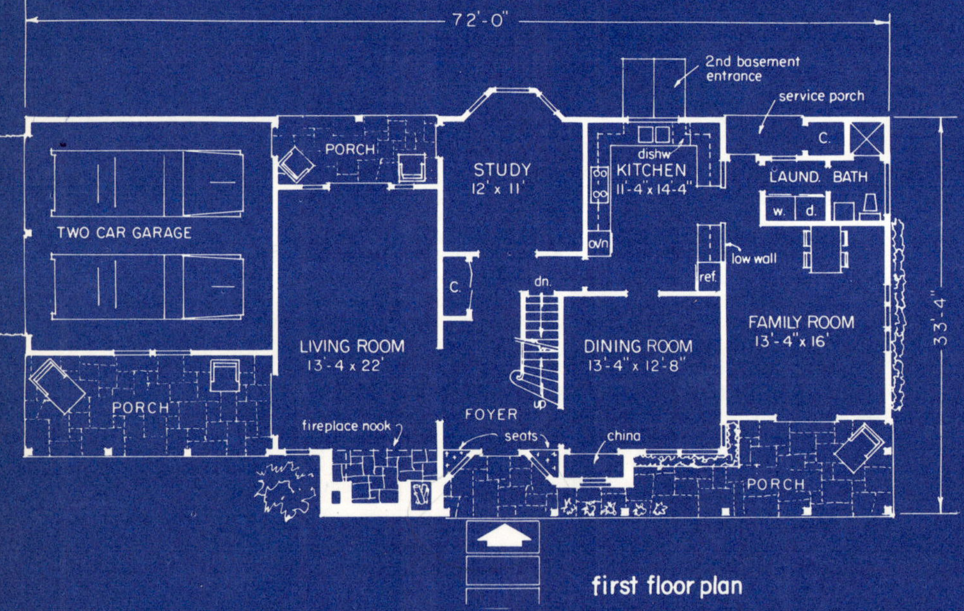

first floor plan

Pitch 12/10

Attic

Bath
Closet
Hall
Bath
8.2

Dining room
Hall
Study
8.2

7.0

Cross section

Cabs
refrig
sink
dishw
Cat
oven
Hood
range top

Kitchen elevations

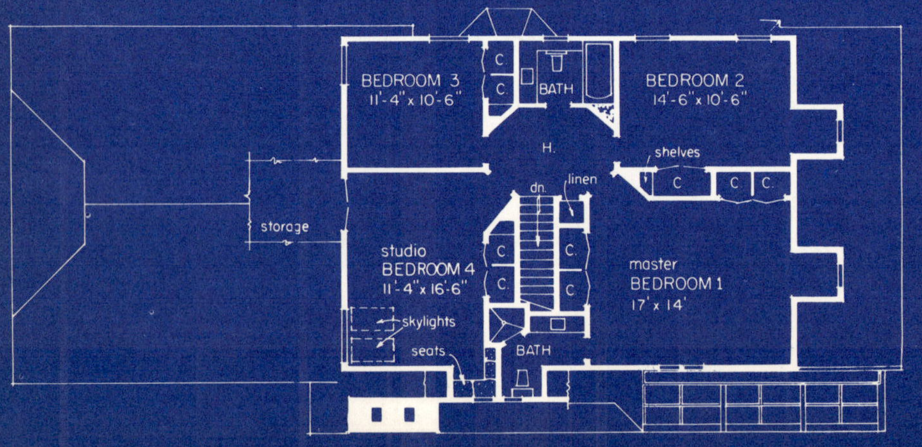

second floor plan

DESIGN S-28

**Size data,**

1,386 square feet on first floor not including garage and porches

1,109 square feet on second floor not including storage

# English Tudor Has Look of Strength

The Tudor design is not new in America, but its return to popularity occurred only recently.

In Tudor, there is a free, unregimented feeling. It rambles by comparison to the formality of some Colonial designs. The heavy atmosphere of timber and masonry gives a feeling of strength and longevity.

The front entrance is located beneath a slate roof. Narrow windows in the walls are of leaded glass. Inside, unusual triangular seats back up the windows in the large foyer. An attractive stairway goes to the second floor, and a wide hall leads to a rear study and the service area. To the left, a spacious living room runs front to rear with exposure in both directions. A huge fireplace nook dominates the front wall. It has a log storage bin and Dutch oven and timber bracing on three sides.

A dining room on the right of the foyer has an interesting built-in closet, with a window in the back wall for the display of fine china. A large kitchen in the rear has low walls on the right which open up to the family room, the laundry and a bath with stall shower. The weather-protected service porch adjoins.

On the second floor is a large octagonal foyer to which all bedrooms connect. There are four rooms with two baths, one of which is private for the master bedroom. One bedroom can be used as a studio, since there are skylight windows in the roof.

## Material List

|  |  |
|---|---|
| **CONCRETE** | |
| Concrete walls, footings, etc. | 2200 cu. ft. |
| Base & garage floors apron | 1112 sq. ft. |
| Porch floor 4" conc. 6 x 6 mesh | 400 sq. ft. |
| **STEEL** | |
| Lally column 3½" x 6'9 | 6 |
| 7" I 15.3 lb. | 60 lin. ft. |
| **MASONRY — CHIMNEY & VENEER** | |
| 4" Brick veneer | 72 sq. ft. |
| 4" Brick veneer chimney | 600 sq. ft. |
| 4" Brick veneer chimney | 100 sq. ft. |
| 4" Stone & mantel — 7200 brick | 25 sq. ft. |
| Fire brick | 120 |
| Flagstone porches | 365 sq. ft. |
| **CARPENTRY** | |
| Plates, joists,2 floors | 5400 BM |
| Joists, garage 2 x 8 | 430 BM |
| Backing — bridging | 600 BM |
| Sub floor ⅝ plyscore 2 floors | 2800 sq. ft. |
| Studs, headers, etc. | 5700 BM |
| Ceiling joists | 1000 BM |
| Sel. fir posts & beams & rails | 1062 BM |
| Rafters | 4700 BM |
| ½" Plyscore roof boards | 3720 sq. ft. |
| ⅜ Ext. Plywood Good 1 side | 2500 sq. ft. |
| Adzed — siding ¾ x 12 | 100 BM |
| 2 x 8 & 2 x 10 for half timber Sel. fir some bevel & some Rbt. | 700 BM |
| 13/16 x 2¼ Cl. red oak floor | 2300 BM |
| ⅝ Plugged plyscore | 450 sq. ft. |
| 15 lb. Saturated felt | 12 Rolls |
| **ROOFING** | |
| Slate roof pre-punched | 39.5 sq. |
| Starter hips & ridges | 450 lin. ft. |
| 15 lb. Saturated felt | 10 rolls |
| **PLASTERBOARD** | |
| ½" for walls | 6400 sq. ft. |
| ⅜ for ceiling | 2400 sq. ft. |

FRONT ELEVATION

RIGHT SIDE ELEVATION

REAR ELEVATION

LEFT SIDE ELEVATION

STATISTICS

BASIC HOUSE........1862 SQ.FT.
MUD RM; GARAGE.....542 SQ.FT.
CELLAR.............672 SQ.FT.

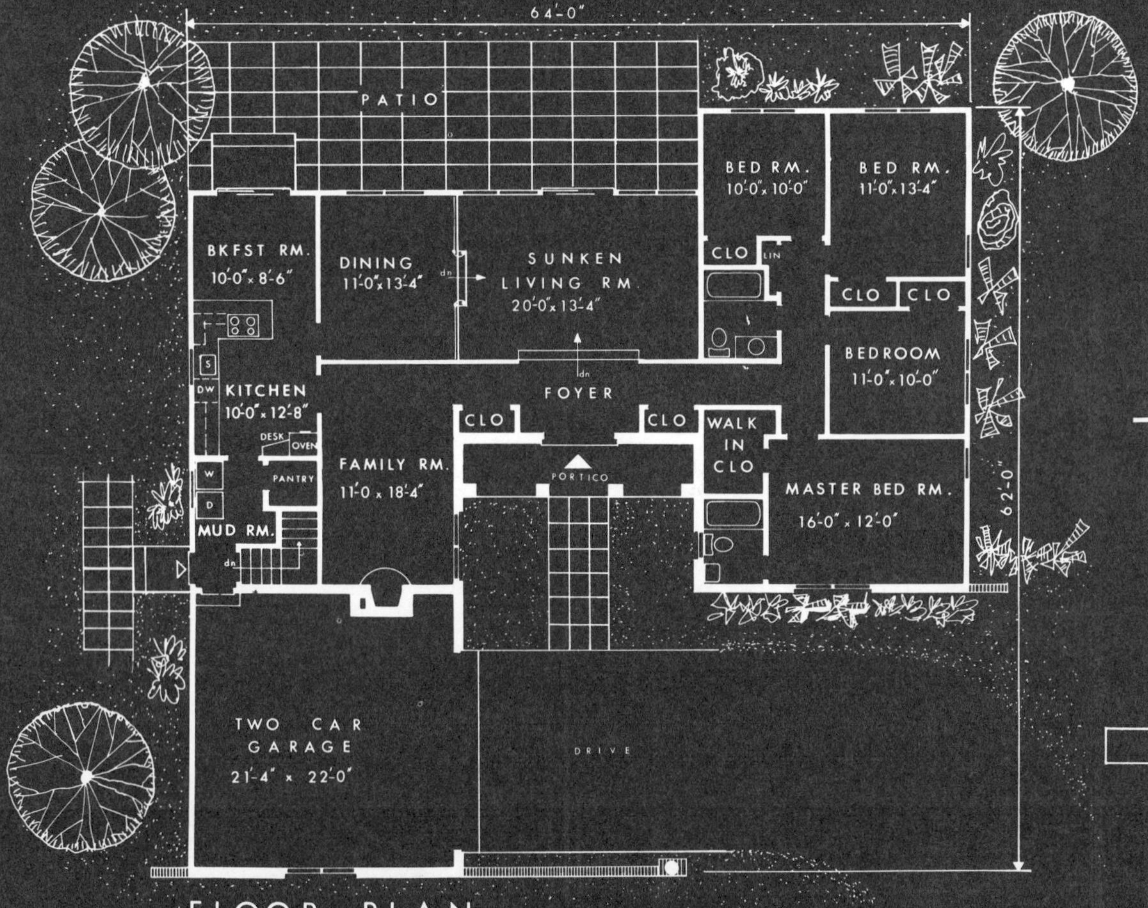

64'-0"
PATIO
BED RM. 10'-0" x 10'-0"
BED RM. 11'-0" x 13'-4"
BKFST RM. 10'-0" x 8'-6"
DINING 11'-0" x 13'-4"
SUNKEN LIVING RM 20'-0" x 13'-4"
CLO
LIN
CLO
CLO
BEDROOM 11'-0" x 10'-0"
KITCHEN 10'-0" x 12'-8"
FOYER
CLO
CLO
WALK IN CLO
DESK
OVEN
PANTRY
FAMILY RM. 11'-0 x 18'-4"
PORTICO
MASTER BED RM. 16'-0" x 12'-0"
MUD RM.
62'-0"
TWO CAR GARAGE 21'-4" x 22'-0"
DRIVE

FLOOR PLAN

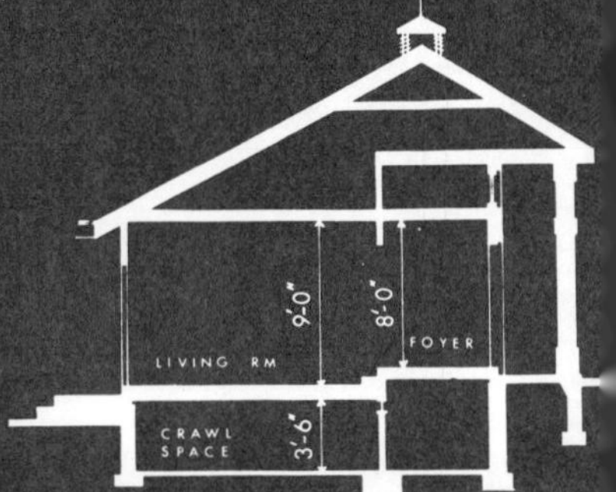

LIVING RM
FOYER
9'-0"
8'-0"
CRAWL SPACE
3'-6"

SECTION

DESIGN S-11

# Ranch with Unusual Exterior Design

This unusual home is a four-bedroom ranch with all the comforts of most well-designed one-story houses, but with the elegance, formality, good zoning and privacy normally found in two-story dwellings.

Note the stately appearance of the outside. A formal courtyard entrance is flanked by two low-roofed wings. In the center is a 12-foot-high columned portico, with a gracious double door entrance. The garage faces the side and the driveway brings guests right to the formal courtyard entrance. The entire front facade is covered in brick and accented with Colonial style windows and shutters.

There is a spacious reception foyer with a closet on each side. Straight ahead is a sunken living room with an attractive window wall overlooking the patio and garden.

Alongside the living room and located up two steps is a balconied dining room, which also faces the rear patio and garden.

To the left of the foyer is the informal activity area of the home, highlighted by a paneled family room with a log-burning brick fireplace. Adjoining this is the kitchen, which stretches out to a breakfast room or dinette for informal meals.

The entire right wing of the house comprises a private and quiet bedroom zone. There are four bedrooms. The master bedroom features an enormous walk-in closet with a full private bath.

## Material List

**CONCRETE WORK**

| | |
|---|---|
| Foundations, footings, slabs, etc. | 88 cu. yds. |

**STEEL**

| | | |
|---|---|---|
| Girder | 8B10 | 26' lin. ft. |
| Girder | 10B11.5 | 84 lin. ft. |
| Lally Cols. | 3½" diam. | 8 pieces |
| Reinforcing Mesh | | 185 sq. ft. |

**FRAMING LUMBER**

| | |
|---|---|
| Total Sills, Joists, Rafters, Studs, Plates, etc. | 13,892 B.F.M. |

**SHEATHING, INSULATION**

| | |
|---|---|
| Sub Flooring | 1950 sq. ft. |
| Wall Sheathing | 2200 sq. ft. |
| Roof Sheathing | 3450 sq. ft. |
| Wall Insulation | 1610 sq. ft. |
| Ceiling Insulation | 1985 sq. ft. |

**FINISHES, INTERIOR**

| | |
|---|---|
| Vinyl Tile | 610 sq. ft. |
| Oak Flooring | 1125 sq. ft. |
| Ceramic Tile Floors | 56 sq. ft. |
| Ceramic Tile Walls | 210 sq. ft. |
| Gypsum Board — house walls | 3800 sq. ft. |
| Gypsum Board — house ceilings | 1855 sq. ft. |
| Gypsum Board — garage | 985 sq. ft. |

**FINISHES, EXTERIOR (other than masonry)**

| | |
|---|---|
| Cedar Shingles | 1080 sq. ft. |
| Asphalt Shingle Roofing | 3450 sq. ft. |
| Plywood Eave & Porch Soffits | 480 sq. ft. |

**DOOR SCHEDULE**

| | |
|---|---|
| Ext. Hardwood, paneled & glazed | 2 units |
| Ext. Hardwood, paneled | 1 |
| Aluminum Sliding | 2 units |
| Int. Hardwood, flush, staingrade | 9 |
| Int. Hardwood, flush, sliding doors | 5 |
| Int. Hardwood, louvered, bi-folding | 1 |
| Metal Clad one side | 1 unit |

FRONT ELEVATION

RIGHT SIDE ELEVATION

REAR ELEVATION

LEFT SIDE ELEVATION

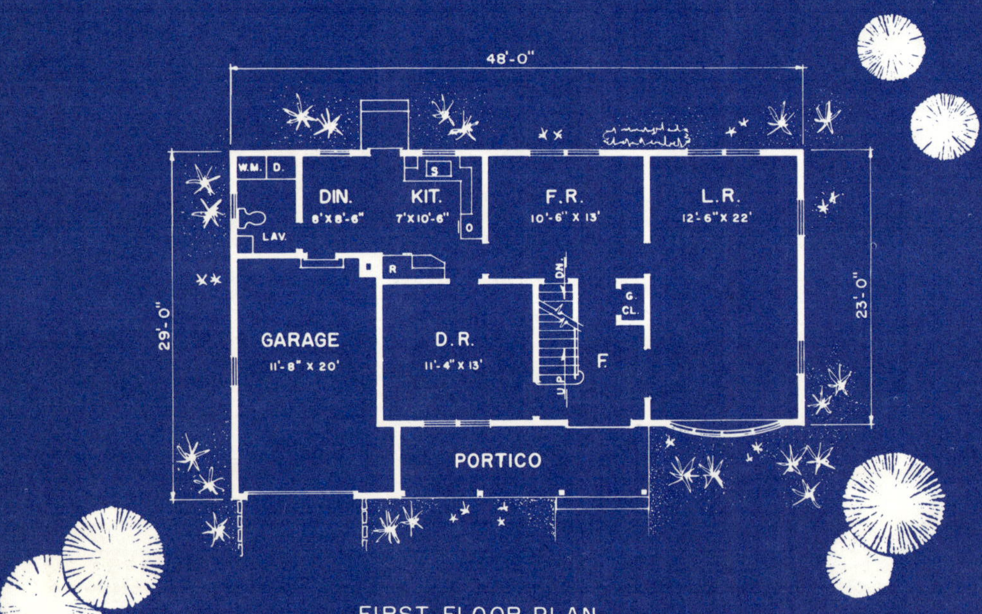


FIRST FLOOR PLAN

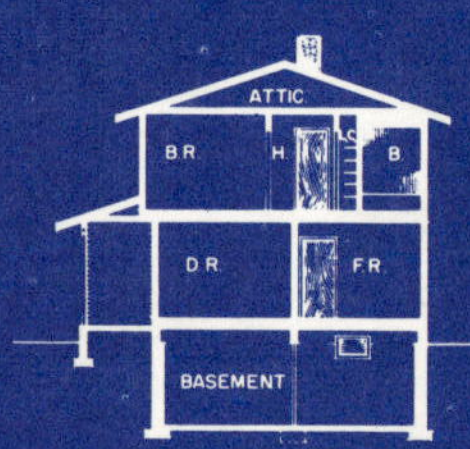


CROSS SECTION

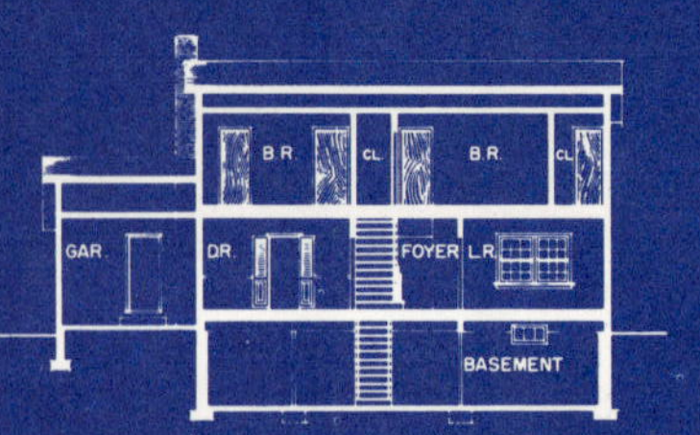


LONGITUDINAL SECTION

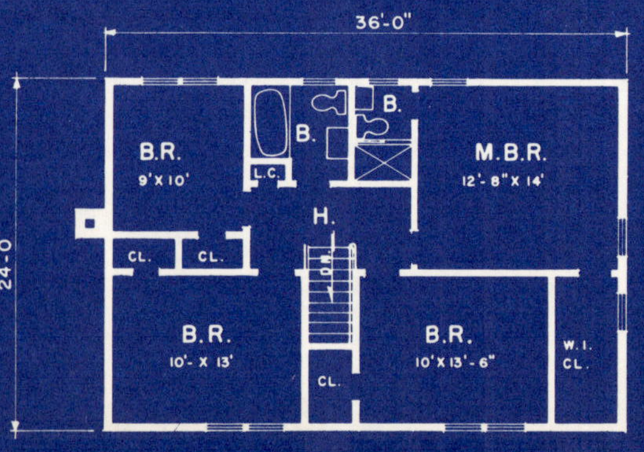


SECOND FLOOR PLAN

## AREA STATISTICS:

FIRST FLOOR . . . . . 925 SQ.FT.
SECOND FLOOR . . . 864 SQ.FT.
BASEMENT . . . . . . 770 SQ.FT.
GARAGE . . . . . . . 240 SQ.FT.

DESIGN L-24

# Two-Story Without Vertical Look

Nobody has yet come up with a more economical way of housing a large family than with a two-story residence.

Building up, rather than sideways, provides more living space for less money. This is especially so when the shape of the house is square, since it helps to cut down construction costs. Design L-24 is almost square, yet has a horizontal look, created by the use of a continous lower roof line.

A dramatic covered entrance leads into a fine open-stair reception area. The dining room is on the left and the living room is on the right with its three windows and decorative bow window at the front. Between the kitchen and the living room the wood-paneled family room is the hub of family and guest entertaining activity. The rear wall can be extended to make the family room larger.

An attractive kitchen with a convenient rear service door includes a dinette area big enough for the entire family.

Equally good planning marks the second floor where all the twin-sized bedrooms are located. The two bathrooms are back-to-back and over the kitchen, minimizing plumbing costs. The rear master bedroom has a private bath and a deep walk-in closet.

With its country-style exterior combining beveled siding and brick veneer, this house has an appearance of comfort paralleling the livable inside plan.

## Material List

**CONCRETE WORK**

| | |
|---|---|
| Footings, floors, etc. | 27 cu. yds. |

**MASONRY**

| | |
|---|---|
| 12" Concrete block | 200 blocks |
| 8" Concrete block | 400 blocks |
| 4" Pancake block | 150 blocks |
| Brick | 450 sq. ft. |

**FRAMING LUMBER**

| | |
|---|---|
| Sills, joists, rafters, studs, plates, etc. | 8000 BFM |

**EXTERIOR SHEATHING**

| | |
|---|---|
| ½" Plyscore wall sheathing | 2000 sq. ft. |
| ½" Plyscore roof sheathing | 1500 sq. ft. |

**DOOR SCHEDULE**

(1) 9-0 x 7-0 wp overhead
(2) 2-8 x 6-8 x 1¾ wp front entrance door
(1) 2-8 x 6-8 x 1¾ wp rear sash door
(1) 2-8 x 6-8 x 1¾ scfp. door
(14) Flush interior veneered doors

**WINDOW SCHEDULE**

(4) 2-8 x ⅛ basement sash
(1) 2-32 - 18 wp awning
(11) 2-8 x 4-2 DH
(3) 2-4 x 3-6 DH
(2) 3-0 x 4-6 DH
(6) 3-0 x 4-2 DH

**EXTERIOR FINISHES**

| | |
|---|---|
| Asphalt strip shingle roofing | 18 squares |
| Clapboard siding or Aluminum | 2600 sq. ft. |

**INTERIOR FINISHES**

| | |
|---|---|
| Ceramic tile floor | 73 sq. ft. |
| Ceramic tile walls | 320 sq. ft. |
| Vinyl tile | 350 sq. ft. |
| Gypsum board walls and ceiling | 8700 sq. ft. |
| Oak flooring | 1300 sq. ft. |

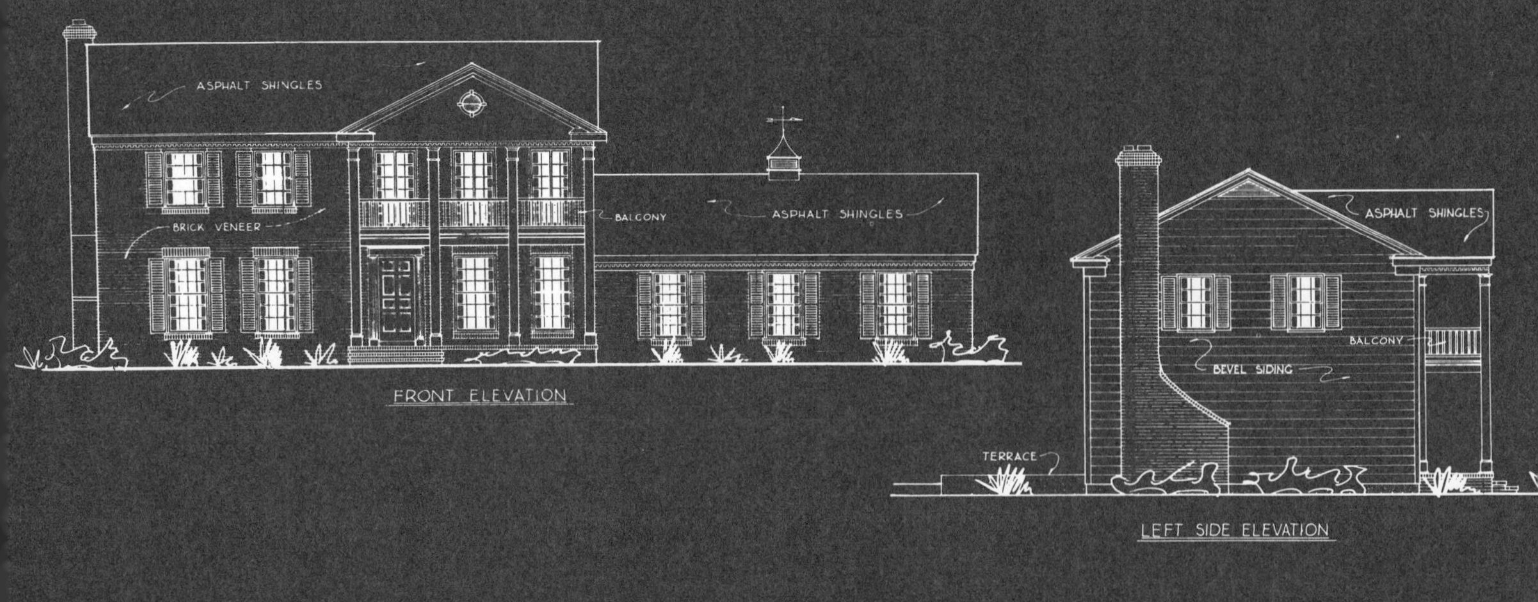

ASPHALT SHINGLES
BRICK VENEER
BALCONY
ASPHALT SHINGLES

FRONT ELEVATION

ASPHALT SHINGLES
BALCONY
BEVEL SIDING
TERRACE

LEFT SIDE ELEVATION

ASPHALT SHINGLES
ASPHALT SHINGLES
BEVEL SIDING
BEVEL SIDING
TERRACE

REAR ELEVATION

ASPHALT SHINGLES
BEVEL SIDING
BALCONY

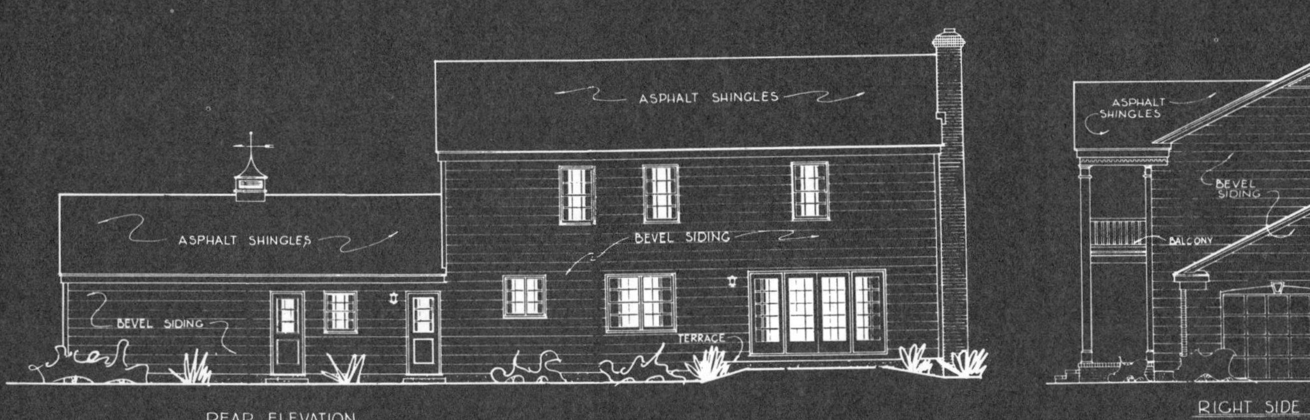

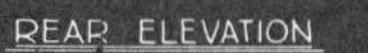

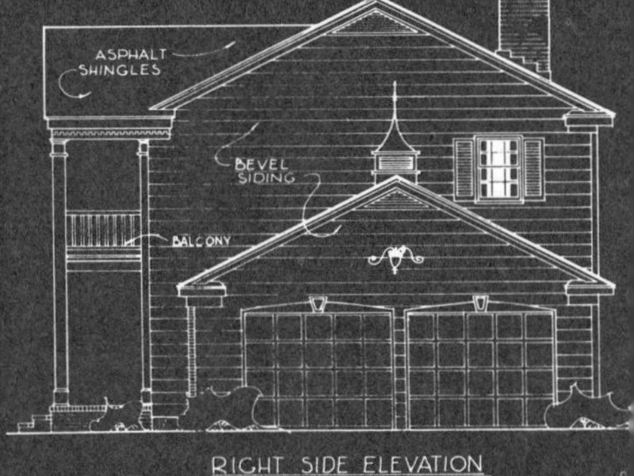

RIGHT SIDE ELEVATION

## HOUSE DATA

| | |
|---|---|
| 1ST FLOOR (OVER CELLAR) | 1163 SQ. FT. |
| 1ST FLOOR (ON SLAB) | 286 SQ. FT. |
| 2ND FLOOR | 1163 SQ. FT. |
| GARAGE | 516 SQ. FT. |

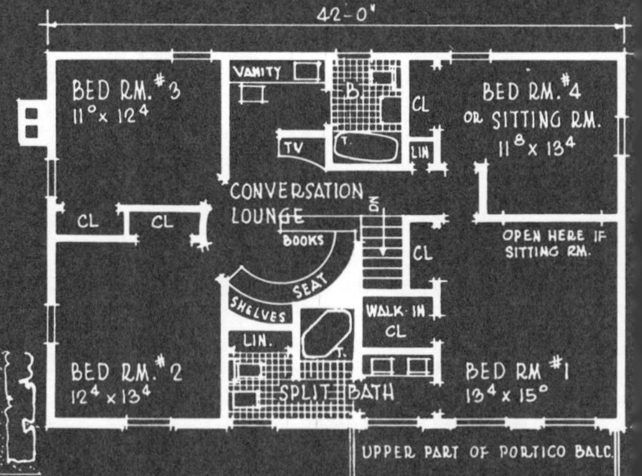

42'-0"
BED RM. #3
11$^0$ x 12$^4$
VANITY
B
CL
LIN
TV
T
BED RM. #4
OR SITTING RM.
11$^8$ x 13$^4$
CONVERSATION
LOUNGE
DN
CL
CL
BOOKS
SEAT
SHELVES
OPEN HERE IF SITTING RM.
CL
WALK-IN
CL
LIN.
T
BED RM. #2
12$^4$ x 13$^4$
SPLIT BATH
BED RM. #1
13$^4$ x 15$^0$
UPPER PART OF PORTICO BALC.

SECOND FLOOR PLAN

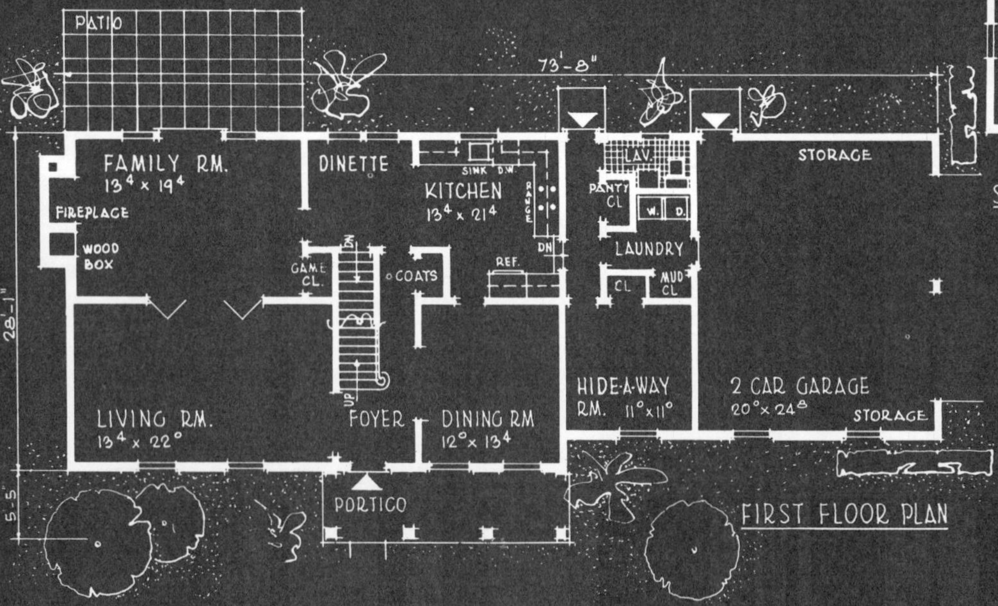

PATIO
73'-8"
FAMILY RM.
13$^4$ x 19$^4$
DINETTE
SINK
D.W.
KITCHEN
13$^4$ x 21$^4$
RANGE
LAV.
PANTRY
CL
W
D
STORAGE
FIREPLACE
WOOD BOX
DN
GAME CL.
COATS
REF.
DN
LAUNDRY
CL
MUD CL
28'-1"
UP
HIDE-A-WAY
RM. 11$^0$ x 11$^0$
2 CAR GARAGE
20$^0$ x 24$^8$
LIVING RM.
13$^4$ x 22$^0$
FOYER
DINING RM
12$^0$ x 13$^4$
STORAGE
5'-5"
PORTICO

FIRST FLOOR PLAN

DESIGN S-59

# Spotlight on Conversation Lounge

Surveys by marketing consultants indicate that many houses of the future will be designed with emphasis on "excitement."

Designed into this four-bedroom, two-story house is an area designated as a conversation lounge. This space is really an enlarged second-floor foyer, circular in shape and intended to produce both glamour and practicality. Furnished with circular bench, bookshelves, stereo, television and works of art it is likely to become a gathering place for younger members of the family.

The architect provides another room in a separate portion of the house planned primarily for the man of the house, as an office or den, so private that its only connection with the rest of the living quarters is through a rear hall.

The rest of the plan is designed to place the principal rooms in their most useful locations. For example, the family room and the kitchen are at the rear to give convenient access from the outdoors and the patio.

On the second floor, the four bedrooms are grouped around the conversation lounge and stair hall. Large closets are featured throughout. The hall bath is of the split variety, with a vanity dressing counter.

The elegance of the two-storied portico, with its second-floor balcony, gives this house a stamp of opulence. The entire front is brick, on a scale reflecting good taste.

## Material List

**MILLWORK**

|  |  |
|---|---|
| Exterior Doors & Frames Compl. | 7 pieces |
| Garage Door Complete Set | 2 sets |
| Interior Doors & Frames Compl. | 13 pieces |
| Sliding Doors | 3 pieces |
| Bi Fold Doors | 11 sets |
| Windows | 24 sets |
| Fascia | 316 lin. ft. |
| Shutters | 20 pieces |
| Base | 505 lin. ft. |

**KITCHEN CABINETS**

|  |  |
|---|---|
| Base Cabinets | 8'-3" long |
| Wall Cabinets | 13'-9" long |

**ROOFING**

|  |  |
|---|---|
| Shingles | 2908 sq. ft. |
| Roofing Paper | 2908 sq. ft. |

**CARPENTRY**

|  |  |
|---|---|
| Framing Lumber | 9550 B.F. |
| Studs | 4736 B.F. |
| Plates | 1420 B.F. |
| Roof Sheathing | 2908 sq. ft. |
| Side Wall Sheathing | 2644 sq. ft. |
| Insulation Walls | 1881 sq. ft. |
| Insulation Ceilings | 1450 sq. ft. |
| Wood Flooring | 2118 sq. ft. |
| Kitchen Plywood | 148 sq. ft. |
| Asphalt Tile Floor | 242 sq. ft. |

**CONCRETE WORK**

|  |  |
|---|---|
| Concrete Walls | 1201 cu. ft. |
| Concrete Footings | 316 cu. ft. |
| Slabs | 745 cu. ft. |

**STRUCTURAL STEEL**

|  |  |  |
|---|---|---|
| Lally Columns | 3½" diam. | 7 pieces |
| Girder | 6" & 8" I Beam | 49 lin. ft. |

**BRICK WORK**

|  |  |  |
|---|---|---|
| Chimney | Brick & Block | 293 cu. ft. |
| Walls | 4" Brick | 869 sq. ft. |

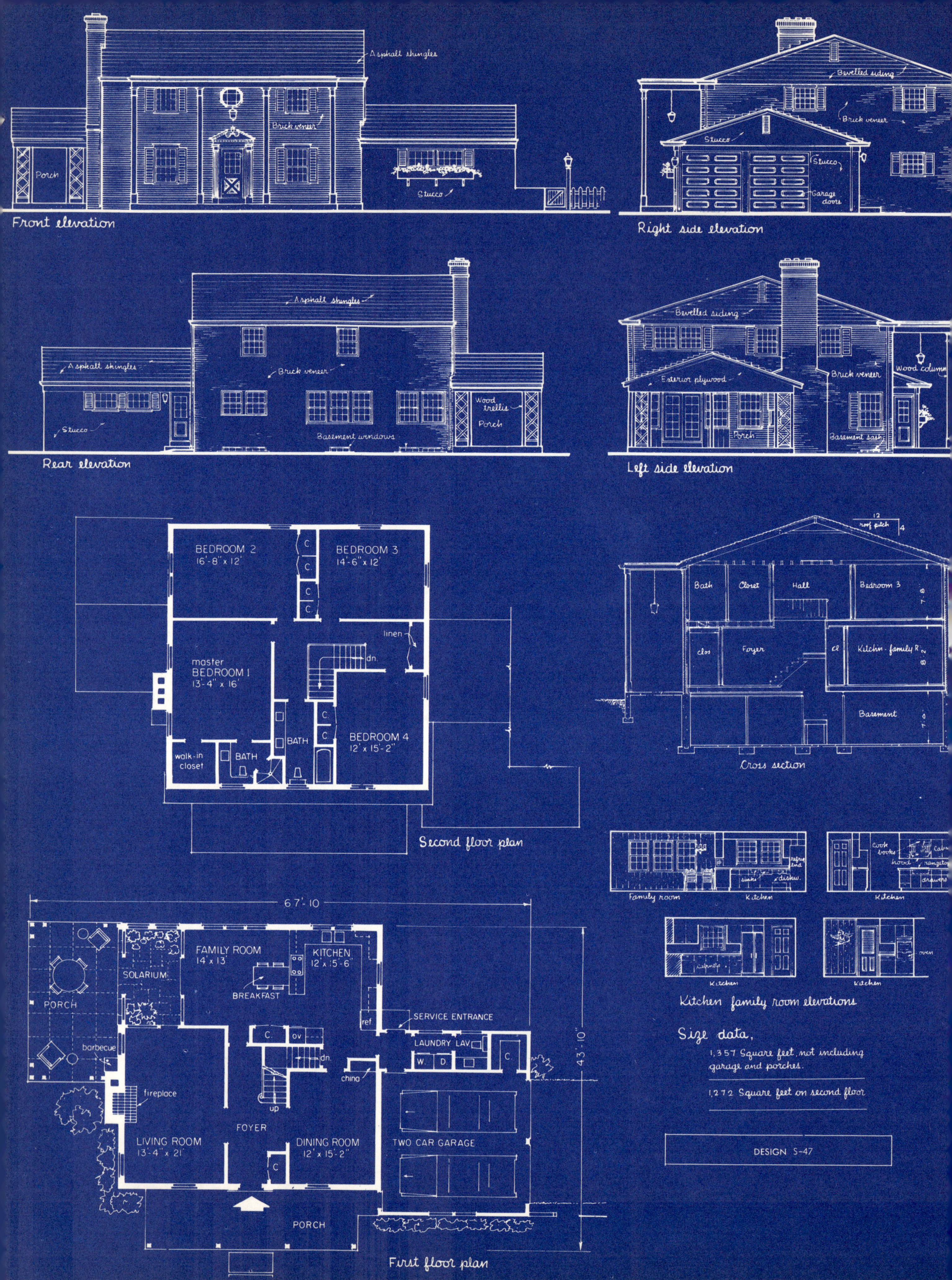

Asphalt shingles
Brick veneer
Porch
Stucco
Front elevation
Bevelled siding
Brick veneer
Stucco
Stucco
Garage doors
Right side elevation
Asphalt shingles
Asphalt shingles
Brick veneer
Stucco
Wood trellis
Porch
Basement windows
Rear elevation
Bevelled siding
Exterior plywood
Brick veneer
Wood column
Porch
Basement sash
Left side elevation
BEDROOM 2
16'-8" x 12'
BEDROOM 3
14'-6" x 12'
linen
dn.
master
BEDROOM 1
13'-4" x 16'
BEDROOM 4
12' x 15'-2"
BATH
walk-in closet
BATH
Second floor plan
Bath
Closet
Hall
Bedroom 3
Foyer
Kitchen-family R.
Basement
Cross section
Family room
Kitchen
Kitchen
Kitchen
Kitchen
Kitchen family room elevations
67'-10
43'-10"
FAMILY ROOM
14' x 13'
KITCHEN
12' x 15'-6"
SOLARIUM
BREAKFAST
PORCH
SERVICE ENTRANCE
ref
LAUNDRY
LAV
W
D
dn.
barbecue
china
fireplace
up
FOYER
LIVING ROOM
13'-4" x 21'
DINING ROOM
12' x 15'-2"
TWO CAR GARAGE
PORCH
First floor plan
Size data,
1,357 Square feet, not including garage and porches.
1,272 Square feet on second floor
DESIGN S-47

# Balanced Colonial With Solarium

There is no thought of a box-like house in looking at this majestic, two-story Colonial. That's because of the placement of an attractive porch at one side of the house and a two-car garage at the other. The squareness of the habitable area is additionally camouflaged by the high-columned front porch, which has always been a dramatic feature of certain Colonial facades.

Inside the front door, a 90-degree turned stairway goes to the second floor. The living and dining rooms are on opposite sides of the foyer. The first glance into the living room brings into immediate view a fireplaced wall flanked by windows.

Across the rear three areas are linked together into kitchen, family room and solarium, with the entire length having an unobstructed vista.

The solarium is both eye-catching and practical — a plant grower's dream and an interesting spot to relax in. It is glazed on all four sides both for the view and to take full advantage of sunlight. Just outside the solarium is the large porch, which can be screened or left open.

Upstairs are four bedrooms, two complete bathrooms and excellent closet space. Design S-47 presents a well-balanced and pleasing appearance. Especially interesting is the antique brick exterior, with black and white bricks interspersed among the red.

## Material List

**CONCRETE**

| | |
|---|---|
| 8"—10" walls — piers, etc. | 1730 cu. ft. |
| Base-garage-floor-apron | 1670 sq. ft. |
| 4" Con. floor 6 x 6 mesh scored | 364 sq. ft. |
| 4" Con. floor 6 x 6 mesh smooth | 250 sq. ft. |
| Slate floor solarium | 110 sq. ft. |

**STEEL**

8 Lally columns 3½ x 6'10
72 Lin. 6' I — 12½ lb.
116 Lin. 4 x 3 —5/16 Angle iron

**MASONRY**

| | |
|---|---|
| 4" Brick veneer | 2370 sq. ft. |
| 50 Press — brick for mantel | |
| Com. brick back up for chimney | 2700 Common |
| 48 Lin. 12 x 13 flue lining | |
| 16 Lin. 8 x 13 flue lining | |

**CARPENTRY**

| | |
|---|---|
| Plates, joists, etc. both floors | 4800 BM |
| Studs, headers, etc. | 5500 BM |
| ⅝ Plyscore sub floor | 2560 sq. ft. |
| Backing — bridging | 460 BM |
| Ceiling joists | 1830 BM |
| Rafters etc. | 3300 BM |
| ½ Plyscore roof boards | 2700 sq. ft. |
| Sheathing — ⅜ plyscore | 3100 sq. ft. |
| 12 Rolls 15 lb. Sat. felt | |
| 13/16 x 2¼ Cl. red Oak floor | 2170 BM |

**ROOFING**

27 sq. 210 lb. thick butt self sealing asphalt shingle
4 rolls starter
7 Rolls 15 lb. Sat. felt
3 sq. 4 ply tar & gravel — 400 lb. gravel per sq.
46 Lin. copper gravel stop

**PLASTERBOARD**

| | |
|---|---|
| ½ for walls | 6700 sq. ft. |
| ⅜ for ceiling | 2400 sq. ft. |

FRONT ELEVATION

RIGHT SIDE ELEVATION

REAR ELEVATION

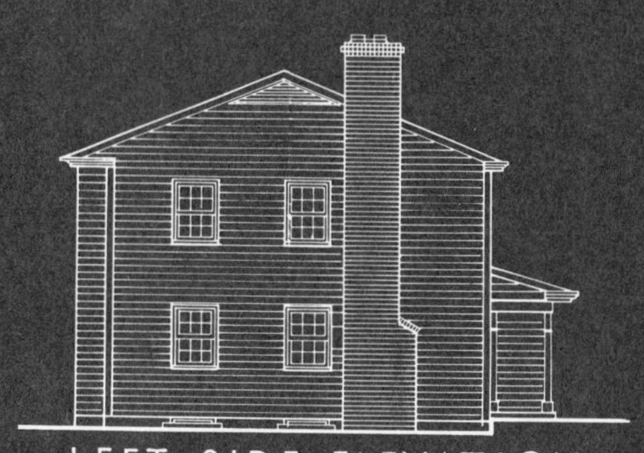

LEFT SIDE ELEVATION

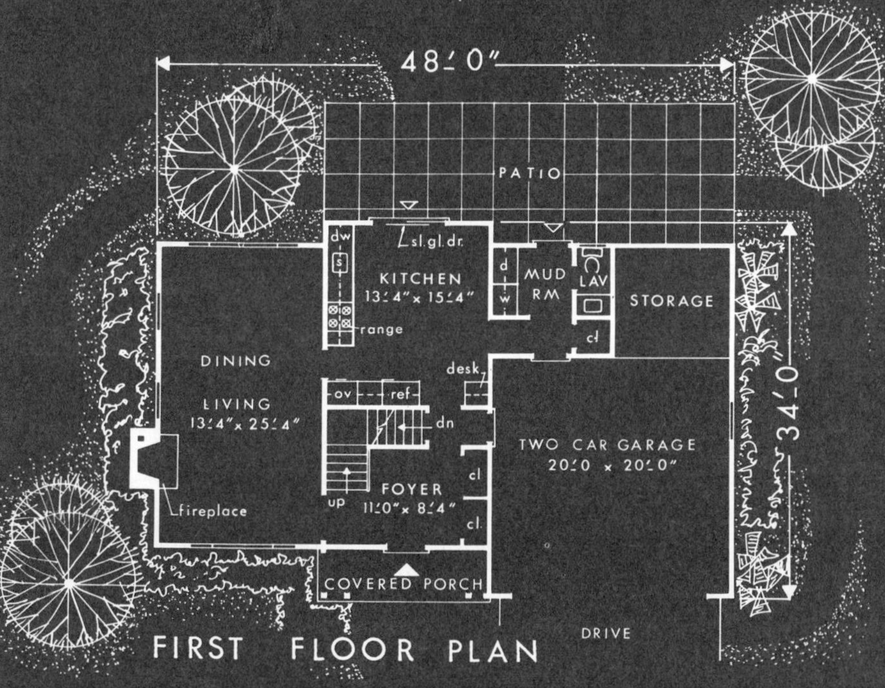


FIRST FLOOR PLAN

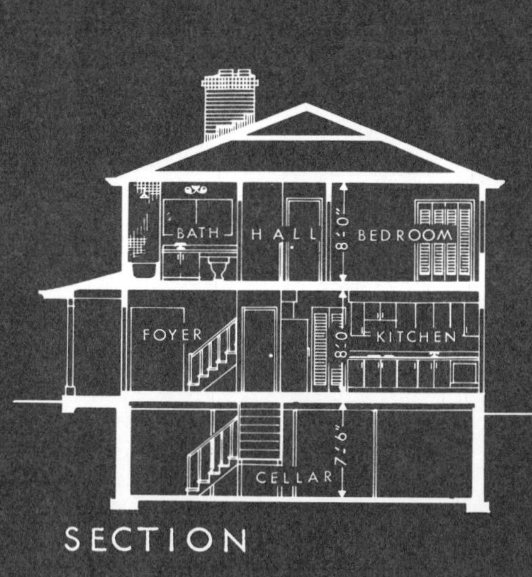


SECTION

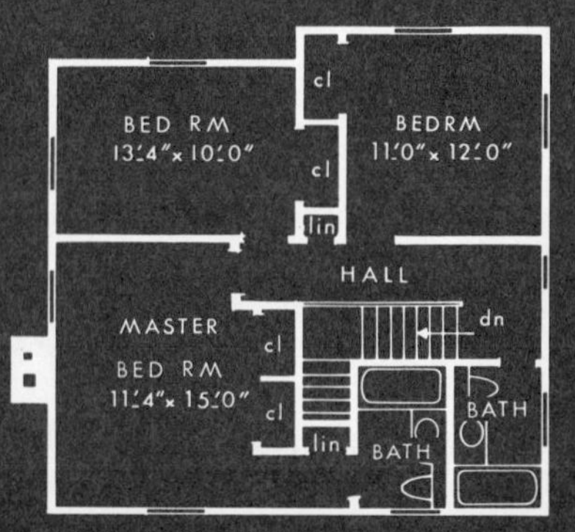


SECOND FLOOR PLAN

AREA STATISTICS :

1st floor............. 902 sq. ft.

2nd floor............ 902 "

mud rm, storage, garage.............. 619 "

cellar................ 902 "

DESIGN L-65

# Solid Design in New England Colonial

With its well-proportioned two-story section, embraced by the one-story garage and portico, this Colonial house has a quiet restraint. Its crisp, narrow horizontal siding, reminiscent of the early New England Colonial, will look equally well whether painted or stained. A red brick chimney on one side provides just enough contrast of material and texture. The gabled roof over the second floor with its large overhang, echoed by the lower roof, maintains the simplicity characteristic of the entire house.

After passing under the welcoming portico, one enters a large foyer with two closets. Access to all rooms on the first floor is off this gracious foyer. The living room-dining room area flows together, giving a feeling of spaciousness enhanced by the large windows and the cozy fireplace. The kitchen is fully equipped and efficient right down to the well-located desk. Sliding glass doors lead to the rear patio, making outdoor meals convenient and easy. There is plenty of space for family dining in this spacious kitchen.

The bedroom area upstairs is reached via the attractive staircase, also off the foyer. The center hall upstairs leads to all bedrooms and to the family bath. The master bedroom is on a corner with double exposure, equipped with its own full bath. The two other bedrooms are also corner rooms with cross ventilation. There is easy access to the second bath.

## Material List

|  |  |
|---|---|
| **CONCRETE WORK** | |
| Foundations, footings, slabs, etc. | 60 cu. yds. |
| **STEEL** | |
| Girder ......... 8B10 | 25 lin. ft. |
| Lally Cols ...... 3½" diam. | 4 pcs. |
| Reinforcing Mesh | 600 sq. ft. |
| **FRAMING LUMBER** | |
| Total sills, joists, rafters, studs, plates, etc. | 8384 B.F.M. |
| **SHEATHING, INSULATION** | |
| Sub Flooring | 1404 sq. ft. |
| Wall Sheathing | 2140 sq. ft. |
| Roof Sheathing | 1576 sq. ft. |
| Wall Insulation | 2000 sq. ft. |
| Ceiling Insulation | 900 sq. ft. |
| **FINISHES, INTERIOR** | |
| Vinyl Tile | 210 sq. ft. |
| Oak Flooring | 1194 sq. ft. |
| Ceramic Tile Floors | 78 sq. ft. |
| Ceramic Tile Walls | 356 sq. ft. |
| Gypsum Board — house | 4854 sq. ft. |
| Gypsum Board — garage | 1140 sq. ft. |
| **FINISHES, EXTERIOR (other than masonry)** | |
| Horizontal siding | 2140 sq. ft. |
| Asphalt Shingle Roofing | 1576 sq. ft. |
| Plywood Eave & Porch Soffits | 100 sq. ft. |
| **WINDOW SCHEDULE** | |
| Wood Casement & Picture Unit | 2 units |
| Wood Double Hung | 12 units |
| Octagonal sash | 1 unit |
| Basement sash | 2 units |
| **DOOR SCHEDULE** | |
| Ext. Hardwood, paneled & glazed | 2 units |
| Aluminum Sliding | 1 unit |
| Int. Hardwood, flush, staingrade | 9 units |
| Int. Hardwood, louvered, bi-folding | 9 units |
| Fireproof self-closing | 2 units |

FRONT ELEVATION

LEFT SIDE ELEVATION

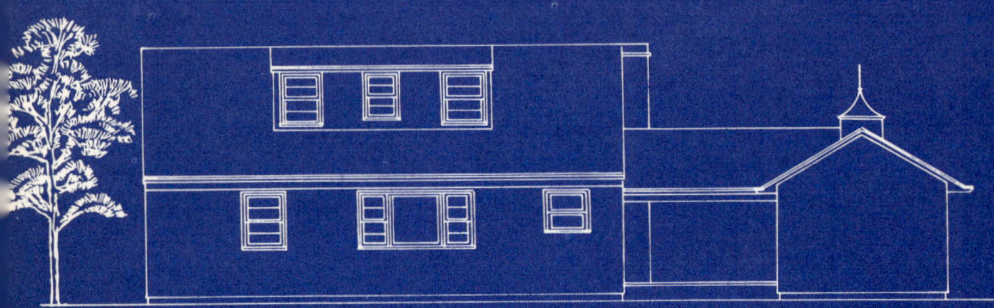

REAR ELEVATION

RIGHT SIDE ELEVATION

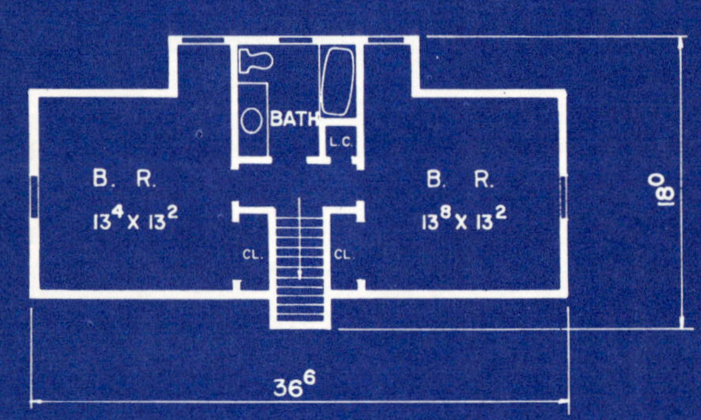


SECOND FLOOR PLAN

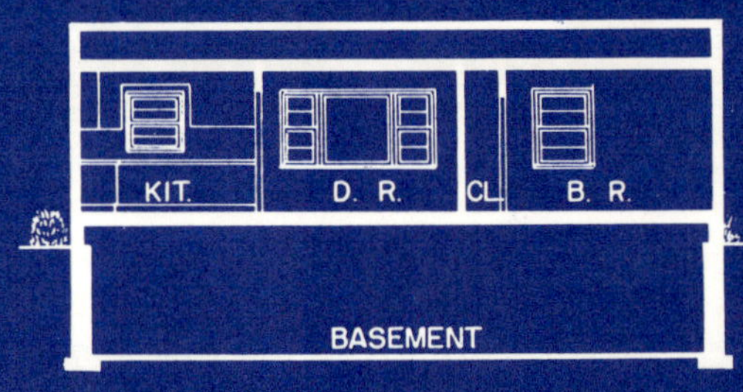


LONGITUDINAL SECTION

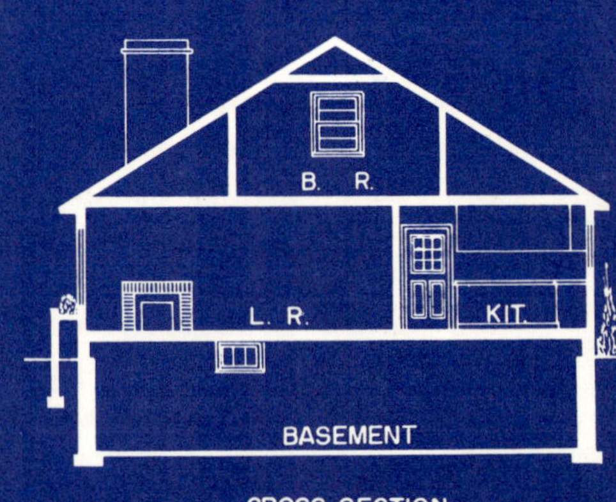


CROSS SECTION

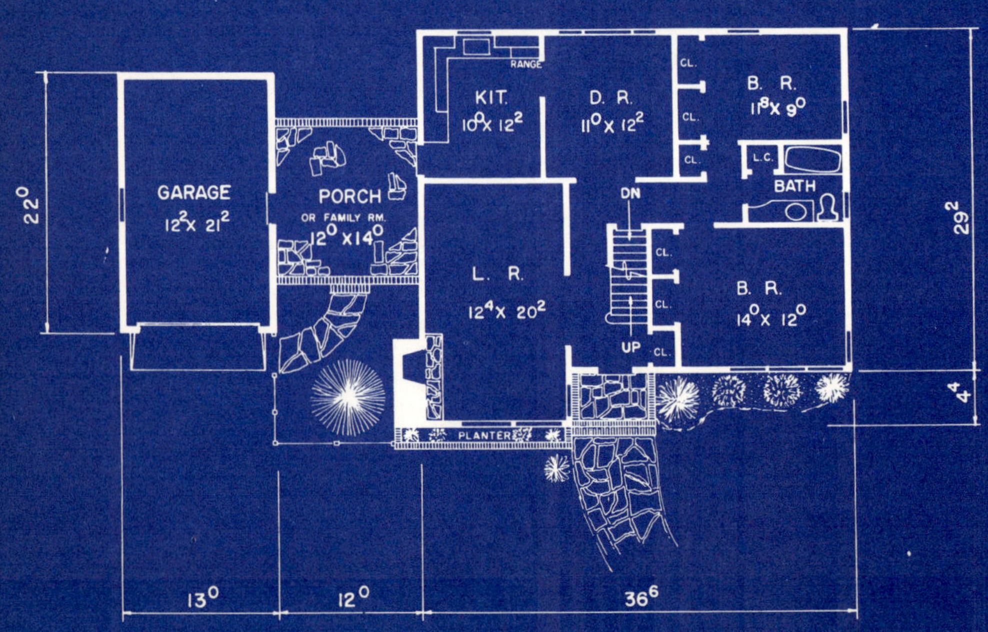


FIRST FLOOR PLAN

AREA STATISTICS

| | |
|---|---|
| FIRST FLOOR | 1104 |
| SECOND FLOOR | 570 |
| BASEMENT | 1110 |
| GARAGE | 286 |

DESIGN S-24

# House To Accommodate Your Budget

Need a small, two-bedroom house with simple charm?

Need a four-bedroom house with the identical dimensions?

If the answer is "yes" to either question, Design S-24 can accommodate your needs. The basic house has a habitable area of 1104 square feet within dimensions of 36′6″ by 33′6″. The expansion space upstairs for two bedrooms and bathroom adds 570 square feet of livability without changing the dimensions. Build the house with a breezeway, porch or family room between house and garage—and the outside figures become 61′ by 33′6″, still requiring only a modest lot.

With all these possibilities, this house can take care of a young couple with a growing family or satisfy older persons who don't want a large home yet would like some extra space for visiting children or friends.

The living room, more than 20′ long, is to the left of the center hall entrance and features an attractive corner fireplace.

Straight ahead from the foyer is the dining room, placed next to the kitchen for easy serving. The kitchen has an L-shaped arrangement, including a built-in range with ventilating hood. To the right of the foyer are two bedrooms separated by a bathroom. There are six closets in this area.

The optional breezeway can be made into an enclosed porch, family room or playroom.

## Material List

**CONCRETE WORK**
Footings, floors, etc. .......... 30 cu. yds.

**MASONRY**
12″ Concrete Block ............ 850 blocks
8″ Concrete Block ............ 920 blocks
4″ Pancake Block ............ 450 blocks
4″ Brick Veneer .............. 265 sq. ft.

**FRAMING LUMBER**
Sills, Joists, Rafters, Studs, Plates, etc. ................ 9000 BFM

**EXTERIOR SHEATHING**
1″ x 10″ fir or N.C. pine wall sheathing 2325 sq. ft.
1″ x 6″ N.C. pine roof sheathing 2200 sq. ft.

**WINDOW SCHEDULE**
(5) 2-8 x 1-8 Basement sash
(1) 2-36-18 wd. awning
(2) 2-4 x 3-6 DH
(4) 3-0 x 3-6 DH
(2) 2-0 x 4-2 DH
(2) 3-0 x 4-2 DH
(5) 3-0 x 3-10 DH
(1) 4-0 x 4-2 Fixed
(2) 4-4 x 5-3 Gliding
(1) 3-4 x 5-3 Gliding

**DOOR SCHEDULE**
(1) 8-0 x 7-0 w.p. overhead
(1) 3-0 x 6-8 x 1¾″ w.p. front entrance door
(2) 2-8 x 6-8 x 1¾″ w.p. sash door
(17) Two-panel 1⅜″ fir int. doors

**EXTERIOR FINISHES**
Asphalt roofing shingles .......... 22 squares
Flush boarding, ½″ ext. plywood. 60 sq. ft.
Red cedar shingles ............ 2225 sq. ft.

**INTERIOR FINISHES**
Ceramic Tile Floor ............ 70 sq. ft.
Ceramic Tile Walls ............ 70 sq. ft.
Vinyl Tile .................... 85 sq. ft.
Gypsum lath and plaster ........ 5800 sq. ft.
Oak Flooring .................. 1420 sq. ft.

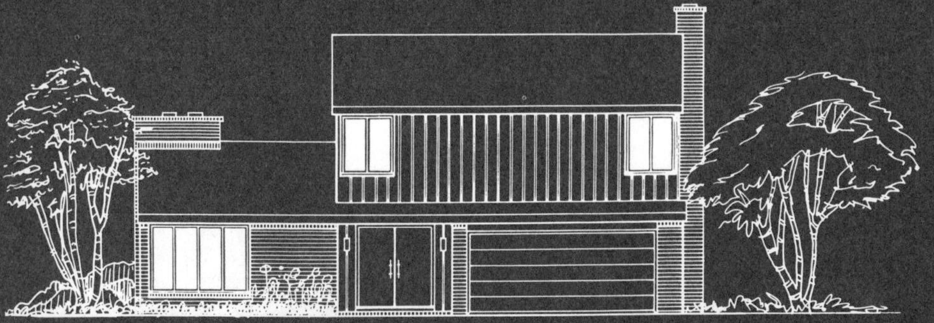

FRONT ELEVATION

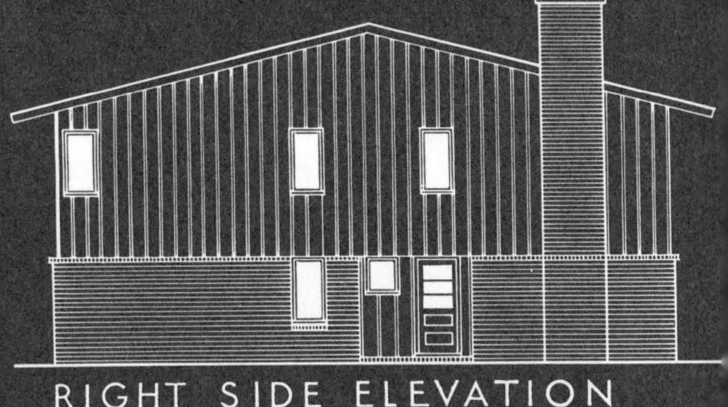

RIGHT SIDE ELEVATION

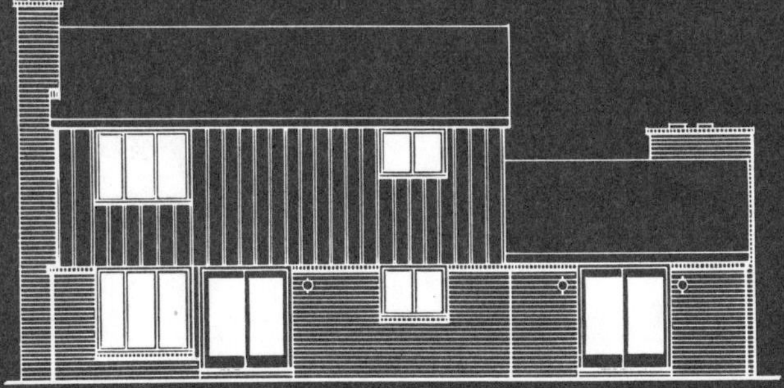

REAR ELEVATION

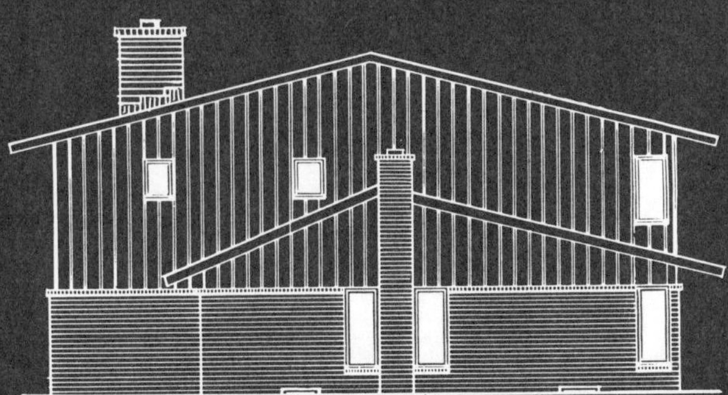

LEFT SIDE ELEVATION

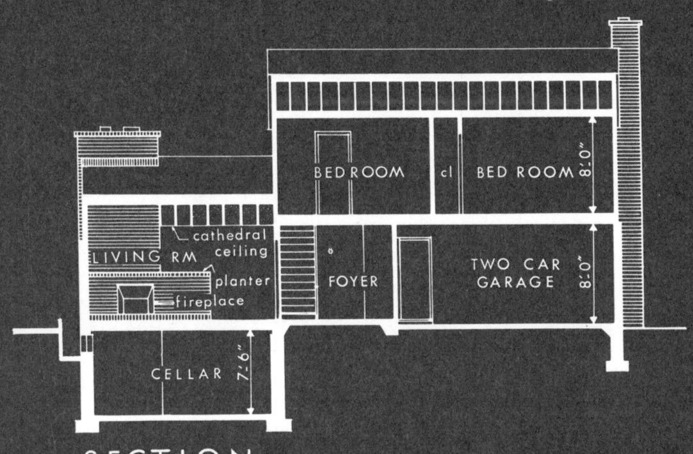


SECTION

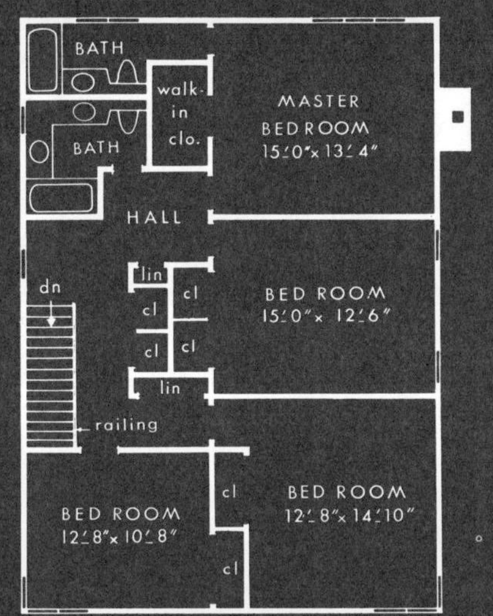


SECOND FLOOR PLAN

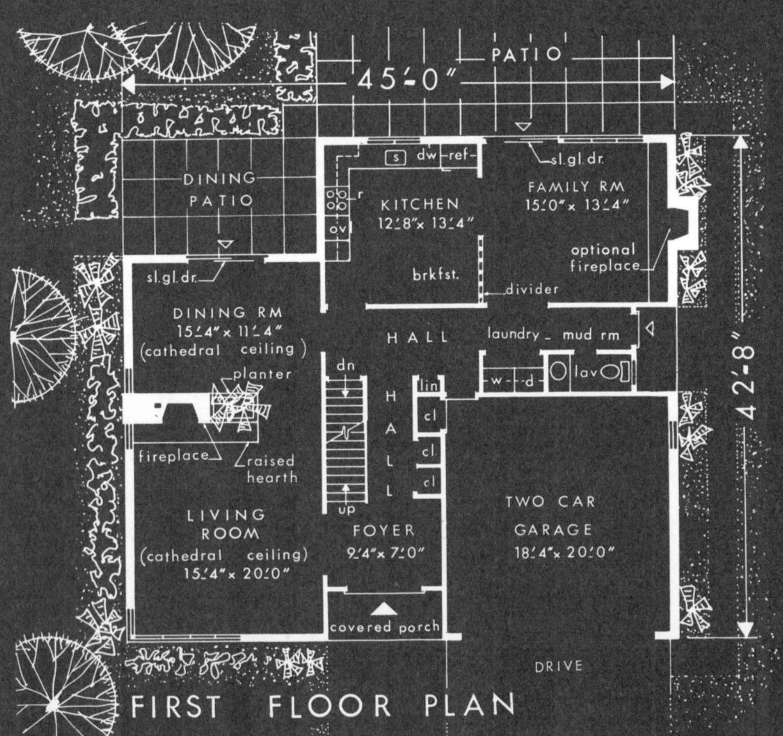


FIRST FLOOR PLAN

## AREA STATISTICS :

| | | |
|---|---|---|
| 1st floor | 1142 | sq.ft. |
| 2nd floor | 1203 | " |
| garage & mud rm | 510 | " |
| partial cellar | 500 | " |

DESIGN L-50

# Some Unusual Features in Contemporary

With its bold design, this contemporary four-bedroom house appeals to the eye in many unusual respects.

The contrast between the horizontal effect of the brick and the vertical wood siding with battens, the corner windows and the treatment of the high and low-pitched roofs combine to create a somewhat dynamic appearance. Although basically a two-story house, the roof of the one-story portion flows beyond it across the entire front in the form of an overhang. A strong horizontal line is thereby achieved.

The indented, covered porch with double-entry doors says "welcome." A roomy foyer leads into a many-windowed living room with beamed cathedral ceiling. The room features a decorative brick fireplace with a raised hearth. Incorporated into the design of the fireplace wall is a low space-dividing planter. On the other side of the fireplace is the dining room, which also has a cathedral ceiling. It opens on to a dining patio.

An L-shaped hall provides easy access to all rooms. Two closets off this hall are close to the entrance foyer. Conveniently, to the right of the hall, are a combined mud room-laundry and a lavatory.

The kitchen includes a breakfast area. It is separated by a perforated divider from the family room which, in turn, has access to a rear patio via sliding glass doors.

## Material List

**CONCRETE WORK**

| | |
|---|---|
| Foundations, footings, slabs, etc. | 80 cu. yds. |

**MASONRY**

| | |
|---|---|
| 4" Brick Veneer | 260 cu. ft. |
| Fireplace — Brick & Block | 274 cu. ft. |

**FRAMING LUMBER**

| | |
|---|---|
| Total Sills, Joists, Rafters, Studs, Plates, etc. | 8650 BFM |

**SHEATHING, INSULATION**

| | |
|---|---|
| Sub Flooring | 1656 sq. ft. |
| Wall Sheathing | 2382 sq. ft. |
| Roof Sheathing | 2090 sq. ft. |
| Wall Insulation | 2000 sq. ft. |
| Ceiling Insulation | 1843 sq. ft. |

**FINISHES, INTERIOR**

| | |
|---|---|
| Flooring | 2470 sq. ft. |
| Ceramic Tile Floors | 95 sq. ft. |
| Ceramic Tile Walls | 350 sq. ft. |
| Gypsum Board — house | 6495 sq. ft. |
| Garage | 436 sq. ft. |

**FINISHES, EXTERIOR (other than masonry)**

| | |
|---|---|
| Vertical Siding — board & batten | 1208 sq. ft. |
| Asphalt Shingle Roofing | 2090 sq. ft. |
| Plywood Eave & Porch Soffits | 360 sq. ft. |

**WINDOW SCHEDULE**

| | |
|---|---|
| Wood Casement | 18 units |
| Basement Windows | 2 units |

**DOOR SCHEDULE**

| | |
|---|---|
| Ext. Hardwood, flush | 2 units |
| Ext. Hardwood, glazed | 1 unit |
| Aluminum Sliding | 2 units |
| Wood Overhead garage, paneled | 8 units |
| Wood Overhead garage, paneled | 1 unit |
| Int. Hardwood, flush, staingrade | 8 units |
| Int. Hardwood, louvered, bi-folding | 14 units |
| Int. Solid core fireproof self-closing | 1 unit |

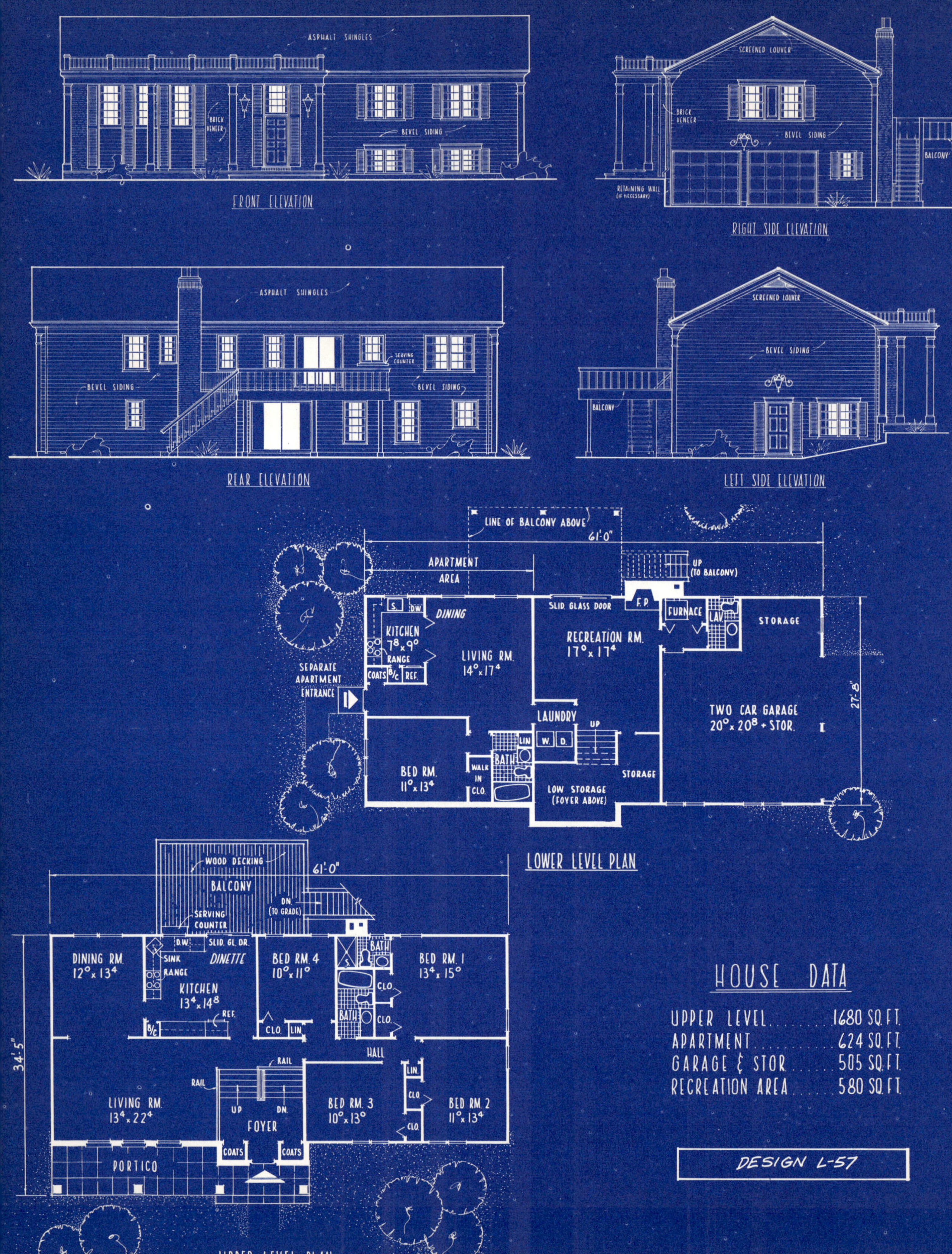
ASPHALT SHINGLES
BRICK VENEER
BEVEL SIDING
FRONT ELEVATION
SCREENED LOUVER
BALCONY
RETAINING WALL (IF NECESSARY)
RIGHT SIDE ELEVATION
SERVING COUNTER
REAR ELEVATION
LEFT SIDE ELEVATION
LINE OF BALCONY ABOVE
61'-0"
APARTMENT AREA
UP (TO BALCONY)
F.P.
SLID. GLASS DOOR
FURNACE
LAV.
STORAGE
DINING
KITCHEN 7⁸ x 9⁰
RANGE
LIVING RM. 14⁰ x 17⁴
RECREATION RM. 17⁰ x 17⁴
SEPARATE APARTMENT ENTRANCE
TWO CAR GARAGE 20⁰ x 20⁸ + STOR.
27'-8"
LAUNDRY
UP
BATH
LIN
BED RM. 11⁰ x 13⁴
WALK IN CLO.
LOW STORAGE (FOYER ABOVE)
LOWER LEVEL PLAN
WOOD DECKING
BALCONY
DN. (TO GRADE)
SERVING COUNTER
SLID. GL. DR.
DINING RM. 12⁰ x 13⁴
SINK
RANGE
DINETTE
KITCHEN 13⁴ x 14⁸
REF.
BED RM. 4 10⁰ x 11⁰
BED RM. 1 13⁴ x 15⁰
CLO.
HALL
34'-5"
RAIL
LIVING RM. 13⁴ x 22⁴
FOYER
COATS
BED RM. 3 10⁰ x 13⁰
BED RM. 2 11⁰ x 13⁴
PORTICO
UPPER LEVEL PLAN
HOUSE DATA
UPPER LEVEL 1680 SQ. FT.
APARTMENT 624 SQ. FT.
GARAGE & STOR 505 SQ. FT.
RECREATION AREA 580 SQ. FT.
DESIGN L-57

# Extra Apartment in Elegant Bi-Level

Of bi-level design (called a high or raised ranch in some regions), this four-bedroom house is unusual in two respects.

Its exterior has been created to resemble an elegant two-story house of the Classic period. And its lower floor contains an apartment, completely private for use either by a relative or as a rentable income-producing unit. Since its structural parts are not controlled by the rooms above, the apartment can be finished immediately or completed at a later time.

Entering through a two-storied porch, one is greeted by a generous foyer with two coat closets at the foyer level. Wide stairways with open railings lead to the upstairs living room or downstairs recreation room. The living room is over 22′ long but appears to be longer because of the wide entry stairway.

The kitchen is planned to serve, through sliding doors, a balcony for outdoor dining.

In the sleeping area we find four huge bedrooms with two baths, one of which, of course, is private for the owners.

On the lower level, there is a large recreation room plus a lavatory and two-car garage.

The apartment adjacent, completely private, can be entered from either the exterior or from the main family area. The architect thus makes it possible for this apartment to be used by either a stranger or by members of the family such as in-laws or grandparents.

## Material List

|  |  |  |
|---|---|---|
| **CONCRETE WORK** | | |
| Concrete Walls | | 628 cu. ft. |
| Foundation Footings | | 166 cu. ft. |
| Slabs | | 553 cu. ft. |
| Misc. Concrete | | 159 cu. ft. |
| **STRUCTURAL STEEL** | | |
| Lally Columns | 3½″ diam. | 5 pieces |
| Girders | 6″ | 30 lin. ft. |
| **BRICK WORK** | | |
| Chimney | Brick & Block | 170 cu. ft. |
| Flue Lining | T. C. | 44 lin. ft. |
| Veneer | Brick | 330 sq. ft. |

|  |  |
|---|---|
| **CARPENTRY** | |
| Framing Lumber | 8970 B.F. |
| Studs | 5333 B.F. |
| Plates | 1500 B.F. |
| Roof Sheathing | 1984 sq. ft. |
| Sub Flooring | 1688 sq. ft. |
| Side Wall Sheathing | 2425 sq. ft. |
| Wall Insulation | 2315 sq. ft. |
| Ceiling Insulation | 2250 sq. ft. |
| Wood Flooring | 1400 sq. ft. |
| Kitchen Plywood | 260 sq. ft. |
| **MILLWORK** | |
| Ext. Doors & Frames Compl. | 2 pieces |

|  |  |  |
|---|---|---|
| Sliding Glass Door Units | | 2 units |
| Garage Door Complete Set | | 2 sets |
| Int. Doors & Frames Compl. | | 23 pieces |
| Bi Fold Units | | 8 units |
| Window Units | | 22 units |
| Fascia | | 214 lin. ft. |
| Base | | 900 lin. ft. |
| Stairs | 7 risers | 1 set |
| Stairs | 6 risers | 1 set |
| **ROOFING** | | |
| Shingles | 235# | 1984 sq. ft. |
| Roofing Paper | 15# felt | 1984 sq. ft. |

FRONT ELEVATION

RIGHT SIDE ELEVATIC

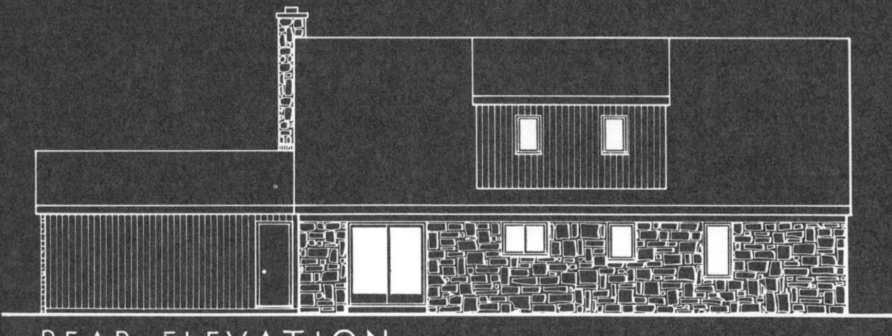

REAR ELEVATION

LEFT SIDE ELEVATION

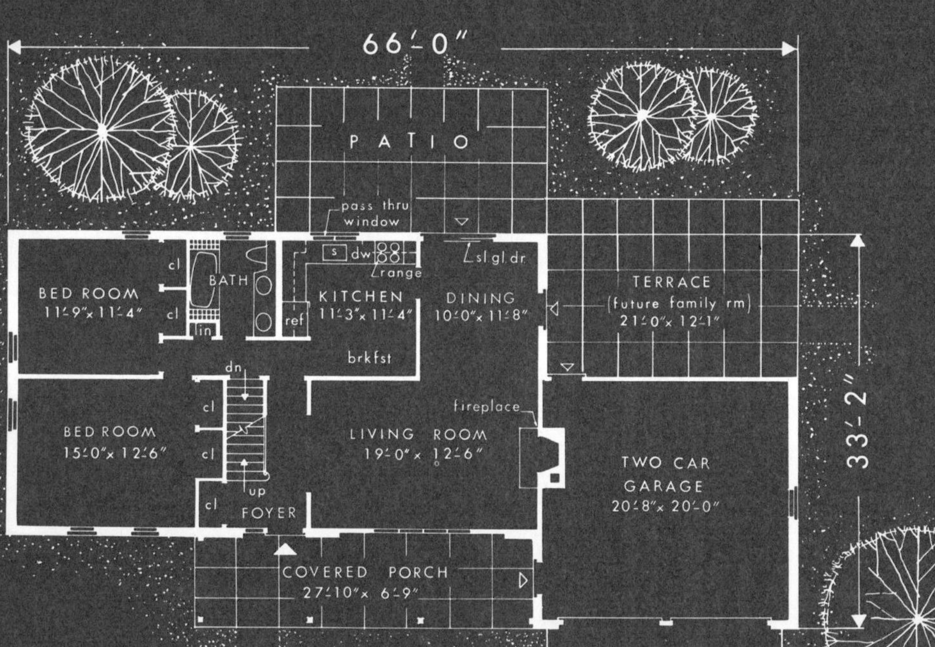


FIRST FLOOR PLAN

## AREA STATISTICS :

| | |
|---|---|
| 1st floor | 1,108 sq.f |
| 2nd floor | 695 " " |
| cellar | 1,108 " " |
| garage | 433 " " |
| covered porch | 196 " " |

BATH
lin
dressing alcove 7'-3"x 6'-8"
dn
BED ROOM 15'-4"x 12'-8"
cl
cl
cl
cl
BED ROOM 20'-8"x 12'-8"

SECOND FLOOR PLAN

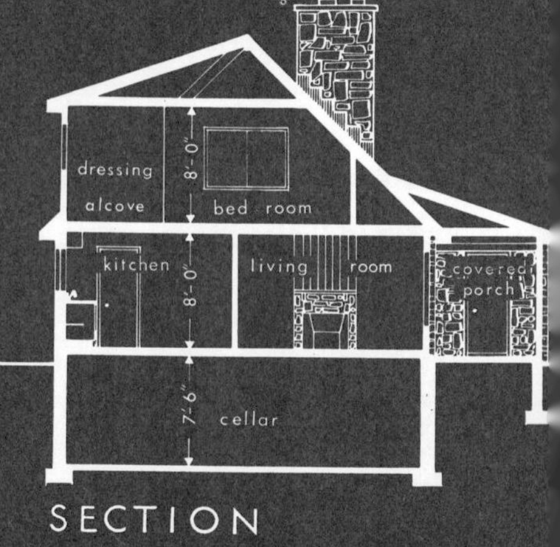


SECTION

DESIGN L-28

# Small House Can Expand If Necessary

Traditional in flavor, but with contemporary features, this house can be categorized as a 1½ story transitional. But the half-story designation is optional, since the structure is planned as a two-bedroom home on a single floor if the partial second floor is not needed.

For those who do not require the two upstairs bedrooms, the basic house contains a living room, dining room, eat-in kitchen, two bedrooms and a bath. And there is a side terrace that can be developed into a family room.

Even with its extra two bedrooms and bath Design L-28 has a charmingly deceptive appearance. It has a long, low horizontal look with a roof line that unifies the covered porch and two-car garage.

Entrance to the house is under a covered porch into the foyer, directly ahead of which is the stair to the second floor. To the right of the foyer is the living room with a decorative fireplace. A large expanse of window faces the front of the house. To the rear, and forming an L, is the dining room with a sliding glass door leading to a rear patio, and another which leads to the side terrace. While the L-shape acts as a divider, a light airy spaciousness is maintained.

The kitchen is next to the dining room and can also be entered through the foyer. The pass-thru window to the patio affords excellent command of outdoor activities and makes serving for barbecues easy.

## Material List

**CONCRETE WORK**

| | |
|---|---|
| Foundations, footings, slabs, etc. | 75 cu. yds. |

**STEEL**

| | |
|---|---|
| Steel Plate | 29 lin. ft. |
| Girder | 44 lin. ft. |
| Lally Columns | 4 pcs. |
| Reinforcing Mesh | 695 sq. ft. |

**MASONRY**

| | |
|---|---|
| Brick Fireplace & Chimney | 166 cu. ft. |
| Stone Chimney | 47 cu. ft. |
| Stone Walls | 170 cu. ft. |

**FRAMING LUMBER**

| | |
|---|---|
| Total Sills, Joists, Rafters, Studs, Plates, etc. | 8301 B.F.M. |

**SHEATHING, INSULATION**

| | |
|---|---|
| Sub Flooring | 1072 sq. ft. |
| Wall Sheathing | 2010 sq. ft. |
| Roof Sheathing | 2456 sq. ft. |
| Wall Insulation | 2010 sq. ft. |
| Ceiling Insulation | 720 sq. ft. |

**FINISHES, INTERIOR**

| | |
|---|---|
| Vinyl Tile | 85 sq. ft. |
| Oak Flooring | 907 sq. ft. |
| Ceramic Tile Floors | 80 sq. ft. |
| Ceramic Tile Walls | 250 sq. ft. |
| Gypsum Board - house walls | 2748 sq. ft. |
| Gypsum Board - house ceilings | 1751 sq. ft. |
| Gypsum Board - garage | 840 sq. ft. |

**DOOR SCHEDULE**

| | |
|---|---|
| Ext. Hardwood, glazed | 2 units |
| Ext. Hardwood, flush | 2 units |
| Wood Overhead garage | 2 units |
| Int. Hardwood, flush, staingrade | 7 units |
| Int. Hardwood, flush, bi-folding | 2 units |
| Int. Hardwood, louvered, bi-folding | 11 units |

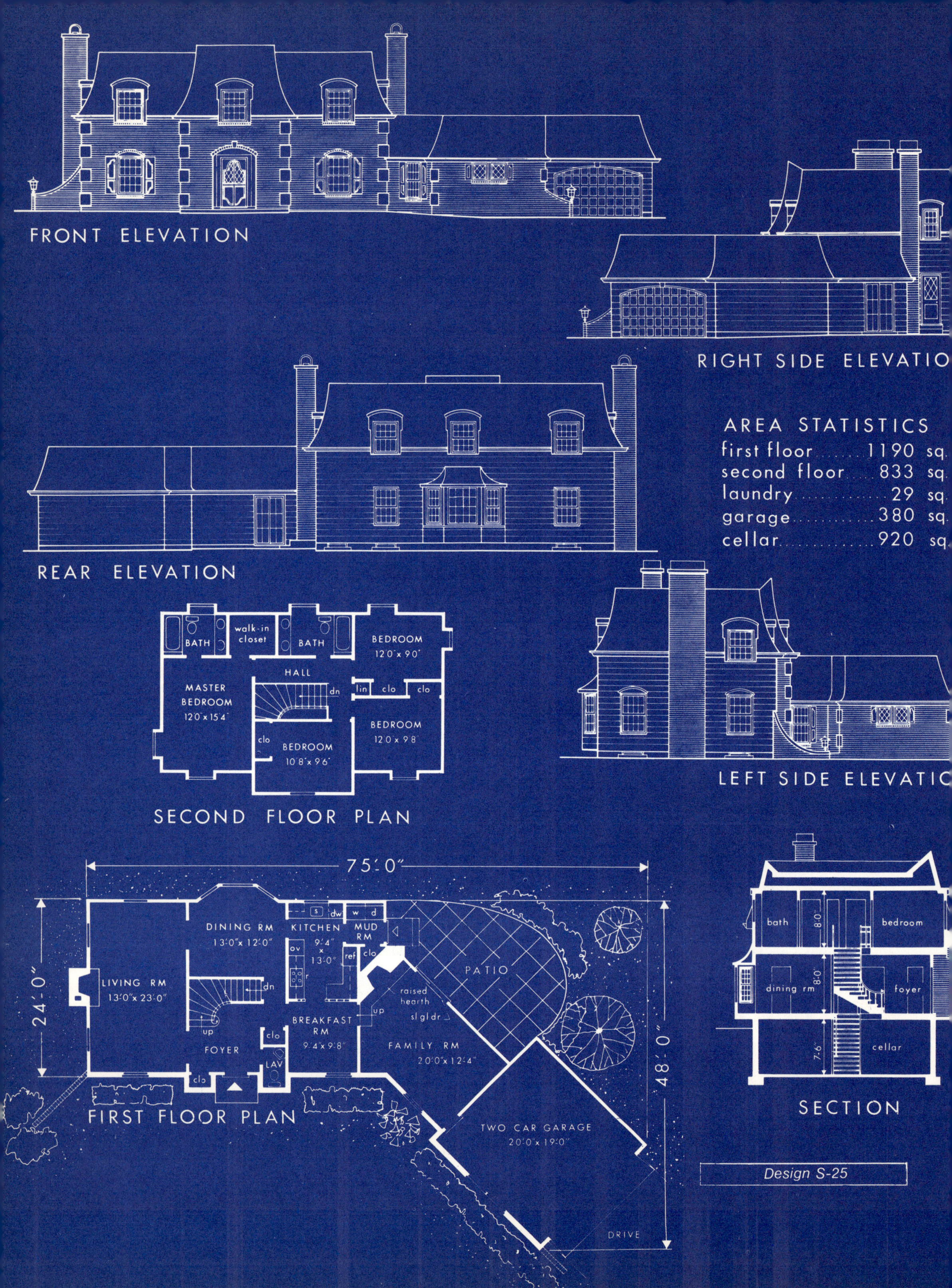
FRONT ELEVATION
RIGHT SIDE ELEVATIO
REAR ELEVATION
AREA STATISTICS
first floor 1190 sq.
second floor 833 sq.
laundry 29 sq.
garage 380 sq.
cellar 920 sq.
LEFT SIDE ELEVATIC
BATH
walk-in closet
BATH
BEDROOM 12'0" x 9'0"
HALL
MASTER BEDROOM 12'0" x 15'4"
dn
lin
clo
clo
BEDROOM 12'0" x 9'8"
clo
BEDROOM 10'8" x 9'6"
SECOND FLOOR PLAN
75'-0"
24'-0"
DINING RM 13'0" x 12'0"
KITCHEN 9'4" x 13'0"
MUD RM
LIVING RM 13'0" x 23'0"
dn
up
FOYER
clo
LAV
clo
BREAKFAST RM 9'4" x 9'8"
raised hearth
sl gl dr
up
PATIO
FAMILY RM 20'0" x 12'4"
48'-0"
FIRST FLOOR PLAN
TWO CAR GARAGE 20'0" x 19'0"
DRIVE
bath
bedroom
dining rm
foyer
cellar
8'-0"
8'-0"
7'-6"
SECTION
Design S-25

# French Chateau Has Elegant Facade

Looking at the facade of this elegant home, one is drawn back in time to another era — the heyday of French nobility. The historic period may be gone, but traces of it remain in the architecture of some French country homes. And the styling is enjoying a resurgence today on both sides of the Atlantic.

Distinguishing features are a high mansard roof curved at its lowest edges, curved brick arches and window dormers, a front entirely of brick, and diamond-paned windows. A one-story angled garage wing and a complementary angled brick wall on the other side add to the elegance of the front facade.

From the covered front portico, the attractive front door leads to a large foyer highlighted by a sweeping, winding staircase. To the left of the foyer is a formal living room 23′ long, with an optional fireplace and mantel. Straight ahead from the foyer is the dining room, featuring a large bay window overlooking the rear garden.

The daily activities areas of the home are all to the right of the foyer. A kitchen, with an abundance of cabinets, is at the rear. Adjacent to it, yet separated by a dwarf partition with turned posts above, is a breakfast room.

Beyond the breakfast room is a paneled family room, with a large brick fireplace and hearth and sliding glass doors to a curved patio. Upstairs are four bedrooms and two bathrooms.

## Material List

**CONCRETE WORK**

| | |
|---|---|
| Foundations, footings, slabs, etc. | 78 cu. yds. |

**STEEL**

| | | |
|---|---|---|
| Girder | 8B10 | 48 lin. ft. |
| Lally Cols. | 3½″ diam. | 5 pcs. |
| Reinforcing Mesh | | 620 sq. ft. |

**MASONRY**

| | |
|---|---|
| Brick Fireplace & Chimney | 300 cu. ft. |
| Brick Patio Walls | 99 cu. ft. |

**FRAMING LUMBER**

| | |
|---|---|
| Total Sills, Joists, Rafters, Studs, Plates, etc. | 18,000 B.F.M. |

**SHEATHING, INSULATION**

| | |
|---|---|
| Sub Flooring | 1800 sq. ft. |
| Wall Sheathing | 2690 sq. ft. |
| Roof Sheathing | 5398 sq. ft. |
| Wall Insulation | 2690 sq. ft. |
| Ceiling Insulation | 833 sq. ft. |

**FINISHES, INTERIOR**

| | |
|---|---|
| Vinyl Tile | 830 sq. ft. |
| Oak Flooring | 1288 sq. ft. |
| Ceramic Tile Floors | 85 sq. ft. |
| Ceramic Tile Walls | 199 sq. ft. |
| Gypsum Board — house walls | 6416 sq. ft. |
| Gypsum Board — House Ceilings | 2052 sq. ft. |

**FINISHES, EXTERIOR (other than masonry)**

| | |
|---|---|
| Horizontal Siding | 1215 sq. ft. |
| Asphalt Shingle Roofing | 2986 sq. ft. |
| Plywood Eave & Porch Soffits | 825 sq. ft. |

**WINDOW SCHEDULE**

| | |
|---|---|
| Wood Double Hung | 16 units |

**DOOR SCHEDULE**

| | |
|---|---|
| Ext. Hardwood, paneled & glazed | 1 unit |
| Ext. Hardwood, paneled | 1 unit |
| Int. Hardwood, flush, staingrade | 9 units |
| Int. Hardwood, louvered, bi-folding | 7 units |

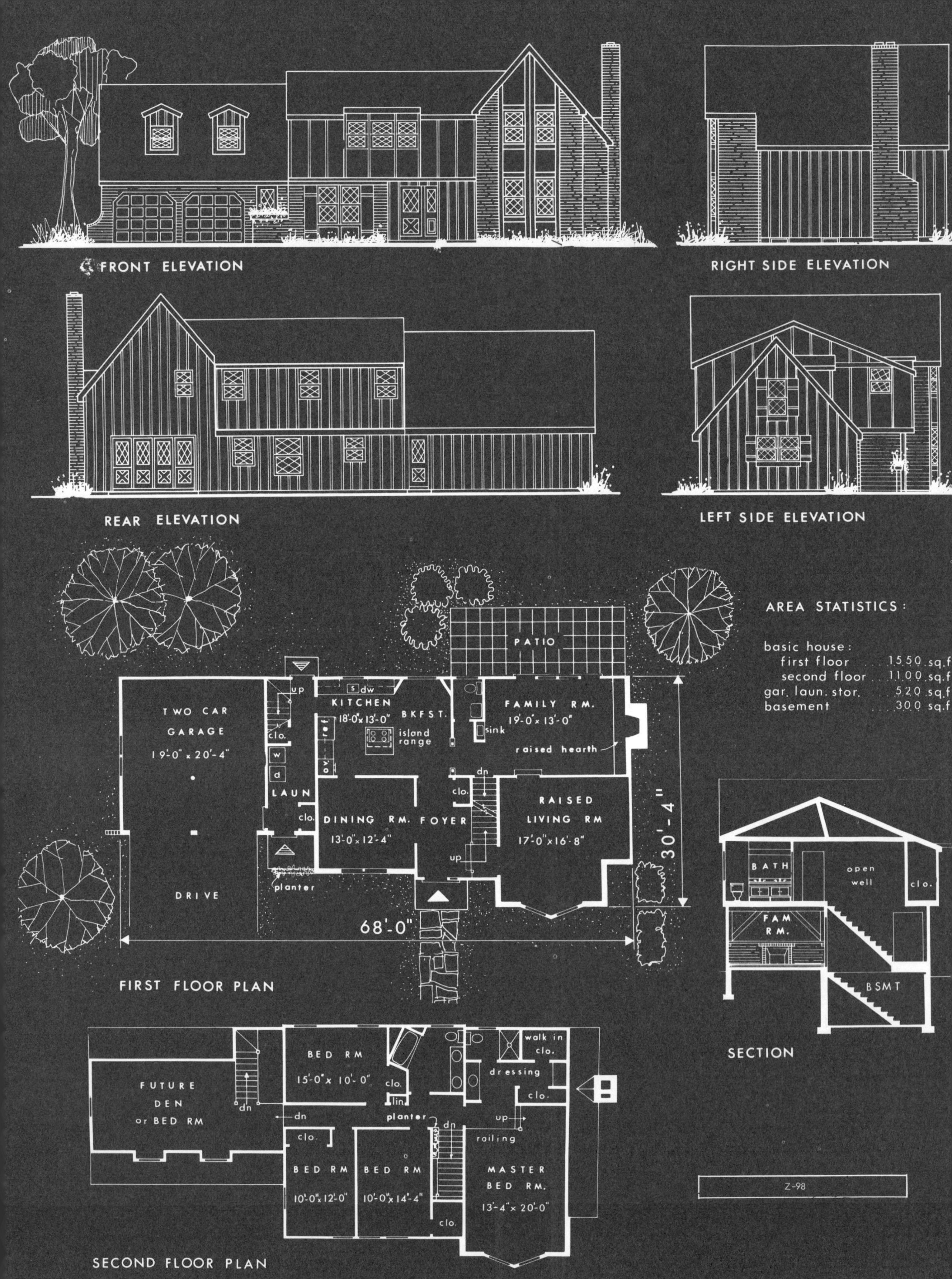
FRONT ELEVATION
RIGHT SIDE ELEVATION
REAR ELEVATION
LEFT SIDE ELEVATION
AREA STATISTICS:
basic house:
first floor 1550 sq.f
second floor 1100 sq.f
gar. laun. stor. 520 sq.f
basement 300 sq.f
PATIO
TWO CAR GARAGE 19'-0" x 20'-4"
KITCHEN 18'-0" x 13'-0"
BKFST.
island range
FAMILY RM. 19'-0" x 13'-0"
sink
raised hearth
LAUN
clo.
DINING RM. 13'-0" x 12'-4"
FOYER
RAISED LIVING RM 17'-0" x 16'-8"
DRIVE
planter
30'-4"
68'-0"
FIRST FLOOR PLAN
BATH
open well
FAM RM.
BSMT
SECTION
FUTURE DEN or BED RM
BED RM 15'-0" x 10'-0"
lin.
walk in clo.
dressing
planter
railing
BED RM 10'-0" x 12'-0"
BED RM 10'-0" x 14'-4"
MASTER BED RM. 13'-4" x 20'-0"
SECOND FLOOR PLAN
Z-98

# Tudor Styling Encloses Modern Home

Tudor home styling combines an informal country atmosphere with an air of stately permanence. In the United States, some of the most fashionable and exclusive homes built 40 or 50 years ago were of Tudor design. With home buyers again becoming style conscious, it is beginning to regain its popularity in many sections of the country.

This house is up-to-date in every respect, packed with many of the charms and nuances typical of Tudor homes. Inside the main entrance foyer the stairway to the second floor features an attractive wood rail. The living room, to the right, gets a private character because it is two steps up. Its shape permits an unusual arrangement of furniture, including a grand piano, if desired, in front of the angled bay window.

The formal dining room is to the left of the foyer, with the kitchen and dinette to the rear. The kitchen features a big center island with a large copper hood above. Splayed walls in the dinette section create a bay.

A curved arch leads from the kitchen to the family room. A rear window wall, with country-style doors, leads to the rear patio. The second floor has four bedrooms, two full baths and an abundance of closets. While the exterior is shown as a combination of brick veneer and vertical board-and-batten siding, other materials might be chosen without destroying the Tudor look.

## Material List

|  |  |
|---|---|
| **CONCRETE WORK** | |
| Foundations, footings, slabs, etc. | 76 cu. yds. |
| **STEEL** | |
| Girder ........ 8B10 ......... | 14 lin. ft. |
| Lally Cols. .... 3½" diam. .... | 2 pcs. |
| Reinforcing Mesh .............. | 400 sq. ft. |
| **MASONRY** | |
| Brick Fireplace & Chimney ...... | 100 cu. ft. |
| Brick Walls .................... | 310 cu. ft. |
| **FRAMING LUMBER** | |
| Total sills, joists, rafters, studs, plates, etc. .......... | 15,780 B.F.M. |
| **SHEATHING, INSULATION** | |
| Sub Flooring .................. | 1400 sq. ft. |
| Wall Sheathing ................ | 2600 sq. ft. |
| Roof Sheathing ................ | 2150 sq. ft. |
| Wall & ceil. insulation ........ | 2700 sq. ft. |
| **FINISHES, INTERIOR** | |
| Vinyl Tile .................... | 300 sq. ft. |
| Oak Flooring .................. | 1600 sq. ft. |
| Ceramic Tile Floors ........... | 112 sq. ft. |
| Ceramic Tile Walls ............ | 241 sq. ft. |
| Gypsum Board .................. | 8044 sq. ft. |
| **FINISHES, EXTERIOR (other than masonry)** | |
| Vertical Siding ............... | 1552 sq. ft. |
| Asphalt Shingle Roofing ........ | 2150 sq. ft. |
| Plywood Eave & Porch Soffits ... | 300 sq. ft. |
| **WINDOW SCHEDULE** | |
| Wood Double Hung with diagonal muntins | 22 units |
| **DOOR SCHEDULE** | |
| Ext. Hardwood, paneled & glazed side lite .................. | 1 unit |
| Ext. Hardwood, paneled (fixed) glazed .................... | 2 units |
| Ext. Hardwood, paneled, glazed .. | 4 units |
| Int. Hardwood, flush, staingrade . | 15 units |
| Int. Hardwood, flush, bi-folding .. | 6 units |

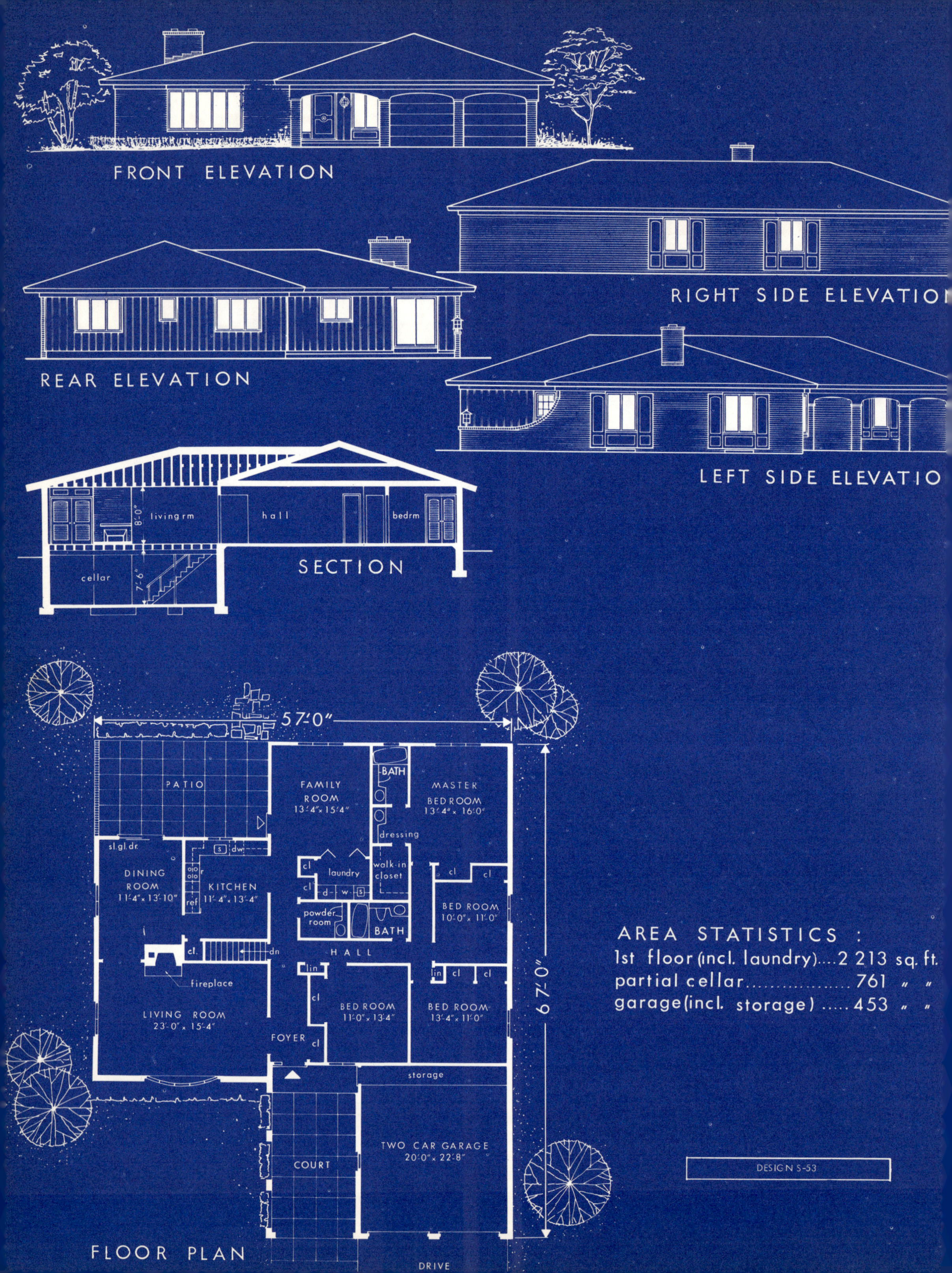
FRONT ELEVATION
RIGHT SIDE ELEVATION
REAR ELEVATION
LEFT SIDE ELEVATION
living rm
8'-0"
hall
bedrm
cellar
7'-6"
SECTION
57'-0"
PATIO
FAMILY ROOM 13'-4" x 15'-4"
BATH
MASTER BED ROOM 13'-4" x 16'-0"
dressing
sl.gl.dr.
s
dw
DINING ROOM 11'-4" x 13'-10"
KITCHEN 11'-4" x 13'-4"
ref
cl
laundry
d
w
walk in closet
BED ROOM 10'-0" x 11'-0"
powder room
BATH
cl.
dn
HALL
fireplace
lin
LIVING ROOM 23'-0" x 15'-4"
BED ROOM 11'-0" x 13'-4"
BED ROOM 13'-4" x 11'-0"
FOYER
67'-0"
storage
COURT
TWO CAR GARAGE 20'-0" x 22'-8"
FLOOR PLAN
DRIVE
AREA STATISTICS :
1st floor (incl. laundry).... 2 213 sq. ft.
partial cellar.................. 761 " "
garage (incl. storage) ..... 453 " "
DESIGN S-53

# Delightful Walkway to Contemporary

This smart, contemporary four-bedroom ranch says "welcome" before you get to the front door. A graceful arched loggia stretches from the front of the garage to the house entrance and is enhanced by the adjoining landscaping.

Once inside, you are greeted by a central foyer from which all the main rooms emanate. The large formal living room to the left features a soft curved bow window and a masonry fireplace. Although bold in character, it is elegant in its simplicity.

To the left of the fireplace is the dining room which can be separated from it by double doors. Sliding doors in the dining room give access to a huge patio.

The kitchen space is divided into two corners. One is occupied by an L-shaped arrangement of counters, cabinets and full equipment. The opposite corner is for informal dining.

The family room, designed for informal living, has access to the patio, which may be termed an outdoor family room.

A powder room off the main foyer is convenient to the kitchen and family room. In this same area is the stairway to a partial basement. Incidentally, this is a house with numerous closets. The bedroom wing consists of four bedrooms and two bathrooms. The master suite, which has its own bathroom, also has a dressing room with a second wash basin and built-in vanity.

## Material List

**CONCRETE WORK**

|  |  |
|---|---|
| Foundations, footings, slabs, etc. | 102 cu. yds. |

**STEEL**

|  |  |
|---|---|
| Girder ....... 8B10 | 23 lin. ft. |
| Reinforcing Mesh | 500 sq. ft. |

**MASONRY**

|  |  |
|---|---|
| Brick Veneer Walls | 361 cu. ft. |
| Brick Fireplace & Chimney | 180 cu. ft. |
| Brick Patio Walls & Piers | 56 cu. ft. |

**FRAMING LUMBER**

|  |  |
|---|---|
| Total Sills, Joists, Rafters, Studs, Plates, etc. | 13,500 B.F.M. |

**SHEATHING, INSULATION**

|  |  |
|---|---|
| Sub Flooring | 736 sq. ft. |
| Wall Sheathing | 1500 sq. ft. |
| Roof Sheathing | 3360 sq. ft. |
| Wall Insulation | 1300 sq. ft. |
| Ceiling Insulation | 2213 sq. ft. |

**FINISHES, INTERIOR**

|  |  |
|---|---|
| Vinyl Tile | 317 sq. ft. |
| Oak Flooring | 1506 sq. ft. |
| Ceramic Tile Floors | 73 sq. ft. |
| Ceramic Tile Walls | 424 sq. ft. |
| Gypsum Board | 5696 sq. ft. |

**FINISHES, EXTERIOR (other than masonry)**

|  |  |
|---|---|
| Vertical Siding (Board & Batten) | 400 sq. ft. |
| Asphalt Shingle Roofing | 3360 sq. ft. |
| Plywood Eave & Porch Soffits | 728 sq. ft. |

**DOOR SCHEDULE**

|  |  |
|---|---|
| Ext. Hardwood, paneled 9 lite | 1 unit |
| Ext. Hardwood, paneled & glazed | 1 unit |
| Ext. Hardwood, paneled | 1 unit |
| Aluminum Sliding | 1 unit |
| Wood Overhead garage, paneled | 2 units |
| Int. Hardwood, flush, staingrade | 8 units |
| Int. Hardwood, louvered, bi-folding | 9 units |
| Int. Hardwood, louvered | 7 units |

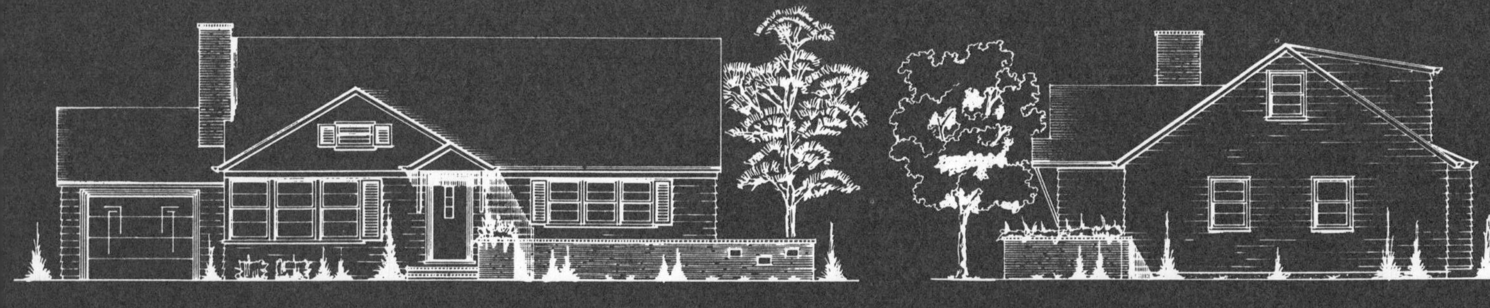

FRONT ELEVATION

RIGHT SIDE ELEVATION

REAR ELEVATION

LEFT SIDE ELEVATION

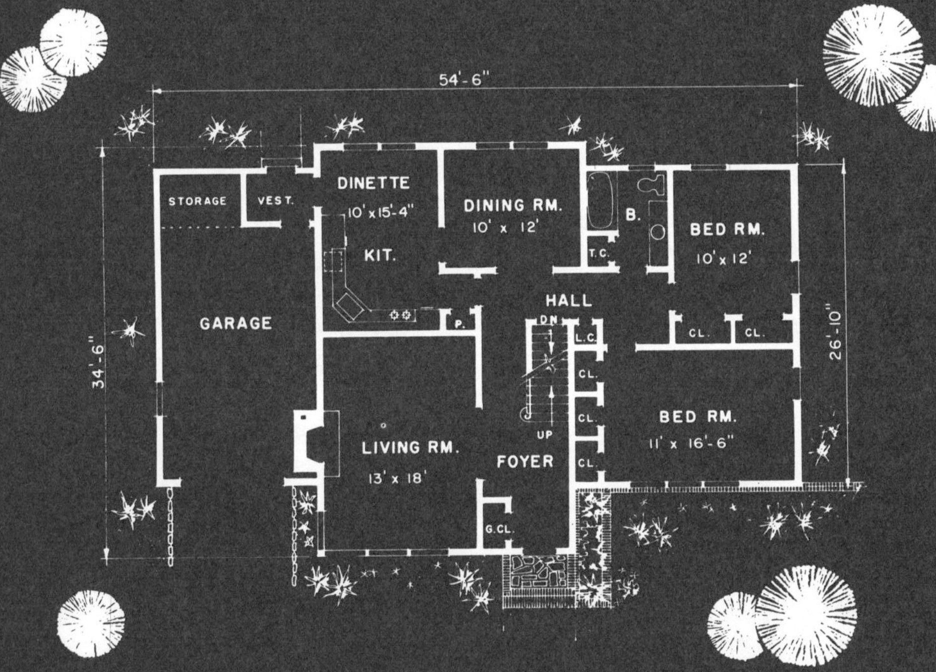


FIRST FLOOR PLAN

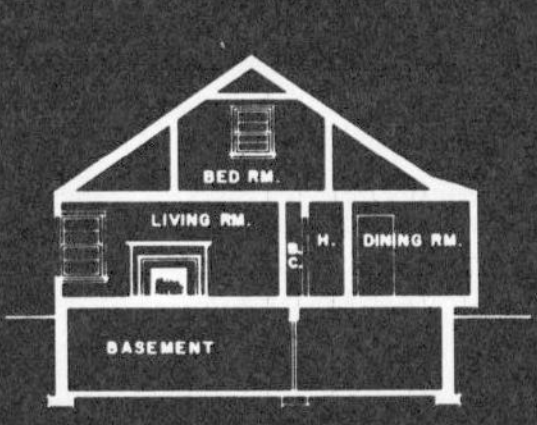


CROSS SECTION

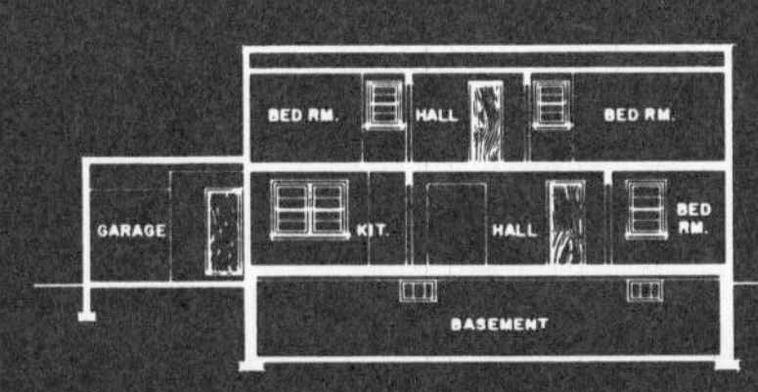


LONGITUDINAL SECTION

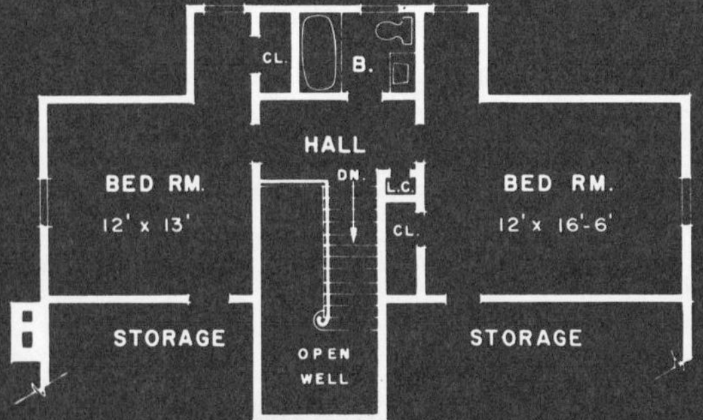


SECOND FLOOR PLAN

## AREA STATISTICS:

FIRST FLOOR . . . . . 1300 SQ. FT.
SECOND FLOOR . . . . . 635 SQ. FT.
BASEMENT . . . . . 1065 SQ. FT.
GARAGE . . . . . . . . . . 300 SQ. FT.

**DESIGN L-39**

# Charm Keynotes This Small Colonial

Styles in architecture change, but the traditional Colonial styling always manages to keep a large and loyal following. Here's a simple design that exudes warmth and hospitality yet does not require a large lot and has an excellent floor plan with no cross traffic from the front foyer.

While it offers complete living on one floor, the plan provides expansion space upstairs for additional living area.

Outside, hand-split wood shingles and horizontally paned windows carry out the updated traditional theme. The front gable, brick planter, sill-height brick-veneered wall, window flower box and shutters are touches that give the small house its charming appearance.

The interior is equally well-planned. To the left of the spacious entrance foyer with its dramatic cathedral ceiling is the living room with a log-burning, brick-faced fireplace and corner picture window.

Straight ahead, the hall leads directly to the dining room, placed next to the kitchen for easy serving. The modern kitchen-dinette has an L-shaped arrangement. To the right of the rear foyer are the bathroom and the two bedrooms; note the seven closets.

This is a comfortable two-bedroom house for a retired couple or one of economy and efficiency for a young couple with one or two children, since it can be expanded from two or four bedrooms as needed.

## Material List

**MASONRY**

| | |
|---|---|
| 8" Block | 300 Blocks |
| 12" Block | 850 Blocks |
| Pancake | 380 Blocks |
| Flagstone | 15 sq. ft. |
| Brick | 280 sq. ft. |

**FRAMING LUMBER**

| | |
|---|---|
| Sills, joists, rafters, studs, plates, etc. | 9412 BFM |

**EXTERIOR SHEATHING**

| | |
|---|---|
| ½" Plyscore wall sheathing (or gyplap) | 1950 sq. ft. |
| ½" Plyscore roof sheathing | 2025 sq. ft. |

**DOOR SCHEDULE**

| | |
|---|---|
| (3) 1-6 x 6-8 | (1) 2-6 x 6-8 Pocket Dr. |
| (8) 2-0 x 6-8 | (1) 2-8 x 6-8 SD |
| (4) 2-4 x 6-8 | (1) 3-0 x 6-8 |
| (6) 2-6 x 6-8 | (1) 9-0 x 7-0 Overhead |
| (1) 2-8 x 6-8 SCFP | |

**WINDOW SCHEDULE**

(4) 2-8 x 1-8 Basement sash

| | |
|---|---|
| (1) 2-0 x 3-6 D.H. | (5) 3-0 x 3-10 D.H. |
| (2) 2-6 x 3-6 D.H. | (1) 2-4 x 3-0 D.H. |
| (8) 3-0 x 4-2 D.H. | (4) 3-4 x 5-2 D.H. |

**INTERIOR FINISHES**

| | |
|---|---|
| Ceramic tile floor | 65 sq. ft. |
| Ceramic tile walls | 220 sq. ft. |
| Sheetrock | 7200 sq. ft. |
| Oak flooring | 1610 sq. ft. |

**EXTERIOR FINISHES**

| | |
|---|---|
| Wood Shingle siding | 1890 sq. ft. |
| Asphalt shingles | 21 squares |

**SUB FLOORING**

| | |
|---|---|
| ⅝" Plyscore | 1800 sq. ft. |

FRONT ELEVATION
RIGHT SIDE ELEVATION
REAR ELEVATION
LEFT SIDE ELEVATION
ATTIC
D.R.
L.R.
F.
BR.
B.R.
GARAGE.
BASEMENT
LONGITUDINAL SECTION
B.R.
M.B.R.
GARAGE
F.R.
CROSS SECTION
TRAY
L.
LAUND
5'-0"x5'-10"
REC. RM.
10'-0"x19'-0"
UP
G.
20'-6"x 23'-10"
ALTERNATE GARAGE
DOOR LOCATION
LOWER LEVEL
61'-4"
M.B.
CL.
DRESS'G RM.
MASTER BED RM.
14' X 16'
BATH.
W.I. CL.
DN
KIT
R. 10'-6"X10'-6"
S.
DINETTE
10'-6"X15'-0"
G. CL.
HALL
CL.
CL.
CL.
REF.
P.
DESK
B.C.
UP
CL.
CL.
L.C.
RAIL
DN
FOYER
DINING RM.
11'-6"X 14'-6"
BED RM.
11'-0"X12'-0"
BED RM.
11'-8"X 15'-0"
32'-4"
34'-4"
LIVING RM.
14'-0"X23'-4"
PORCH
(OPTIONAL)
12'X14'
DEN.
11'-6"X 14'-0"
UPPER LEVELS
AREA STATISTICS
UPPER & LOWER LIVING AREA
....2005 SQ.FT.
GARAGE....... 485 SQ.FT.
BASEMENT... 1000 SQ.FT.
DESIGN L-53

# Sunken Living Room Feature of Split-Level

Advocates of the split-level design will find much that is pleasing in this three-bedroom house.

Of special interest is the sunken living room, more than 23′ long and featuring a brick-faced fireplace with flagstone hearth, a sloping exposed-beam ceiling, corner windows and wrought iron railings.

A brick planter separates the foyer from the living room. Adjacent to the living room is a den to which, as the floor plan shows, a porch could be added if desired.

More than 25 feet across the back, the kitchen-dinette area is a model of convenience, providing sufficient accommodations for the entire family.

Seven steps below the dinette, and under the bedrooms, is the lower level with a two-car garage, lavatory, laundry and a wood-paneled recreation room that has sliding glass doors to the backyard.

Five graceful curving steps lead to the upper level with its three bedrooms and two complete baths. The master bedroom, facing the rear, has corner windows and three closets, one a walk-in. It has a private bathroom with a dressing area. The main bath has a full-length mirrored vanity and a bathtub with overhead shower.

The excellent grouping of the large horizontal window areas enhances the graceful lines of this fine split-level.

## Material List

**CONCRETE WORK**

Footings, floors, etc. .......... 38 cu. yds.

**MASONRY**

12″ Concrete Block ............ 580 blocks
8″ Concrete Block ............ 1650 blocks
4″ Pancake Block ............ 180 blocks
4″ Brick Veneer .............. 500 sq. ft.

**FRAMING LUMBER**

Sills, joists, rafters, studs, plates, etc. ................. 15,800 BFM

**EXTERIOR SHEATHING**

½″ Exterior plywood .......... 2400 sq. ft.
⅝″ Exterior plywood .......... 2500 sq. ft.

**INSULATION**

Rockwool semi-thick ........... 2200 sq. ft.
Rockwool full-thick ............ 2300 sq. ft.

**DOOR SCHEDULE**

(2) 8-0 x 6-6 w.p. overhead type
(1) 3-0 x 6-8 x 1¾″ front door
(2) 2-8 x 6-8 x 1¾″ sash door
(1) 2-6 x 6-8 x 1⅜″ S.C.F.P. door
(7) 2-6 x 6-8 Int. flush doors
(3) 2-4 x 6-8 Int. flush doors
(11) 2-0 x 6-8 Int. flush doors
(2) 1-6 x 6-8 Int. flush doors

**EXTERIOR FINISHES**

Asphalt roofing shingles ........ 25 squares
WD shingles .................. 2100 sq. ft.
Brick Veneer .................. 500 sq. ft.

**INTERIOR FINISHES**

Ceramic Tile Floor ............ 96 sq. ft.
Ceramic Tile Walls ............ 250 sq. ft.
Vinyl Tile .................... 160 sq. ft.
Sheetrock .................... 8000 sq. ft.
Oak Flooring .................. 2000 sq. ft.

FRONT ELEVATION

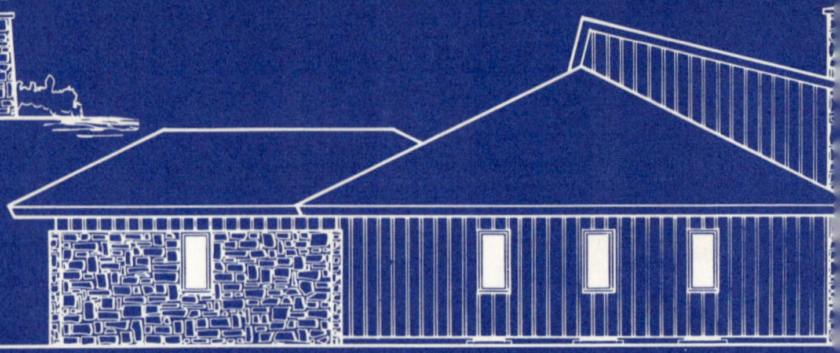

RIGHT SIDE ELEVATION

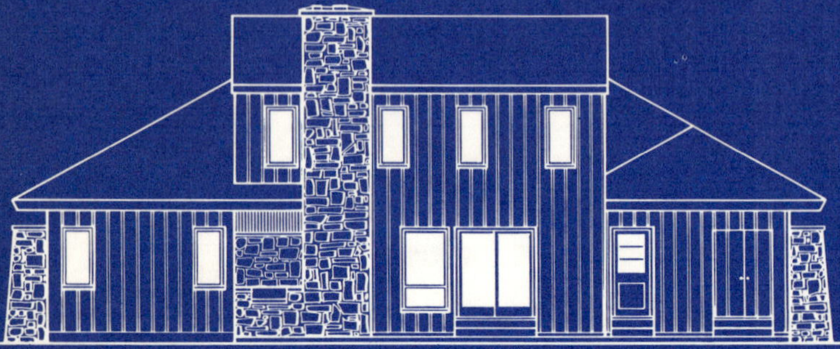

REAR ELEVATION

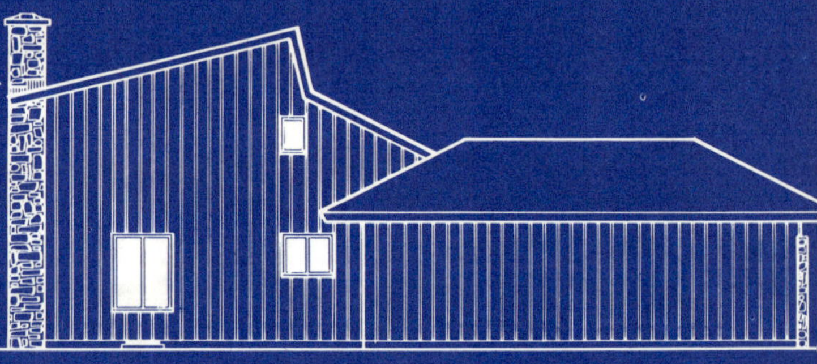

LEFT SIDE ELEVATION

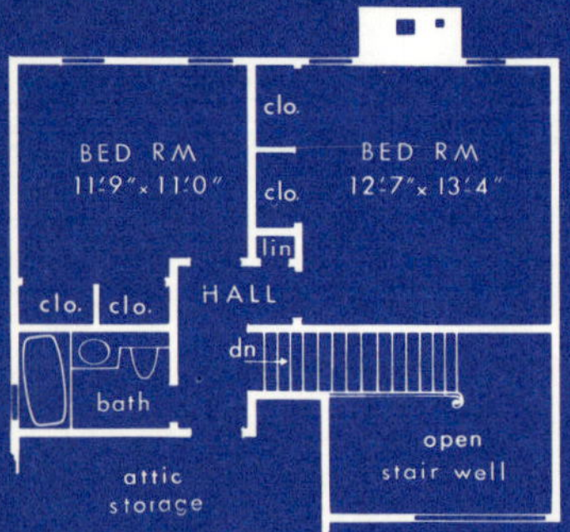

BED RM 11'9"x 11'0"
BED RM 12'7"x 13'4"
clo. clo. lin
HALL
clo. clo.
bath
dn
open stair well
attic storage

SECOND FLOOR

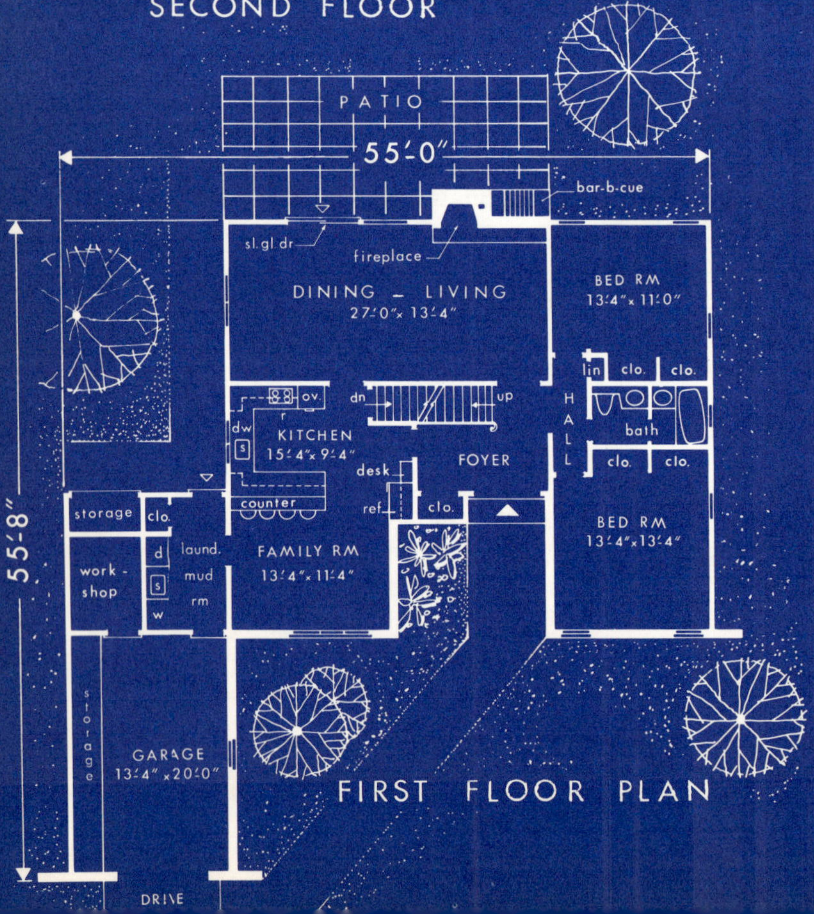

PATIO
55'-0"
bar-b-cue
sl. gl. dr.
fireplace
DINING – LIVING 27'0"x 13'4"
BED RM 13'4"x 11'0"
lin clo. clo.
dn up
HALL
bath
KITCHEN 15'4"x 9'4"
ov. dw s
desk
FOYER
clo. clo.
counter
ref.
clo.
BED RM 13'4"x 13'4"
storage clo.
laund. mud rm
d s w
work-shop
FAMILY RM 13'4"x 11'4"
55'-8"
storage
GARAGE 13'4"x 20'0"
DRIVE

FIRST FLOOR PLAN

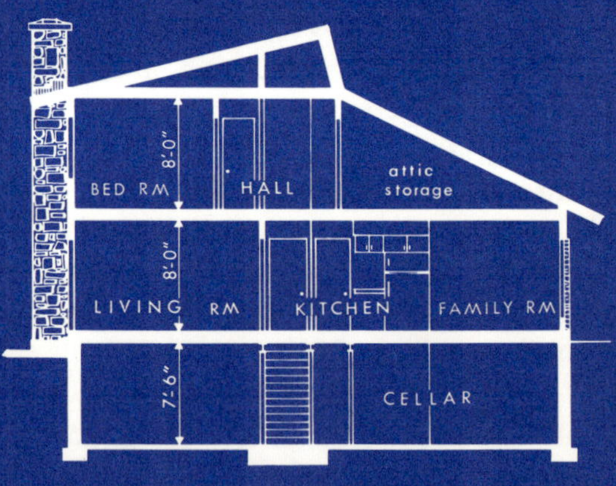


SECTION

AREA STATISTICS

| | | |
|---|---|---|
| 1st floor | 1,312 | sq. ft. |
| 2nd floor | 469 | " |
| garage, storage | 346 | " |
| laundry mud rm | 80 | " |
| full cellar | 1,312 | " |

DESIGN S-74

# Stone Exterior On Attractive Contemporary

This house bears a resemblance to the midwestern prairie ranch, but its styling is strictly contemporary. And it's not a pure ranch because it has a partial second floor which accommodates two bedrooms and a bath, making it a four-bedroom house within the modest overall dimensions of 55′ by 55′8″.

The front entrance doors are approached through a welcoming garden and a small exterior court which acquaints one with the character and spirit of the home. The entrance itself, composed of double doors over which is a large glass area, leads into the main two-story dramatic foyer. All the rooms and a decorative stair to the second floor emanate from this space.

Looking beyond the foyer and through the living room, one is confronted with a handsome stone fireplace. The living room combines with the dining room to merge into one large space.

To the left of the foyer are an efficient U-shaped kitchen and a family room. To the right of the foyer is the bedroom wing with two bedrooms and a compartmentalized bathroom.

The second floor has two bedrooms and a bath, with storage space comprising the remainder of the floor. The large open stairwell which joins the first and second floors vertically is well lighted by the windows over the entrance doors.

## Material List

**CONCRETE WORK**
Foundations, footings, slabs, etc. 78 cu. yds.

**STEEL**
Girder ......... 8B10 ........ 66 lin. ft.
Lally Cols. ..... 3½″ diam. ... 9 pcs.
Reinforcing Mesh .............. 500 sq. ft.

**MASONRY**
Stone Walls (veneer) ........... 147 cu. ft.
Stone Fireplace & Chimney ...... 240 cu. ft.

**FRAMING LUMBER**
Total Sills, Joists, Rafters, Studs, Plates, etc. ..........12,000 B.F.M.

**SHEATHING**
Sub Flooring .................. 1210 sq. ft.
Wall Sheathing ................ 300 sq. ft.
Roof Sheathing ............... 2700 sq. ft.

**FINISHES, INTERIOR**
Vinyl Tile .................... 514 sq. ft.
Oak Flooring .................. 1093 sq. ft.
Ceramic Tile Floors ........... 70 sq. ft.
Ceramic Tile Walls ............ 480 sq. ft.
Gypsum Board .................. 6763 sq. ft.

**FINISHES, EXTERIOR (other than masonry)**
Vertical Siding ............... 1586 sq. ft.
Asphalt Shingle Roofing ........ 2700 sq. ft.
Plywood Eave & Porch Soffits ... 677 sq. ft.

**WINDOW SCHEDULE**
Wood Casement ............... 14 units
Flexivent Awning .............. 1 unit
Fixed Glazed .................. 3 units
Gliding ....................... 1 unit

**DOOR SCHEDULE**
Ext. Hardwood, paneled ........ 4 units
Ext. Hardwood, paneled & glazed . 1 unit
Int. Hardwood, flush, staingrade . 11 units
Int. Hardwood, louvered, bi-folding 13 units

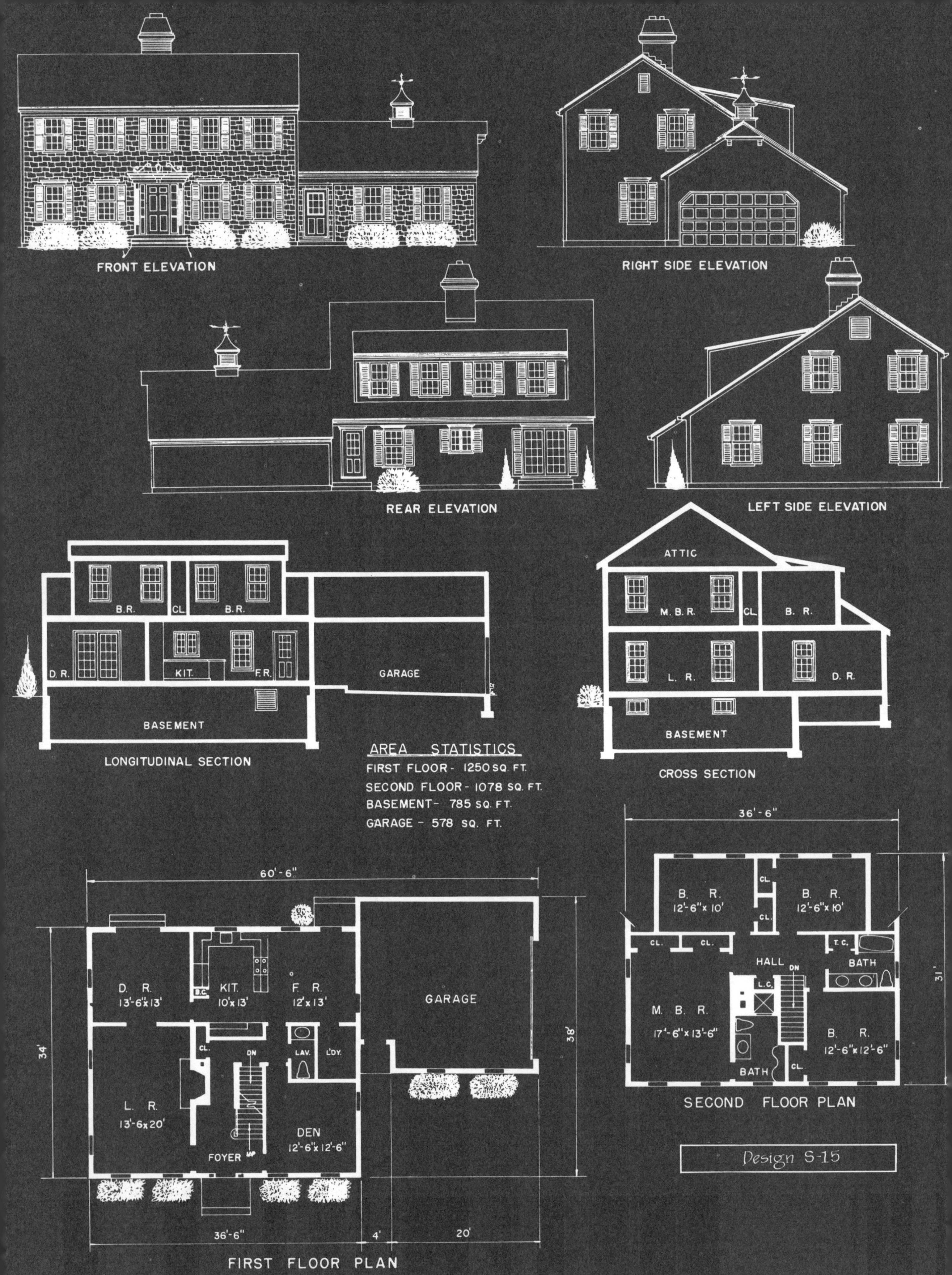
FRONT ELEVATION
RIGHT SIDE ELEVATION
REAR ELEVATION
LEFT SIDE ELEVATION
B.R.
CL
B.R.
D. R.
KIT.
F.R.
GARAGE
BASEMENT
LONGITUDINAL SECTION
ATTIC
M. B. R.
CL
B. R.
L. R.
D. R.
BASEMENT
CROSS SECTION
AREA STATISTICS
FIRST FLOOR - 1250 SQ. FT.
SECOND FLOOR - 1078 SQ. FT.
BASEMENT - 785 SQ. FT.
GARAGE - 578 SQ. FT.
60'-6"
D. R.
13'-6"x13'
KIT.
10'x13'
F. R.
12'x13'
GARAGE
CL.
LAV.
L'DY.
L. R.
13'-6x20'
FOYER
DEN
12'-6"x12'-6"
34'
38'
36'-6"
4'
20'
FIRST FLOOR PLAN
36'-6"
B. R.
12'-6"x10'
CL.
B. R.
12'-6"x10'
CL.
CL.
T. C.
HALL
DN
BATH
L.C.
M. B. R.
17'-6"x13'-6"
B. R.
12'-6"x12'-6"
BATH
CL.
31'
SECOND FLOOR PLAN
Design S-15

# Authentic Saltbox with Four Bedrooms

The simple, tasteful exterior of the Early American Colonial house has enabled it to maintain its popularity throughout the centuries. One of the most interesting variations is the New England Saltbox, distinguished by its rear roof lines, which extend downward much lower than the roof lines at the front of the house.

This Saltbox, authentic in its exterior styling, has a modern interior, with a center hall entrance, a practical room arrangement on the first floor and four bedrooms and two bathrooms upstairs.

The living room has plenty of wall space plus a brick-faced log-burning fireplace. On the other side of the foyer is a wood paneled den. A separate dining room adjacent to the kitchen with twin French glass doors takes full advantage of the view and provides access to the rear patio. A kitchen-family room affords enough space for informal meals and relaxation.

A staircase leads from the entrance foyer to the bedrooms. Three bedrooms are close to the family bathroom, which has a clear plastic, sliding bathtub enclosure and a full-length mirrored double vanity. The master bedroom has two sliding door closets, four windows and a full bath with a glass-enclosed tiled shower and full-length mirrored dressing vanity.

## Material List

**CONCRETE WORK**

| | |
|---|---|
| Footings, floor, etc. | 30 cu. yds. |

**STEEL**

| | |
|---|---|
| 4"0 Lally cols. | 2 pieces |
| Metal areaways | 3 pieces |

**MASONRY**

| | |
|---|---|
| Flagstone | 30 sq. ft. |
| 8" Concrete Block | 400 Blocks |
| 12" Concrete Block | 1000 Blocks |
| Brick | 200 sq. ft. |
| 4" Pancake Block | 110 Blocks |

**FRAMING LUMBER**

| | |
|---|---|
| Sills, joists, rafters, studs, plates, etc. | 10,000 B.F.M. |

**EXTERIOR SHEATHING**

| | |
|---|---|
| ½" Plyscore wall sheathing | 2800 sq. ft. |
| ½" Plyscore roof sheathing | 2300 sq. ft. |

**SUBFLOORING**

| | |
|---|---|
| ⅝" Plyscore | 2400 sq. ft. |

**DOOR SCHEDULE**

(1) 3-0 x 6-8 x 1¾ wp front entrance door
(2) 3-0 x 6-8 1¾ wp. sash doors
(1) 2-8 x 6-8 x 1⅜ scfp. door
(2) 3-0 x 1¾ french doors
(24) Flush interior veneered doors

**WINDOW SCHEDULE**

(3) 2-8 x 1-8 Basement sash
(13) 2-8 x 4-6 DH
(11) 2-8 x 5-2 DH
(1) 3-0 x 3-2 casement

**EXTERIOR FINISHES**

| | |
|---|---|
| Red cedar shingles | 2150 sq. ft. |
| Asphalt shingles | 21 Squares |

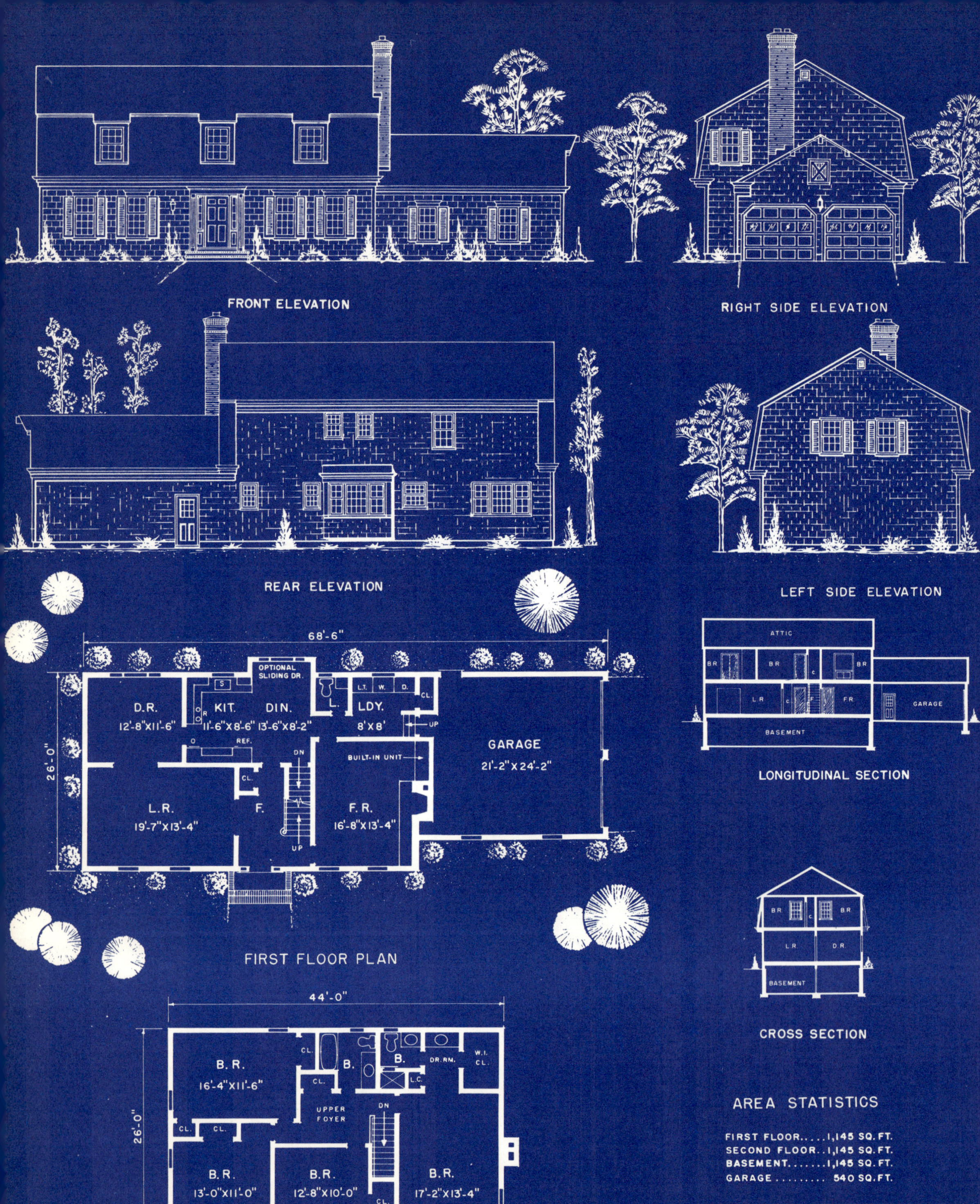


Design L-29

SECOND FLOOR PLAN

# Gambrel Roof Has Purpose in Two-Story

Comfort and serenity seem to emerge from the exterior of this two-story traditional.

Much of the appearance of warmth comes from the gambrel roof, a feature borrowed from the Dutch Colonial design. A variation of the gable, the gambrel has two slopes on either side of the center ridge.

There are four bedrooms and two bathrooms upstairs. An attractive staircase leads from the entrance foyer to a spacious rectangular second-floor foyer. The master bedroom has a walk-in closet, a dressing area and a full bath.

The living room, to the left of the foyer, has three windows, yet still has plenty of wall space. As the upper part of the L combination, the dining room is accessible to the kitchen as well as the living room.

A large kitchen includes a boxed-bay mullion window in the dinette area to provide room for the entire family. Close at hand is the stairway to the full basement.

To the right of the kitchen is a lavatory and laundry complex with space for the washing and drying appliances and a door leading to the garage. The family room can be reached from the kitchen, laundry area and the front foyer without crossing any rooms. It has two front windows and, since more actual living is likely to be done in the family room, the stone-faced log-burning fireplace with raised hearth is located there rather than in the living room.

## Material List

**CONCRETE WORK**

| | |
|---|---|
| Footings, floors, etc. | 30 cu. yds. |

**MASONRY**

| | |
|---|---|
| Flagstone | 30 sq. ft. |
| 8" Concrete Block | 400 blocks |
| 12" Concrete Block | 1000 blocks |
| Brick | 300 sq. ft. |
| 4" Pancake Block | 110 blocks |

**FRAMING LUMBER**

| | |
|---|---|
| Sills, Joists, Rafters, Studs, Plates, etc. | 10,000 BFM |

**EXTERIOR SHEATHING**

| | |
|---|---|
| ½" Plyscore wall sheathing | 2800 sq. ft. |
| ½" Plyscore roof sheathing | 2300 sq. ft. |

**INSULATION**

| | |
|---|---|
| Rockwool Semi-thick | 2300 sq. ft. |
| Rockwool Full-thick | 1150 sq. ft. |

**DOOR SCHEDULE**

(2) 8-0 x 7-0 wp overhead
(1) 3-0 x 6-8 x 1¾ wp front entrance door
(1) 2-8 x 6-8 x 1¾ wp rear sash door
(1) 2-8 x 6-8 x 1⅜ scfp. door
(23) Flush interior veneered doors

**EXTERIOR FINISHES**

| | |
|---|---|
| Red cedar shingles | 1700 sq. ft. |
| Asphalt shingles | 25 squares |
| Clapboards | 700 sq. ft. |

**INTERIOR FINISHES**

| | |
|---|---|
| Ceramic Tile Floor | 125 sq. ft. |
| Ceramic Tile Walls | 400 sq. ft. |
| Vinyl Tile | 300 sq. ft. |
| Gypsum board walls & ceiling | 10,000 sq. ft. |
| Oak flooring | 2000 sq. ft. |

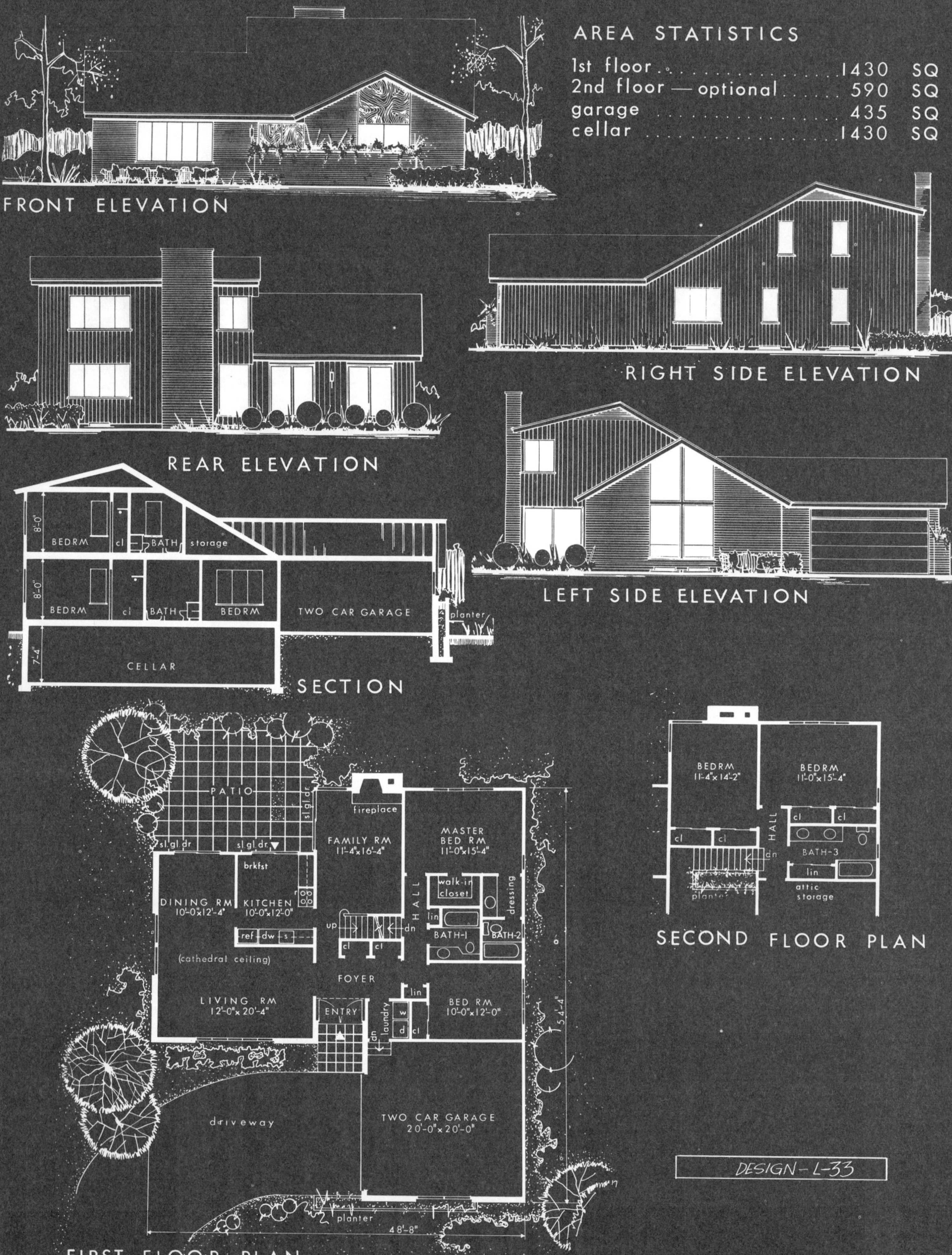

AREA STATISTICS
1st floor .... 1430 SQ F
2nd floor — optional .... 590 SQ F
garage .... 435 SQ F
cellar .... 1430 SQ F
FRONT ELEVATION
RIGHT SIDE ELEVATION
REAR ELEVATION
LEFT SIDE ELEVATION
8'-0"
BEDRM
cl
BATH
storage
8'-0"
BEDRM
cl
BATH
BEDRM
TWO CAR GARAGE
planter
7'-4"
CELLAR
SECTION
PATIO
sl.gl.dr.
fireplace
FAMILY RM
11'-4"x16'-4"
MASTER
BED RM
11'-0"x15'-4"
brkfst
DINING RM
10'-0"x12'-4"
KITCHEN
10'-0"x12'-0"
walk-in
closet
dressing
HALL
lin
BATH-1
BATH-2
up
dn
ref
dw
s
(cathedral ceiling)
FOYER
LIVING RM
12'-0"x20'-4"
ENTRY
laundry
w
d
cl
lin
BED RM
10'-0"x12'-0"
54'-4"
driveway
TWO CAR GARAGE
20'-0"x20'-0"
planter
48'-8"
FIRST FLOOR PLAN
BEDRM
11'-4"x14'-2"
BEDRM
11'-0"x15'-4"
HALL
cl
dn
BATH-3
lin
planter
attic
storage
SECOND FLOOR PLAN
DESIGN-L-33

# Contemporary for Small or Large Family

There's plenty of exterior sparkle in this contemporary house with its large overhangs, recessed front entry, spacious windows, soaring roof line and tasteful blend of brick veneer and vertical siding.

This is an "expansion" home that need be expanded only if necessary. It can serve as a one-floor, two-bedroom house for a small family or retired couple; a flexible unit that can grow along with a family; as a four-bedroom home right from the start.

Inside the double-door front entry, there is a feeling of elegance. A large foyer with a two-story-high ceiling and a built-in planter above a pair of closets are definitely not "plain" features. From there, circulation is easy to all parts of the house.

To the left is a cathedral-ceilinged living room adjoining a dining room with the same type of ceiling and with sliding glass doors to a rear patio. Immediately adjacent is an eat-in kitchen, also with sliding glass doors to the rear. And behind the foyer is a charming family room featuring a brick fireplace and, once again, sliding glass doors to the patio.

For those who do not need the upstairs area for bedrooms, the first floor has two such rooms. The master bedroom is a suite unto itself. Also on the first floor is a separate laundry room next to the bedroom wing and which also serves as a mud room entry from the garage.

## Material List

|  |  |
|---|---|
| **CONCRETE WORK** | |
| Total cu. yds. | 67 cu. yds. |
| **STEEL** | |
| Lally Cols. | 4 |
| Girder | 75 lin. ft. |
| **MASONRY** | |
| Brick Veneer | 610 sq. ft. |
| **LUMBER** | |
| Total Sills, Beams, Rafters, Studs, Plates, etc. | 13,632 BFM |
| **DRYWALL TILE** | |
| Ceramic Tile | 482 sq. ft. |
| ½" Gypsum Board | 7500 sq. ft. |
| **SHEATHING, FLOORING, INSULATION, ETC.** | |
| Wall Sheathing | 2200 sq. ft. |
| Sub Flooring | 1920 sq. ft. |
| Roof Sheathing | 2440 sq. ft. |
| Vertical Siding | 1630 sq. ft. |
| Oak Flooring | 1575 sq. ft. |
| Wall Insulation | 1860 sq. ft. |

**WINDOWS, DOORS**
10 Wood Casement Units
3 Wood Awning Units
6 Steel Basement Windows
2 Exterior Solid core Doors
3 Aluminum Sliding door units
14 Interior hollow-core doors
2 Interior bi-fold units
6 Interior sliding units

FRONT ELEVATION

RIGHT SIDE ELEVATION

REAR ELEVATION

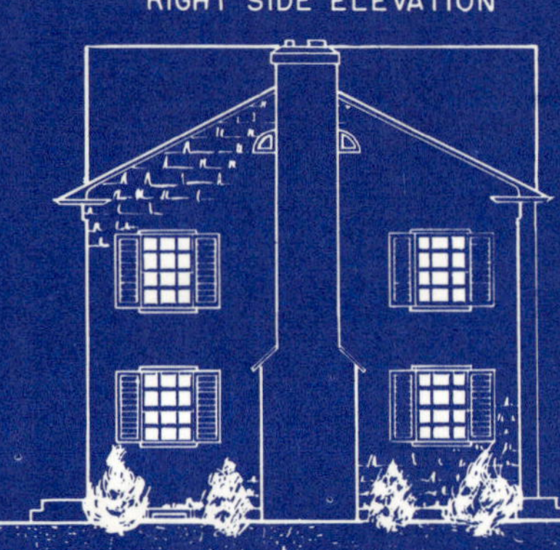

LEFT SIDE ELEVATION

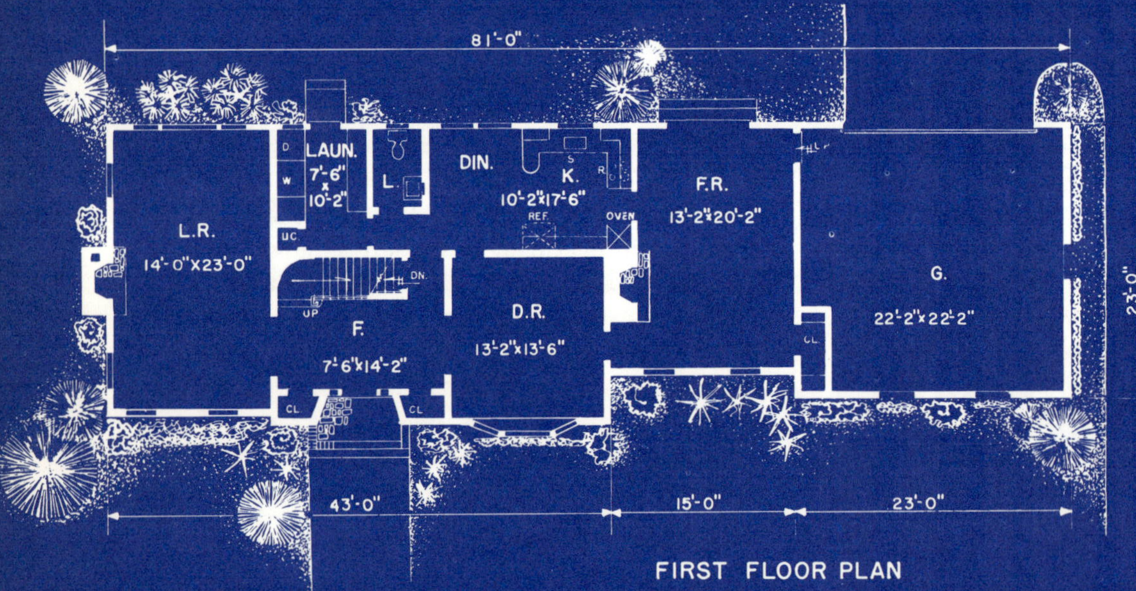


FIRST FLOOR PLAN

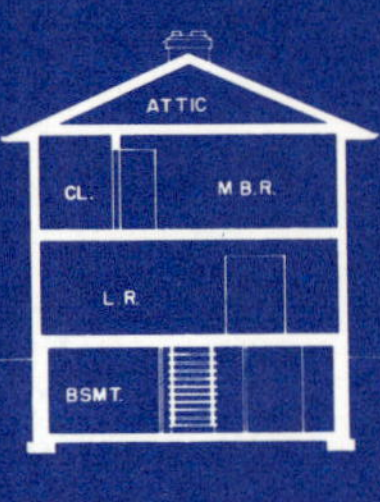


CROSS SECTION

## AREA STATISTICS

| | |
|---|---|
| FIRST FLOOR | 1365 SQ.FT. |
| SECOND FLOOR | 1050 SQ.FT. |
| BASEMENT | 1053 SQ.FT. |
| GARAGE | 529 SQ.FT. |

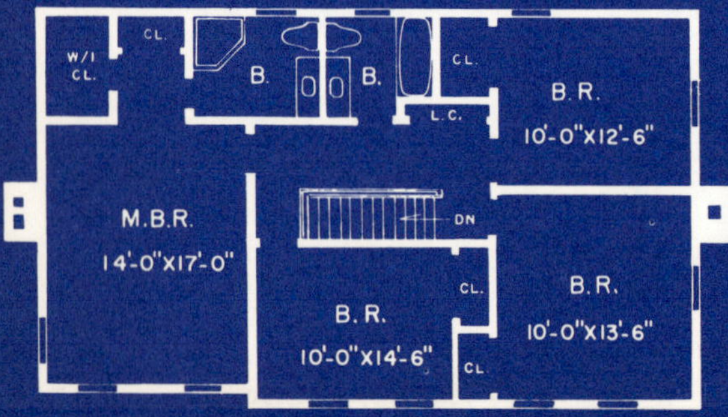


SECOND FLOOR PLAN

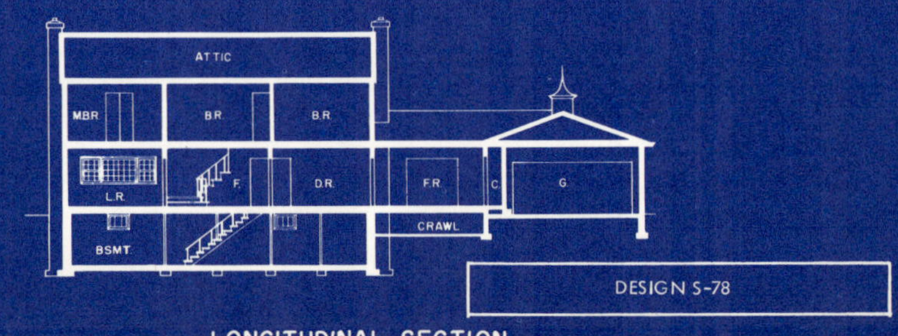


LONGITUDINAL SECTION

# Modern Layout in Colonial Exterior

The 18th and 20th centuries are pleasantly combined in this enchanting two-story house.

Colonial features give the outside its warm, hospitable appearance. The architect has used narrow clapboard siding, small-paned windows, big chimneys, a paneled entrance door and a cupola on the garage.

An excellent floor plan and present-day comforts are highlights of the modern interior. To the right of the foyer is the dining room, accessible both to the kitchen-dinette and the family room. The U-shaped kitchen provides a maximum of counter tops and cabinets.

A mullion window in the dinette section and a window over the kitchen sink assure plenty of light and air as well as a clear view of the backyard. Only a few steps away is the rear service entry, laundry room, lavatory and utility closet. Traffic is effectively distributed through the first floor and, by means of an attractive staircase, directly to the second floor. The lavatory is well-located.

The appealing family room has all the attributes for informal entertaining — accessibility to the kitchen; a fireplace; wood paneled sidewalls for a feeling of warmth; floor-to-ceiling sliding glass doors that lead to the outdoors. It also has a spacious closet.

The master bedroom has two closets, one of which is a walk-in. A private full bath is equipped with a built-in mirrored vanitory and tiled stall shower.

## Material List

|  |  |
|---|---|
| **CONCRETE WORK** | |
| Footings, floors, etc. | 37 cu. yds. |
| **STEEL** | |
| 4″ Lally columns | 6 pieces |
| Metal areaways | 5 pieces |
| **MASONRY** | |
| Quarry tile | 120 sq. ft. |
| 8″ block | 1500 blocks |
| 12″ block | 450 blocks |
| Pancakes | 160 blocks |
| **FRAMING LUMBER** | |
| Sills, joists, rafters, studs, plates, etc. | 10,000 BFM |
| **EXTERIOR SHEATHING** | |
| ½″ plyscore wall sheathing (or gyplap) | 3400 sq. ft. |
| ½″ plyscore roof sheathing | 2800 sq. ft. |
| **SUB FLOORING** | |
| ⅝″ plyscore | 2400 sq. ft. |
| **INSULATION** | |
| Rockwool semi-thick | 3500 sq. ft. |
| Rockwool full-thick | 1300 sq. ft. |
| **DOOR SCHEDULE** | |
| (1) 2-8 x 6-8 SD | |
| (1) 3-0 x 6-8 SD | |
| (29) flush interior veneered doors | |
| **EXTERIOR FINISHES** | |
| Wood shingles siding | 3400 sq. ft. |
| Asphalt shingles | 28 squares |
| **INTERIOR FINISHES** | |
| Ceramic tile floors | 110 sq. ft. |
| Ceramic tile walls | 300 sq. ft. |
| Gypsum wall and ceiling boards | 10,000 sq. ft. |
| Oak flooring | 1900 sq. ft. |

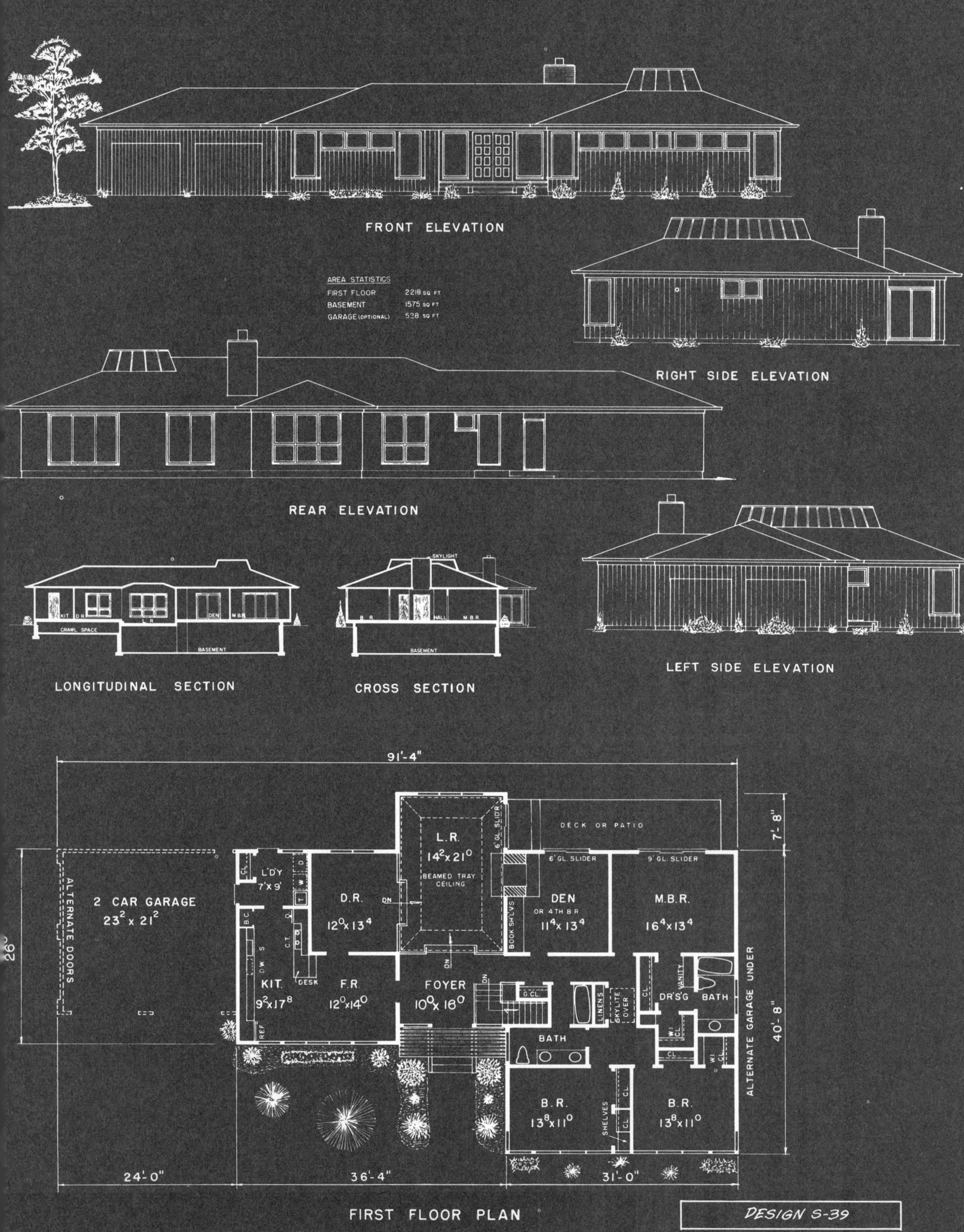

FRONT ELEVATION
AREA STATISTICS
FIRST FLOOR 2218 SQ FT
BASEMENT 1575 SQ FT
GARAGE (OPTIONAL) 528 SQ FT
RIGHT SIDE ELEVATION
REAR ELEVATION
LEFT SIDE ELEVATION
LONGITUDINAL SECTION
CROSS SECTION
FIRST FLOOR PLAN
91'-4"
2 CAR GARAGE 23² x 21²
ALTERNATE DOORS
L'DY 7'x9'
KIT. 9²x17⁸
D.R. 12⁰x13⁴
F.R. 12⁰x14⁰
L.R. 14²x21⁰
BEAMED TRAY CEILING
FOYER 10⁰x 16⁰
DECK OR PATIO
DEN OR 4TH B.R 11⁴x 13⁴
M.B.R. 16⁴x13⁴
DR'S'G
BATH
B.R. 13⁸x11⁰
B.R. 13⁸x11⁰
ALTERNATE GARAGE UNDER
40'-8"
7'-8"
24'-0"
36'-4"
31'-0"
DESIGN S-39

# Interesting Roof on Contemporary Ranch

Continuity of design is evident in this crisp-looking contemporary ranch, intended for a family requiring three or four bedrooms and for placement on a fairly sizable lot.

Among the eye-catching features of the exterior are a trellised roof which casts interesting shadows over the double-door entrance and a tapered skylight roof over the central foyer of the bedroom wing.

Directly behind the spacious foyer is a sunken living room with a tray-beamed sloped ceiling and a two-way stone fireplace. On the foyer and dining room sides of the living room are wrought iron railings. Sliding glass doors lead to a patio area.

Modern in every respect is the combined kitchen-family room to the left of the foyer.

To the right of the living room is a den or a fourth bedroom. Like the master bedroom and the living room, it has sliding glass doors to the rear patio. The master bedroom suite includes a dressing area with two closets and a bath with a full-width sliding door enclosure and recessed vanity. The hall bath is a divided room with double-basin vanity and over-tub shower. Two bedrooms are at the front.

With its natural V-joint siding, long sweeping silhouette hip roof and tapered skylight, large front and continuous ribbon windows, Design S-39 has a pleasant contemporary appearance matched by the modern, spacious look on the inside.

## Material List

**CONCRETE WORK**
Footings, floors, etc. .......... 50 cu. yds.

**STEEL**
4" Lally columns .............. 7 pieces
Metal areaways ............... 4 pieces

**MASONRY**
12" Concrete Block ............1400 blocks
8" Concrete Block ............1700 blocks

**FRAMING LUMBER**
Sills, joists, rafters, studs, plates, etc. .................9000 B.F.M.

**EXTERIOR SHEATHING**
1" x 10" fir or N.C. pine wall sheathing ..................2000 sq. ft.
1" x 6" N.C. pine roof sheathing .3300 sq. ft.

**SUB FLOORING**
1" x 6" N.C. pine .............2000 sq. ft.

**WINDOW SCHEDULE**
(8) 2-8 x 1-8 basement sash
(3) 30-24 awning
(12) 36-24 awning
(1) 48-24 awning
(2) 36-24 fixed
(5) 36-48 fixed
(6) 34-72 fixed
(2) 40-72 fix-sidelights

**DOOR SCHEDULE**
(2) 9-0 x 7-0 w.p. overhead
(2) 2-8 x 6-8 x 1¾" w.p. front entrance door
(1) 2-8 x 6-8 x 1¾" w.p. flush doors
(1) 2-8 x 6-8 x 1⅜" S.C.F.P. door
(28) Flush 1⅜" fir interior doors

**EXTERIOR FINISHES**
Asphalt roofing shingles ........ 33 squares
Vertical V-joint boarding .......2000 sq. ft.

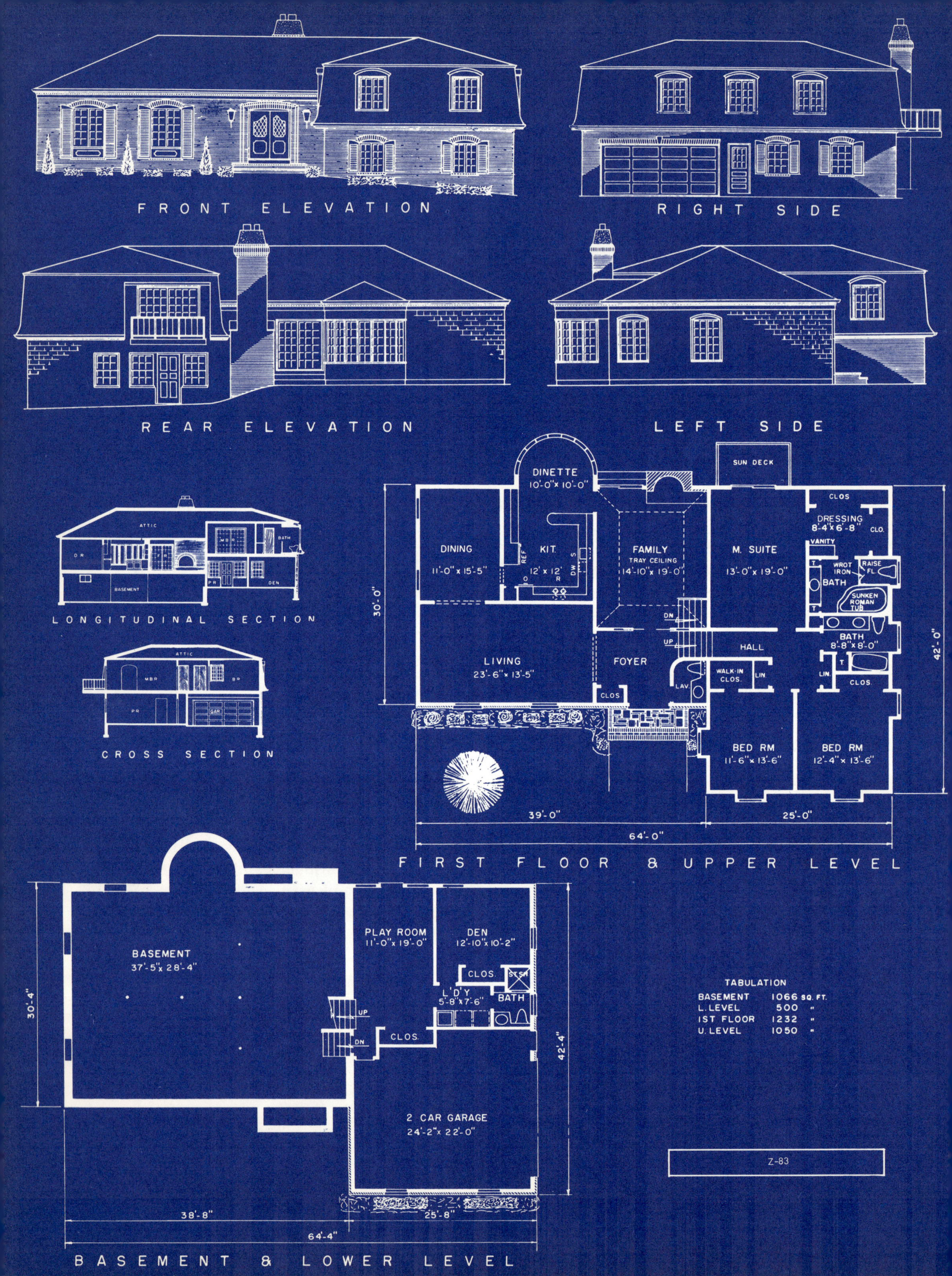
FRONT ELEVATION
RIGHT SIDE
REAR ELEVATION
LEFT SIDE
LONGITUDINAL SECTION
CROSS SECTION
FIRST FLOOR & UPPER LEVEL
DINETTE 10'-0" x 10'-0"
SUN DECK
DINING 11'-0" x 15'-5"
KIT. 12' x 12'
FAMILY TRAY CEILING 14'-10" x 19'-0"
M. SUITE 13'-0" x 19'-0"
DRESSING 8'-4" x 6'-8"
BATH
SUNKEN ROMAN TUB
LIVING 23'-6" x 13'-5"
FOYER
HALL
WALK-IN CLOS.
BATH 8'-8" x 8'-0"
BED RM 11'-6" x 13'-6"
BED RM 12'-4" x 13'-6"
39'-0"
25'-0"
64'-0"
30'-0"
42'-0"
BASEMENT & LOWER LEVEL
BASEMENT 37'-5" x 28'-4"
PLAY ROOM 11'-0" x 19'-0"
DEN 12'-10" x 10'-2"
L'D'Y 5'-8" x 7'-6"
BATH
2 CAR GARAGE 24'-2" x 22'-0"
38'-8"
25'-8"
64'-4"
30'-4"
42'-4"
TABULATION
BASEMENT 1066 SQ. FT.
L. LEVEL 500 "
1ST FLOOR 1232 "
U. LEVEL 1050 "
Z-83

# French Mansard Roof on Split-Level

The split-level became popular after World War II, but has been dressed in many traditional architectural styles in the ensuing years.

Here's one that could very well be nestling in a French countryside. Its outstanding exterior feature, which has a distinct interior advantage, is a mansard roof. Because its lower slope is very slight — almost vertical — the dormered top floor has virtually the same amount of headroom as the area below.

The mansard roof in this case is directly over a spacious bedroom wing, highlighted by a luxurious master bedroom suite. The sleeping quarters, with long expanses of wall space for easy furniture placement, stretch from the bedroom hall to the rear of the house, with glass doors leading to a private balcony.

Two other bedrooms, each with double exposure, are served by a sizable bathroom.

The living room is directly to the left of a large foyer and has two front windows. The adjacent dining room is accessible from the living room and the kitchen. The kitchen is enhanced by an imposing dinette area in the form of a semicircular, floor-to-ceiling bay window with nine casement windows.

Off the kitchen is the family room, set off by a striking arched fireplace in a brick wall.

The level underneath the bedrooms has a two-car garage, a playroom opening to the backyard, a den or fourth bedroom and a bath, complete with stall shower.

## Material List

**CONCRETE WORK**

| | |
|---|---|
| Footings, floors, etc. | 32 cu. yds. |

**STEEL**

| | |
|---|---|
| 4" Lally cols. | 7 pieces |
| Metal areaways | 3 pieces |
| 3½ x 3½ x 5/16 angle lintel | 1 piece |
| 12" Channel 20.7 w/ ½ x 7 PL Welded | 1 piece |
| 12 WF 27 | 1 piece |

**MASONRY**

| | |
|---|---|
| 8" Concrete Block | 300 blocks |
| 12" Concrete Block | 1100 blocks |
| Brick | 1200 sq. ft. |

**FRAMING LUMBER**

| | |
|---|---|
| Sills, joists, rafters, studs, plates, etc. | 10,000 BFM |

**EXTERIOR SHEATHING**

| | |
|---|---|
| ½" Plyscore wall sheathing | 1800 sq. ft. |
| ½" Plyscore roof sheathing | 3250 sq. ft. |

**DOOR SCHEDULE**

(2) 2-8 x 6-8 x 1¾ wp front entrance door
(1) 3-0 x 6-8 x 1¾ wp rear sash door
(1) 2-8 x 6-8 x 1¾ wp side sash door
(2) 9" x 6-8 x 1⅛ wp louvered door
(29) Flush interior veneered doors
(2) 4-0 x 6-8 x 1⅜ wood folding doors
(1) 6-0 x 6-8 wood sliding glass door

**INTERIOR FINISHES**

| | |
|---|---|
| Ceramic tile floor | 150 sq. ft. |
| Ceramic tile walls | 220 sq. ft. |
| Vinyl tile | 560 sq. ft. |
| Gypsum board walls & ceilings | 9800 sq. ft. |
| Oak flooring | 1800 sq. ft. |

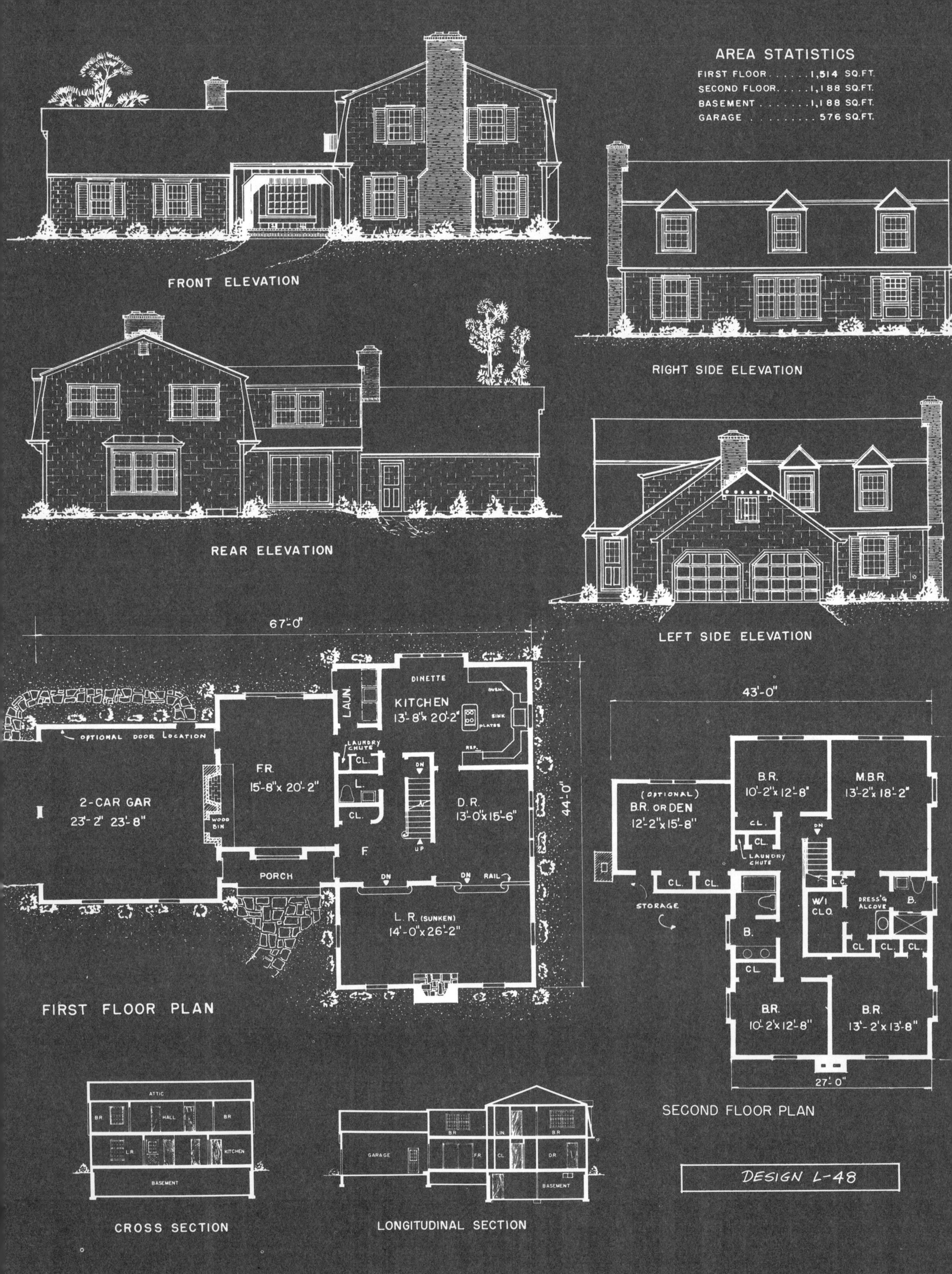
AREA STATISTICS
FIRST FLOOR . . . . . . 1,514 SQ.FT.
SECOND FLOOR . . . . . 1,188 SQ.FT.
BASEMENT . . . . . . . 1,188 SQ.FT.
GARAGE . . . . . . . . . 576 SQ.FT.
FRONT ELEVATION
RIGHT SIDE ELEVATION
REAR ELEVATION
LEFT SIDE ELEVATION
67'-0"
DINETTE
KITCHEN
13'-8" x 20'-2"
LAUN.
OVEN
SINK
LAUNDRY CHUTE
CL.
REF.
OPTIONAL DOOR LOCATION
F.R.
15'-8" x 20'-2"
2-CAR GAR
23'-2" 23'-8"
WOOD BIN
L.
CL.
DN
UP
D.R.
13'-0" x 15'-6"
44'-0"
F.
PORCH
DN
DN
RAIL
L.R. (SUNKEN)
14'-0" x 26'-2"
FIRST FLOOR PLAN
43'-0"
B.R.
10'-2" x 12'-8"
M.B.R.
13'-2" x 18'-2"
(OPTIONAL)
B.R. OR DEN
12'-2" x 15'-8"
CL.
CL.
LAUNDRY CHUTE
DN
CL.
CL.
STORAGE
W/I CLO.
DRESS'G ALCOVE
B.
B.
CL.
CL.
CL.
CL.
B.R.
10'-2" x 12'-8"
B.R.
13'-2" x 13'-8"
27'-0"
SECOND FLOOR PLAN
ATTIC
B.R.
HALL
B.R.
L.R.
KITCHEN
BASEMENT
CROSS SECTION
B.R.
LIN.
B.R.
GARAGE
F.R.
CL.
D.R.
BASEMENT
LONGITUDINAL SECTION
DESIGN L-48

# Queen Anne Provides Royal Living

An interesting variation of the popular Dutch Colonial is the Queen Anne, which also features the space-creating gambrel roof.

The principal difference between the two styles is that the Queen Anne has dormer windows in the lower of the two roof slopes characteristic of the gambrel. The traditional flavor of the exterior is accentuated by the symmetry of the shuttered multi-paned windows, the large chimney and the hand-split cedar shingles.

The family room is wood-paneled and has a brick fireplace with a raised hearth. Sliding glass doors lead to the rear patio or garden.

For formal entertaining, the living room is something special. It is more than 26′ long, has three exterior exposures, a Colonial brick fireplace and wall areas well suited to various kinds of pleasing furniture arrangements. Decorative wrought iron railings produce a balcony effect to the dining room.

More than 20′ across the back of the house, the combined kitchen-dinette area is a model of convenience.

On the second floor, three bedrooms are close to the family bathroom. The master bedroom suite has a dressing alcove with a sink vanity, two closets, one a walk-in, and a private bath with tiled shower stall. Entrance into this area is through louvered doors. An optional additional room on the second floor could be a den, sewing room or bedroom.

## Material List

**CONCRETE WORK**

| | |
|---|---|
| Footings, Floors, etc. | 37 cu. yds. |

**MASONRY**

| | |
|---|---|
| Slate | 10 sq. ft. |
| 8″ block | 1500 blocks |
| 12″ block | 450 blocks |
| Pancake | 160 blocks |
| Flagstone | 300 blocks |
| 4″ Brick Veneer | 450 sq. ft. |

**FRAMING LUMBER**

| | |
|---|---|
| Sills, joists, rafters, studs, plates, etc. | 10,000 BFM |

**EXTERIOR SHEATHING**

| | |
|---|---|
| ½″ plyscore wall sheathing (or gyplap) | 3400 sq. ft. |
| ½″ plyscore roof sheathing | 2800 sq. ft. |

**SUB FLOORING**

| | |
|---|---|
| ⅝″ plyscore | 2400 sq. ft. |

**INSULATION**

| | |
|---|---|
| Rockwool semi-thick | 1300 sq. ft. |
| Rockwool full-thick | 3500 sq. ft. |

**WINDOW SCHEDULE**

(5) 2-8 x 1-8 Basement sash

| | |
|---|---|
| (2) 2-8 x 4-2 DH | (6) 3-0 x 4-6 DH |
| (6) 2-8 x 4-6 DH | (7) 3-4 x 4-6 DH |
| (2) 3-0 x 4-2 DH | (4) 3-4 x 5-6 DH |

**INTERIOR FINISHES**

| | |
|---|---|
| Ceramic tile floor | 156 sq. ft. |
| Ceramic tile walls | 622 sq. ft. |
| Rocklath | 10,000 sq. ft. |
| Oak flooring | 1600 sq. ft. |

**EXTERIOR FINISHES**

| | |
|---|---|
| Wood shingles siding | 3400 sq. ft. |
| Vertical "V" joint siding | 140 sq. ft. |

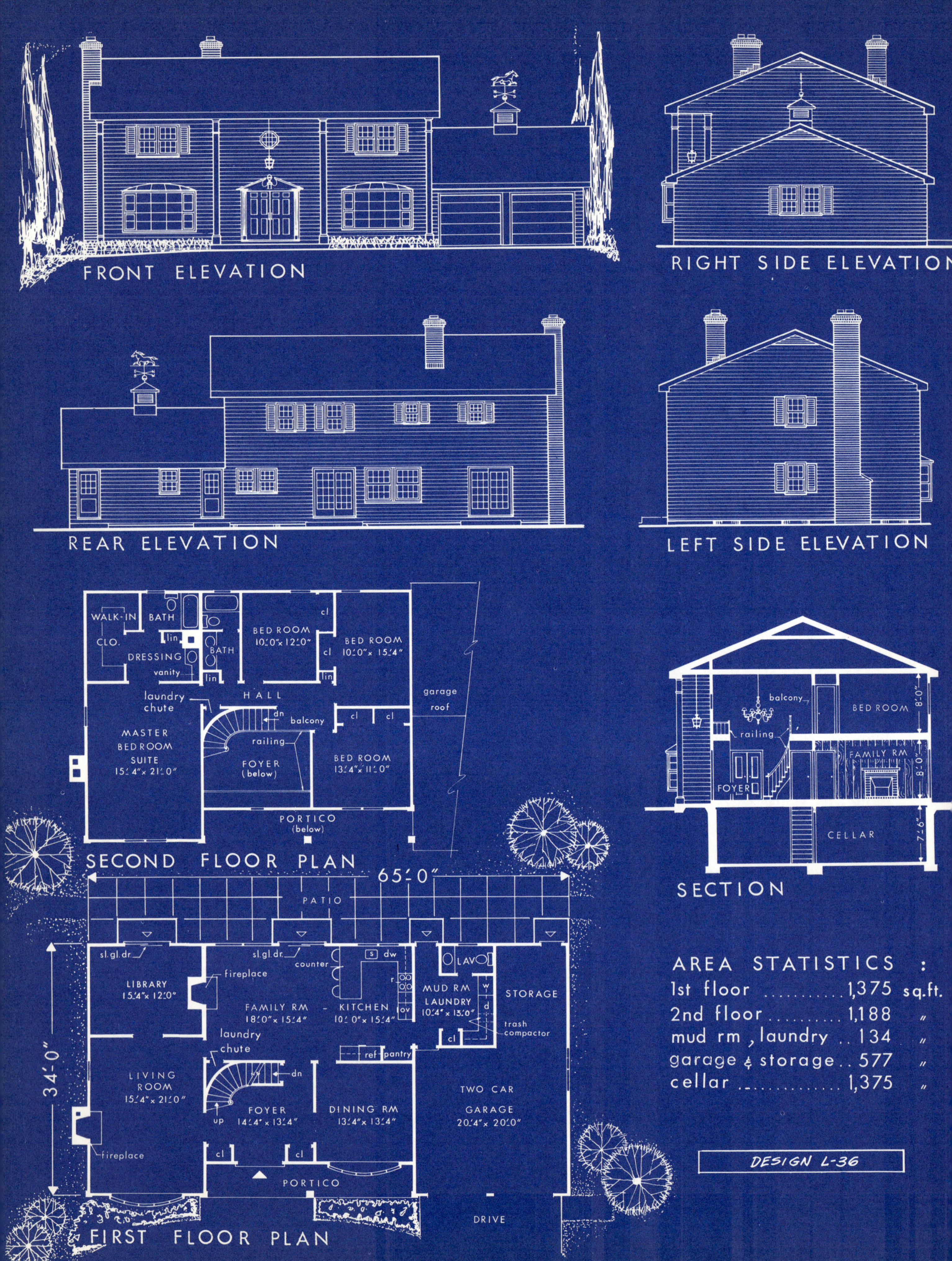

FRONT ELEVATION
RIGHT SIDE ELEVATION
REAR ELEVATION
LEFT SIDE ELEVATION
WALK-IN CLO.
BATH
lin
DRESSING
vanity
BATH
lin
BED ROOM 10'0" x 12'0"
cl
BED ROOM 10'0" x 15'4"
cl
lin
garage roof
laundry chute
HALL
dn
balcony
cl
cl
MASTER BED ROOM SUITE 15'4" x 21'0"
railing
FOYER (below)
BED ROOM 13'4" x 11'0"
PORTICO (below)
SECOND FLOOR PLAN
balcony
BED ROOM
8'0"
railing
FAMILY RM
8'0"
FOYER
CELLAR
7'6"
SECTION
65'0"
PATIO
sl.gl.dr.
sl.gl.dr.
counter
dw
LAV
LIBRARY 15'4" x 12'0"
fireplace
MUD RM LAUNDRY 10'4" x 13'0"
STORAGE
FAMILY RM 18'0" x 15'4"
KITCHEN 10'0" x 15'4"
trash compactor
cl
laundry chute
ref
pantry
34'0"
LIVING ROOM 15'4" x 21'0"
dn
up
FOYER 14'4" x 13'4"
DINING RM 13'4" x 13'4"
TWO CAR GARAGE 20'4" x 20'0"
fireplace
cl
cl
PORTICO
DRIVE
FIRST FLOOR PLAN
AREA STATISTICS :
1st floor .......... 1,375 sq.ft.
2nd floor .......... 1,188 "
mud rm, laundry .. 134 "
garage & storage .. 577 "
cellar .............. 1,375 "
DESIGN L-36

# Columned Colonial Has Luxury Features

Distinctiveness and luxury are immediately evident in the facade of this two-story house.

Design L-36 is reminiscent of a Colonial mansion with its high colonnade, recessed portico, window shutters, brick chimneys, cupola on the garage roof and the hanging light fixture suspended over the main entry doors. A narrow horizontal siding used for the entire exterior reduces the overall height and unifies the house with one basic material.

Inside, two stories high, is a reception foyer which features a circular stair leading to a second-floor balcony. A large hanging chandelier adds to the elegance.

To the left of the reception foyer, and facing the front, is a large living room with bow window. Immediately visible on entrance is the fireplace, always an attractive focal feature. Toward the rear of the house and beyond the living room is the library.

Directly to the rear of the foyer and facing the patio are the family room and the kitchen. An almost solid wall of glass faces the rear patio. A sliding glass door in the center defines the kitchen from the family room.

In the bedroom wing, reached by the main circular stair, all four bedrooms are off a central hall. The huge master suite — 33′ by 15′4″—includes the bedroom, a dressing room, walk-in closet and full bath, with a sink vanity in the dressing portion. The bath serving the other 3 bedrooms is "split."

## Material List

**CONCRETE WORK**

| | |
|---|---|
| Foundations, footings, slabs, etc. | 85 cu. yds. |

**MASONRY**

| | |
|---|---|
| Brick Fireplace & Chimney | 190 cu. ft. |

**FRAMING LUMBER**

| | |
|---|---|
| Total Sills, Joists, Rafters, Studs, Plates, etc. | 12,594 B.F.M. |

**SHEATHING, INSULATION**

| | |
|---|---|
| Sub Flooring | 2750 sq. ft. |
| Wall Sheathing | 2328 sq. ft. |
| Roof Sheathing | 4640 sq. ft. |
| Wall Insulation | 2000 sq. ft. |
| Ceiling Insulation | 1509 sq. ft. |

**FINISHES, INTERIOR**

| | |
|---|---|
| Vinyl Tile | 440 sq. ft. |
| Oak Flooring | 1200 sq. ft. |
| Ceramic Tile Floors | 95 sq. ft. |
| Ceramic Tile Walls | 292 sq. ft. |
| Gypsum Board - house walls | 5736 sq. ft. |
| Gypsum Board - house ceilings | 2884 sq. ft. |
| Gypsum Board - garage | 1313 sq. ft. |

**FINISHES, EXTERIOR (other than masonry)**

| | |
|---|---|
| Horizontal Siding (wood or alum.) | 2328 sq. ft. |
| Asphalt Shingle Roofing | 4640 sq. ft. |
| Plywood Eave & Porch Soffits | 1380 sq. ft. |

**WINDOW SCHEDULE**

| | |
|---|---|
| Octagonal Sash | 1 unit |
| Wood Double Hung | 17 units |
| Bow Window | 2 units |
| Basement Hopper | 5 units |

**DOOR SCHEDULE**

| | |
|---|---|
| Ext. Hardwood, paneled | 2 units |
| Ext. Hardwood, glazed | 2 units |
| Aluminum Sliding | 2 units |
| Int. Hardwood, flush, staingrade | 18 units |
| Int. Hardwood, louvered, bi-folding | 8 units |

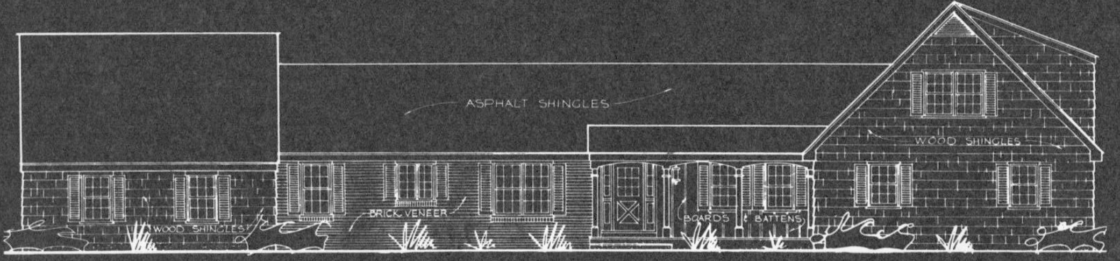


FRONT ELEVATION

ASPHALT SHINGLES

WOOD SHINGLES

RIGHT SIDE ELEVATION

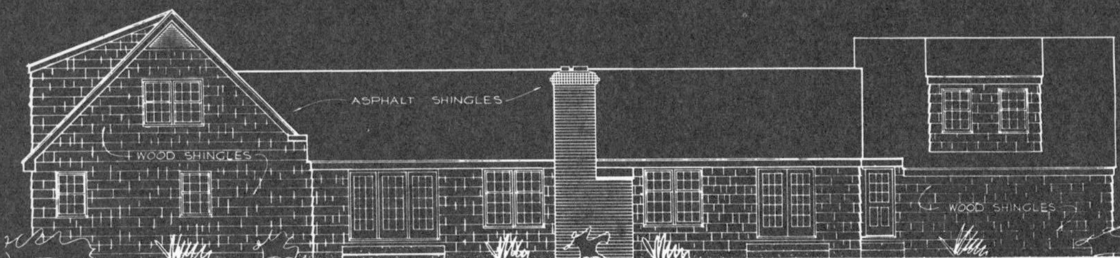


REAR ELEVATION

LEFT SIDE ELEVATION

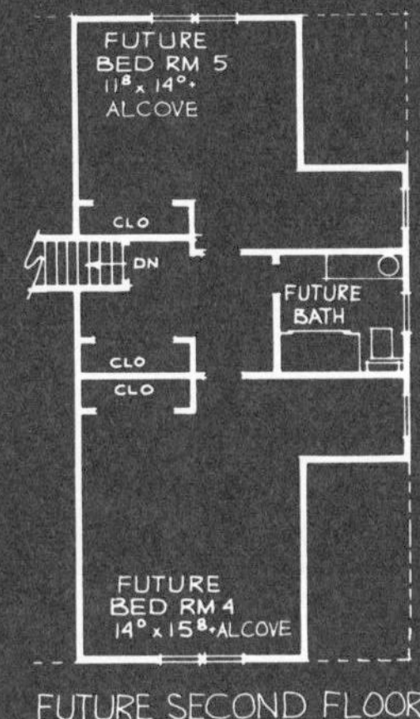


FUTURE SECOND FLOOR

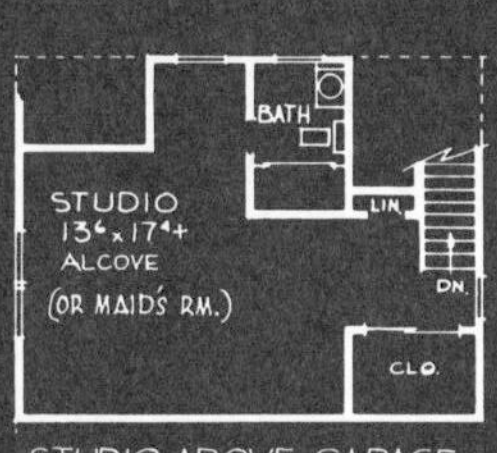


STUDIO ABOVE GARAGE

HOUSE DATA

| | |
|---|---|
| FIRST FLOOR | 2844 SQ.FT. |
| SECOND FL | 738 SQ.FT. |
| STUDIO RM. | 393 SQ.FT. |
| GARAGE | 593 SQ.FT. |

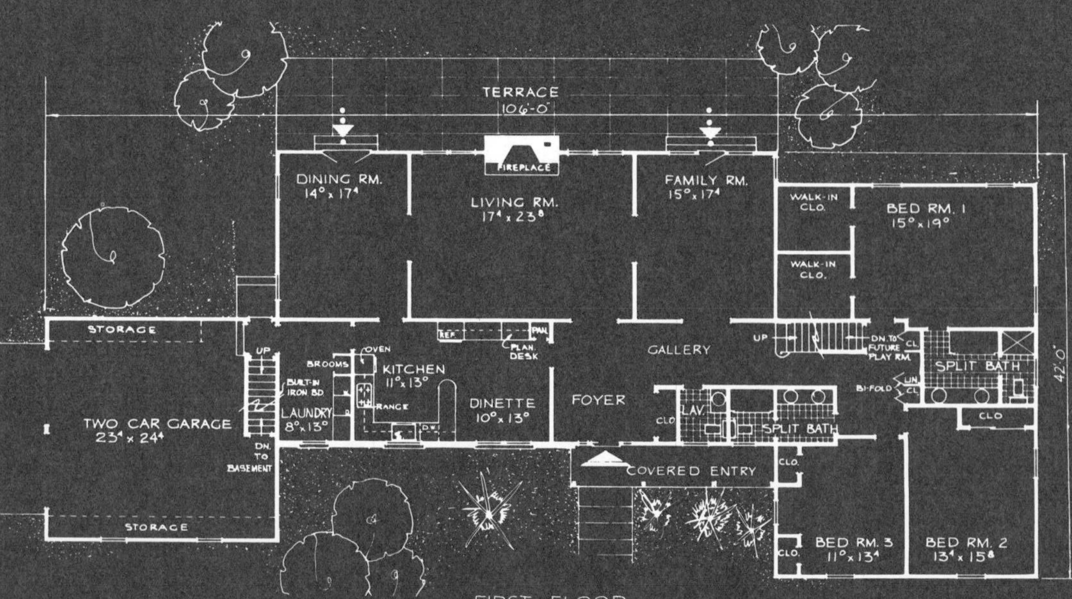


FIRST FLOOR

DESIGN S-45

# Gracious Living For Large Family

This is a spacious home intended for a large family, with emphasis on gracious living and areas of privacy.

In addition to the three bedrooms on the first floor (one of them is a sizable 19′ by 15′), there is an upstairs expansion space for two extra bedrooms or for individual pursuits. And there's a studio over the garage, separate from other living areas.

Entering the house through a huge foyer and gallery section, one can move directly to the living room, the family room, the kitchen or the bedroom wing. Everything is generously conceived.

The living room, dining room and family room all face onto the rear terrace, which is directly accessible from the dining room and family room. While the family room is next to the bedroom wing, it is sound isolated from it by two large walk-in closets that are part of the master bedroom. The bathroom serving this bedroom and the one in the bedroom hall are both of the "split" variety, which gives greater flexibility.

There is a traditional exterior with gabled roof and shed dormers. A small area of brick is used on a portion of the front facade. The other walls are finished in wood shingles and vertical boards and battens. The architect suggests the shingles be left natural or stained, with accent color achieved in the painted front door and shutters.

## Material List

**CONCRETE WORK**

| | | |
|---|---|---|
| Concrete Walls | | 1789 cu. ft. |
| Foundation Footings | | 414 cu. ft. |
| Slabs | | 1138 cu. ft. |
| Misc. Concrete | | 240 cu. ft. |

**STRUCTURAL STEEL**

| | | |
|---|---|---|
| Lally Columns | 3½″ diam. | 15 pieces |
| Girder | 7″ I Beam | 118 lin. ft. |

**BRICK WORK**

| | | |
|---|---|---|
| Chimney | Brick | 229 cu. ft. |
| Flue Lining | T. C. | 40 lin. ft. |
| Veneer | Brick | 230 sq. ft. |

**CARPENTRY**

| | |
|---|---|
| Framing Lumber | 16,644 B.F. |
| Studs | 4667 B.F. |
| Plates | 1400 B.F. |
| Roof Sheathing | 5000 sq. ft. |
| Sub Flooring | 3231 sq. ft. |
| Side Wall Sheathing | 4014 sq. ft. |
| Wall Insulation | 2850 sq. ft. |
| Ceiling Insulation | 3436 sq. ft. |
| Wood Floors | 2637 sq. ft. |
| Kitchen Floor | 273 sq. ft. |

**MILLWORK**

| | |
|---|---|
| Exterior Doors & Frames Compl. | 7 units |
| Garage Door Complete Set | 2 units |
| Interior Doors & Frames Compl. | 25 units |
| Sliding Doors | 2 units |
| Bi Fold Doors | 3 units |
| Windows | 38 units |
| Fascia | 480 lin. ft. |

**KITCHEN CABINETS**

| | |
|---|---|
| Counters | 27 lin. ft. |
| Hangers | 24 lin. ft. |

**ROOFING**

| | | |
|---|---|---|
| Shingles | 235# asphalt | 5000 sq. ft. |
| Roofing Paper | 15# felt | 5000 sq. ft. |

FRONT ELEVATION

RIGHT SIDE ELEVATION

REAR ELEVATION

LEFT SIDE ELEVATION

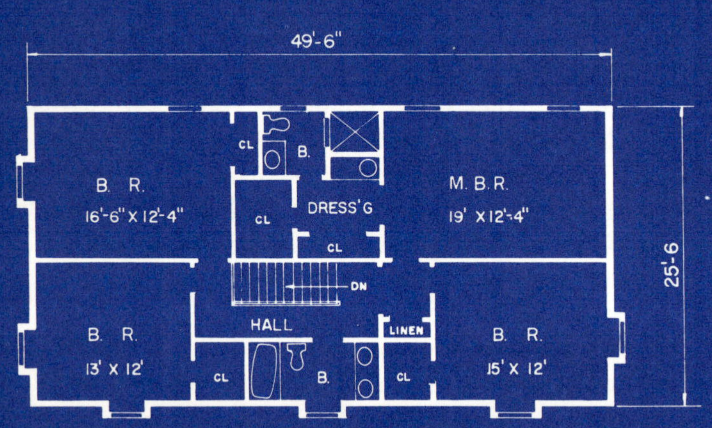


SECOND FLOOR PLAN

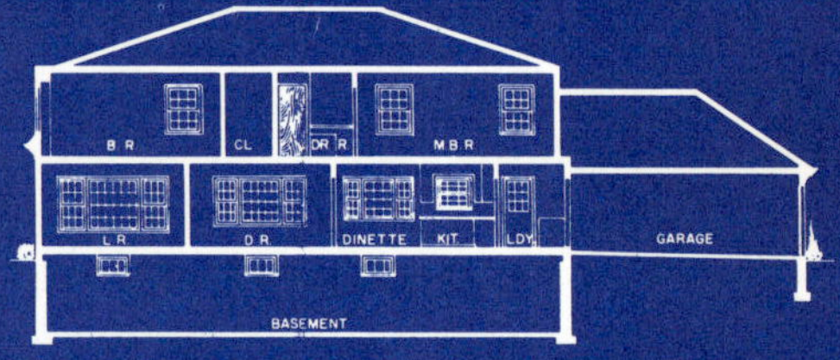


LONGITUDINAL SECTION

B.R.
B.R.
L.R.
BASEMENT

CROSS SECTION

74'-6"

D. R.
14' x 12'-4"

DINETTE
8' x 12'-4"

KIT.
7' x 12'-4"

LDY.

CL.

GARAGE
22'-6" x 21'

22'

26'-6"

L. R.
14' x 25'-8"

DN

UP

CL.

STUDY
16' x 13'

FOYER
9'-6" x 13'-6"

LAV.

4'-6"

51'-6"

23'

FIRST FLOOR PLAN

AREA STATISTICS

| | |
|---|---|
| FIRST FLOOR | 1364 SQ FT |
| SECOND FLOOR | 1264 |
| BASEMENT | 1364 |
| GARAGE | 506 |

DESIGN S-34

# More Space Upstairs with Mansard Roof

The capacity of a 17th century French mansard roof to provide extra space on the second floor is clearly illustrated in this pleasant traditional house.

Because the lower roof slope is very slight — almost vertical — the floor area upstairs is almost the same as the first floor; that is, 1264 square feet to 1364 square feet. As a result the architect has been able to place four good-sized bedrooms and two bathrooms on the second floor.

On the first floor, the exceptionally large entrance foyer is virtually a room itself. To the left is a living room 25′8″ long from the front of the house to the rear. Centered on the long wall is a brick-faced fireplace with a window on each side.

The kitchen-dinette is large enough to accommodate the whole family, thus preserving the regular dining room for formal occasions. Two windows, one over the sink and a rectangular-bayed one in the dinette, provide natural light and a view of the rear yard.

What the architect has designated as a study, but can be used as a family room or even an extra bedroom, is the right front corner of the house. It is separated from the foyer by a lavatory and a clothes closet. Because it is a "dead end" room, with only a single entrance, it can be utilized in any possible way with no possibility that it will be used as a passageway.

## Material List

**CONCRETE WORK**
Footings, floors, etc. .......... 50 cu. yds.

**MASONRY**
12″ Concrete Block ............ 1300 blocks
8″ Concrete Block ............ 400 blocks
4″ Pancake Block ............ 118 blocks
4″ Brick Veneer ............ 460 sq. ft.

**FRAMING LUMBER**
Sills, joists, rafters, studs, plates, etc. ................ 10,000 BFM

**EXTERIOR SHEATHING**
1″ x 10″ fir or N.C. pine wall sheathing ............ 1700 sq. ft.
1″ x 6″ N.C. pine roof sheathing . 3500 sq. ft.

**SUB FLOORING**
1″ x 6″ N.C. pine ............ 2600 sq. ft.

**INSULATION**
Rockwool semi-thick .......... 3000 sq. ft.
Rockwool full-thick ............ 1300 sq. ft.

**DOOR SCHEDULE**
(2) 9-0 x 7-0 w.p. overhead
(1) 3-6 x 6-8 x 1¾″ w.p. front entrance door
(1) 2-8 x 6-8 x 1¾″ w.p. sash door
(1) 2-6 x 6-8 x 1⅜″ S.C.F.P. door
(25) Two-panel 1⅜″ fir interior doors

**EXTERIOR FINISHES**
Wood roofing shingles ........ 35 squares
Flush boarding ½″ ext. plywood. 34 sq. ft.
Red Cedar Shingles ............ 1120 sq. ft.

**INTERIOR FINISHES**
Ceramic Tile Floor ............ 100 sq. ft.
Ceramic Tile Walls ............ 300 sq. ft.
Vinyl Tile ................... 220 sq. ft.
Gypsum Board walls & ceiling ....10,000 sq. ft.
Oak Flooring ................ 2000 sq. ft.

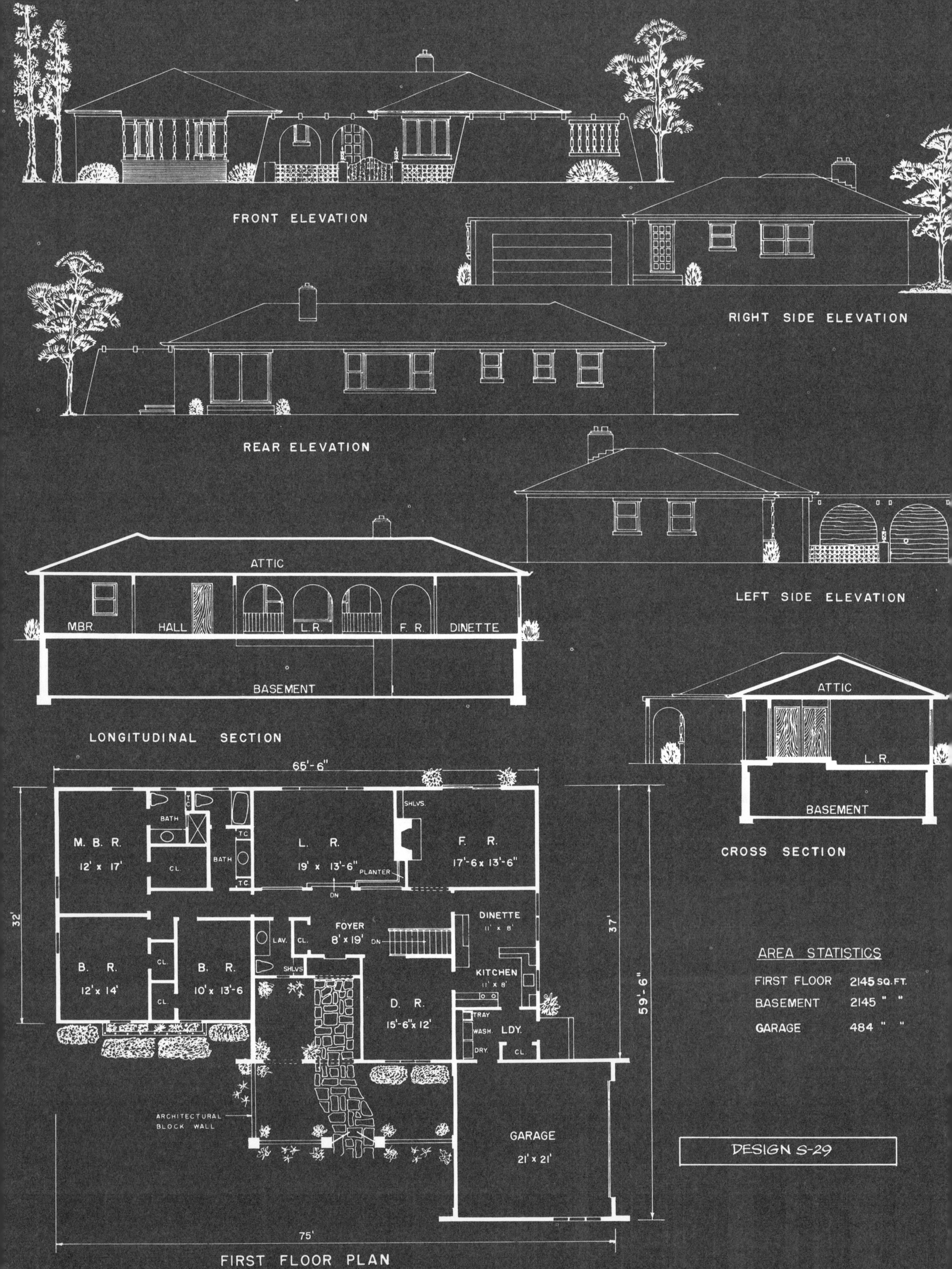
FRONT ELEVATION
RIGHT SIDE ELEVATION
REAR ELEVATION
LEFT SIDE ELEVATION
ATTIC
M.B.R.
HALL
L.R.
F. R.
DINETTE
BASEMENT
LONGITUDINAL SECTION
ATTIC
L. R.
BASEMENT
CROSS SECTION
65'-6"
M. B. R.
12' x 17'
BATH
T.C.
CL.
BATH
L. R.
19' x 13'-6"
SHLVS.
PLANTER
DN
F. R.
17'-6 x 13'-6"
DINETTE
11' x 8'
FOYER
8' x 19'
DN
LAV.
CL.
SHLVS
B. R.
12' x 14'
CL.
CL.
B. R.
10' x 13'-6
KITCHEN
11' x 8'
D. R.
15'-6" x 12'
TRAY
WASH.
DRY.
LDY.
CL.
32'
37'
59'-6"
ARCHITECTURAL
BLOCK WALL
GARAGE
21' x 21'
75'
FIRST FLOOR PLAN
AREA STATISTICS
FIRST FLOOR 2145 SQ. FT.
BASEMENT 2145 " "
GARAGE 484 " "
DESIGN S-29

# Authentic Spanish Styling in Ranchero

A basic three-bedroom, one-story house has been given the old ranchero treatment in this interesting design.

The exterior of rough stucco, turned wood posts, tapered walls and projecting beams is in the popular Spanish tradition. While the exterior architectural details and some of those on the inside adhere to the Spanish motif, the floor plan is arranged for present-day living, a central foyer leading directly to all rooms in the house.

The entrance is impressive with wrought iron gates leading to the front door through a private courtyard. Straight ahead from the foyer is a sunken living room with a ceiling 9′ 4″ high and with a full-height picture window overlooking the rear.

To the right of the living room and also directly accessible from the foyer is the family room with a stone-faced fireplace, bookshelves, wood-beamed ceiling and a sliding door unit to the rear.

A somewhat unusual design touch is the location of the dining room, which is at the front of the house and across the foyer from the living room. There are two windows in the kitchen-dinette, one over the sink and a mullion type in the eating section.

Three bedrooms and two bathrooms comprise the left wing of the house. The master bedroom bath has a tiled shower stall and a full-length mirrored vanity.

## Material List

**CONCRETE WORK**

| | |
|---|---|
| Footings, floors, etc. | 42 cu. yds. |

**MASONRY**

| | |
|---|---|
| 12″ Concrete Block | 1300 blocks |
| 8″ Concrete Block | 850 blocks |

**FRAMING LUMBER**

| | |
|---|---|
| Sills, joists, rafters, studs, plates, etc. | 10,000 BFM |

**EXTERIOR SHEATHING**

| | |
|---|---|
| 1″ x 10″ fir or N.C. Pine Wall | 1820 sq. ft. |
| 1″ x 6″ N.C. Pine Roof sheathing | 2940 sq. ft. |

**SUB FLOORING**

| | |
|---|---|
| 1″ x 6″ N.C. Pine | 2140 sq. ft. |

**WINDOW SCHEDULE**

| | |
|---|---|
| (2) 2-4 x 3-6 DH | (1) 6-4 x 5-2 Fixed |
| (3) 3-0 x 4-2 DH | (1) 16-36 csmt |
| (2) 3-4 x 4-6 DH | (1) 3-24-40 csmt |
| (2) 2-4 x 5-2 DH | (1) 3-24-60 csmt |
| | (2) 2-18-54 csmt |

**DOOR SHEDULE**

(2) 16-0 x 7-0 w.p. overhead
(1) 3-0 x 6-8 x 1¾″ w.p. front entrance door
(1) 2-8 x 6-8 x 1¾″ w.p. door
(1) 2-8 x 6-8 x 1⅜″ S.C.F.P. door
(23) Two-panel 1⅜″ fir int. doors

**EXTERIOR FINISHES**

| | |
|---|---|
| Asphalt roofing shingles | 29.4 squares |
| Stucco | 1820 sq. ft. |

**INTEROR FINISHES**

| | |
|---|---|
| Ceramic Tile floor | 70 sq. ft. |
| Ceramic Tile walls | 220 sq. ft. |
| Vinyl Tile | 200 sq. ft. |
| ½″ Sheetrock | 5000 sq. ft. |
| Oak Flooring | 1540 sq. ft. |

FRONT ELEVATION

RIGHT SIDE ELEVATION

REAR ELEVATION

LEFT SIDE ELEVATION

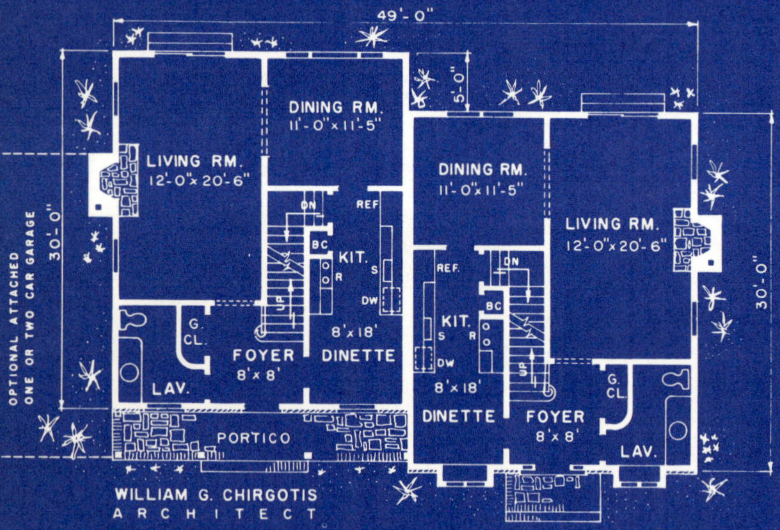


FIRST FLOOR PLAN

## AREA STATISTICS

BASEMENT....1470 SQ. FT.
FIRST FL. .....1470 SQ. FT.
SECOND FL....1568 SQ. FT.

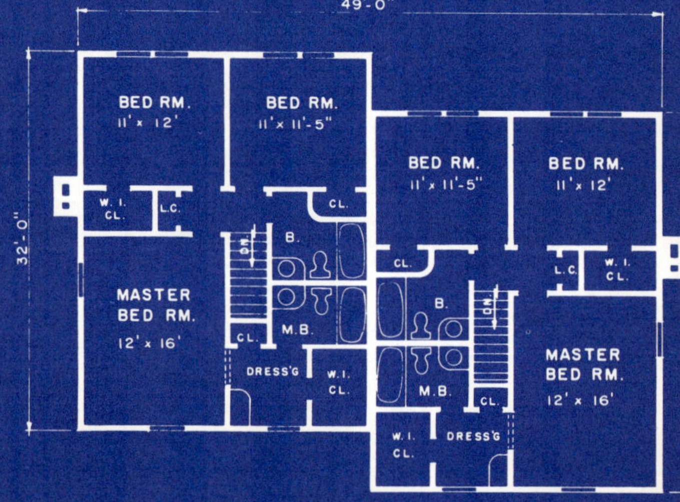


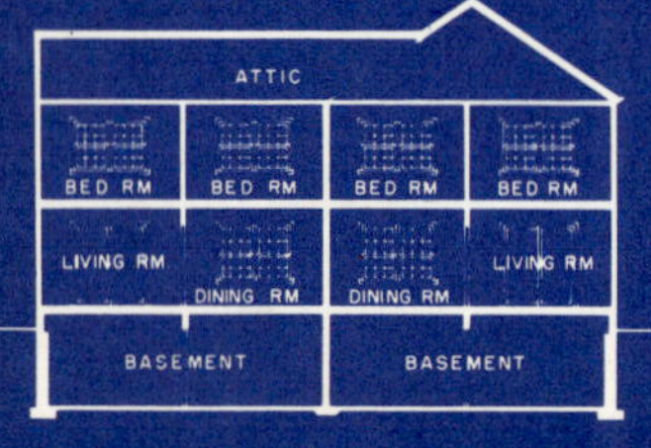


LONGITUDINAL SECTION

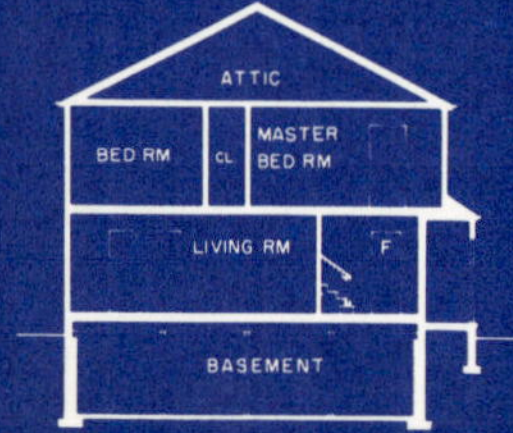


CROSS SECTION

SECOND FLOOR PLAN

**DESIGN S-50**

# Duplex That Doesn't Appear To Be Duplex

You can look at the outside of this house without suspecting that it is a duplex.

Although it has two front doors, it is difficult to guess that each leads to a separate unit. The architect has accomplished this by setting back half the house several feet, with each half having distinctively different Colonial facades.

Each apartment is identical in size and has a habitable area of 1470 square feet, consisting of living room, dining room, kitchen-dinette, foyer and lavatory downstairs and three bedrooms and two baths upstairs.

When entering either foyer, there is an immediate view of the living room with its all-brick fireplace and sliding door leading to the rear yard or patio.

Few women could resist the 18-foot-long kitchen-dinette with its full complement of appliances, counter top, broom closet and an eating area large enough to accommodate the family.

With its shingles, narrow beveled siding, corner boards, multi-paned windows, moulded cornices, shutters and doors in the tradition of the Colonial period, Design S-50 will fit in with any surrounding where duplex construction is permitted.

And then, of course, this type of house permits the owners to live well and to obtain an income — without having a tenant family on the floor above or the floor below.

## Material List

**CONCRETE WORK**
Footings, floors, etc. .......... 34 cu. yds.

**STEEL**
4" Lally columns .............. 8 pieces
Metal areaways ............... 6 pieces

**MASONRY**
8" Block .................... 600 blocks
12" Block .................... 1400 blocks
Brick ....................... 1000 sq. ft.

**FRAMING LUMBER**
Sills, joists, rafters, studs, plates, etc. ................ 14,000 BFM

**EXTERIOR SHEATHING**
½" Plyscore wall sheathing (or gyplap) ................. 2000 sq. ft.
½" Plyscore roof sheathing .... 2300 sq. ft.

**SUB FLOORING**
⅝" Plyscore .................. 3150 sq. ft.

**INSULATION**
Rockwool semi-thick ........... 3000 sq. ft.
Rockwool full-thick ............ 1600 sq. ft.

**DOOR SCHEDULE**
(2) 3-0 x 6-8 x 1¾" W.P. front entrance doors
(2) 6-0 unit plate glass alum. sliding glass door units
(40) Panel interior doors

**WINDOW SCHEDULE**
(6) 2-8 x 1-8 basement sash
(8) 3-0 x 4-2 DH
(16) 3-0 x 4-6 DH
(2) 3-5 x 4-6 /Csmt.

**EXTERIOR FINISHES**
Hand-Split Shakes ............ 140 sq. ft.
Beveled Siding ................ 300 sq. ft.
Asphalt shingles .............. 23 squares
Wood shingles ................ 2800 sq. ft.

# Chapter Five
# NEW HOUSE MAINTENANCE

## 1. Importance of Exterior Checkups

You are not going to have the same kind of repair and maintenance chores in your new house as you would have had if you purchased an old house. A considerable part of your attention will be given to preventing trouble before it occurs or, at the very least, while it is still minor and easily handled.

Given time, small openings in the outside of your house become large openings. And large openings permit the entrance of water that can cause hundreds of dollars worth of damage to walls, ceilings and furniture. That stitch-in-time proverb was not invented to apply to caulking gaps in the exterior of homes, but it certainly fits the situation.

Caulking by the do-it-yourselfer has been made comparatively easy in recent years by the use of cartridges filled with the repair material. The cartridge is placed in the frame of an inexpensive so-called cartridge gun. When the trigger is pressed the compound is forced through the nozzle into the house opening. It's a lot simpler than applying bulk compound with a putty knife or a cylinder gun.

When should caulking be done? As soon as an opening is spotted. Since small gaps are often difficult to detect via a casual examination, from ground level, close inspections should be made twice a year — in the spring and fall. These inspections should be made from a ladder, permitting a close look at possible trouble spots.

Gaps usually occur around windows and doors and wherever dissimilar materials come together; in short, wherever there are joints. Besides blocking the entrance of water, caulking compound keeps out drafts and insects. Its big virtue is its flexibility, which allows it to retain its position even when the materials around it are expanding and contracting.

It is not enough merely to apply the compound wherever there is an opening. The area must be cleared of all pieces of old compound and anything else that is loose to the touch. This can be done with an old screwdriver, awl or similar tool.

While the patch need not be painted, most compounds will stay in place much longer (usually many years) if they are painted about an hour after being applied. In purchasing caulking compound, read the label on the cartridge to determine whether it can be used over a painted surface.

## 2. Interior Walls

If you notice a few small cracks in the plaster or wallboard of your interior walls a few months after you have moved into your new house, don't be dismayed. This sometimes happens while the house is settling and is not serious.

Such cracks can be filled with a spackling compound, sanded and then repainted to match the rest of the wall. However, you don't want to do this and find that the cracks have lengthened a few days later. To avoid this, delay the spackling until you are sure the house settling will not affect the walls any further. When the cracks (most of the time merely hairline openings) make an appearance, place tiny pencil or chalk marks at both ends of each crack. All you have to do then is to wait a few weeks to see whether the cracks have extended beyond the marks. If they haven't, go ahead with the repair.

## 3. Floors

While squeaky floors generally are synonymous with older houses, they occasionally make an appearance in a new residence. Whether the contractor will make an effort to correct the condition during your first year of occupancy depends on the terms of your agreement with him. It is well for you to know how to silence the squeaks in any case, since they often show up after expiration of your guarantee.

First, you need an assistant who need merely possess the ability to stand and walk. If the squeaks are coming from the first floor and there is an exposed basement ceiling, go downstairs and get yourself a step-ladder or chair, a piece of chalk and a flashlight. While your helper walks across the floor following your directions, listen carefully and make chalk marks on the underside of the floor wherever squeaks are heard. Have your helper repeat the performance as often as necessary. This time, using a flashlight, see what is causing the squeaks.

When the subfloor or under-floor moves up and down slightly over a joist, the most frequent cause of noise, the remedy is a wedge driven between the floor and the joist. When the noise is coming from the flooring between the joists, it is likely that the top or finish floor is loose and must be brought back to the under-floor. This is done by having your assistant stand on the offending board while you drive a screw upward through the subfloor into the finish floor. Use a screw one and one-quarter inches in length.

A less likely but possible source of squeaks is loose bridging, those strips of wood used as supports between joists. If you find one of them not secured, hammer a couple of nails into it until there is no movement. Or you can buy metal bridging strips made especially for that purpose.

If the squeaks are coming from an upper floor, they cannot be attacked from the underside and must be handled from the top. After locating a noisy spot, again by the walking method, drive two nails into the floor there. They should be about two inches apart and driven at an angle so that their ends meet or almost meet inside the wood. When possible, place the nails so that they go through the two layers of flooring into a joist.

Use pilot holes for the nails if the floor is made of oak or some other hardwood. These are drilled holes smaller in diameter than the nails being used, which should be long and of the so-called finishing type. Countersink the nails slightly below the surface so that they can be covered with wood putty or plastic wood.

## 4. Plumbing, Heating and Air Conditioning

Plumbing, heating and air conditioning are not likely to give you much trouble during the first years in your new house. Nearly all utility equipment, however, is guaranteed to perform satisfactorily for certain periods of time, some for as long as 10 years.

But there is one thing you can do which will be of help to you and your family. Buy a couple of dozen tags, the kind that can be attached to objects with cord or wire.

Attach each of these tags to everything possible, but most especially to the various water valves. It is not enough for you to know where they are. What happens when you aren't there? Do the other members of the family know where the main shut-off valve is located? Do they know which valve to close when the cold water must be turned off without affecting the hot water? And vice versa?

The answer to the latter three questions is probably "no." The best way to avoid the trouble that can ensue when the water can't be shut off properly is to place tags on all the key valves. Each tag should tell what it controls and how the valve is turned to halt the flow of water. And each responsible member of the family should be advised that all such valves work the same way: when one is turned clockwise, the valve closes and the water is halted; when it is turned counter-clockwise, the valve opens.

In addition to the main valves, there are separate shutoffs at some fixtures, especially in newer houses, so that the water can be cut off from a particular sink or fixture without disturbing the flow to other parts of the house.

## 5. Landscaping Tips

Landscaping is a custom job which takes into consideration the needs — and pocketbooks — of each individual family.

Obviously a family with several growing children will find it impractical to specialize in a velvet-smooth, weed-free lawn. A hillside house will have problems never encountered with a house on level ground. This family of adults will go in for flower gardening; that family of adults, to whom grounds keeping is a chore and not a hobby, will want to keep outside work to a minimum.

Unfortunately few men and women buying their first house really know just how much outside work they want. Much better to start modestly with a carefully prepared lawn, basic tree and shrub plantings and a minimum of flower and vegetable gardens. It is easy enough to add.

Ideally, a landscaping plan should be developed as carefully as house-building plans. More often landscaping in the broad sense turns out to be something added after the house is finished.

The first move of the home owner is to make a plan on paper. Every family will have a different answer, for each family has its own particular needs.

If there is need for privacy, there should be a screen. This can be a closely woven fence, a hedge of evergreens or a long planting of shrubs.

If the family enjoys outdoor living in season, provisions should be made for a sunny patio, a shady spot out of summer heat, and an outdoor grill for family fun.

Gardeners will want to mark out their future flower beds, taking into consideration the amount of sun which will reach it in growing season.

Everybody with grounds has to consider grass. And, no matter what you've been told, grass is a plant, which appreciates good soil, adequate drainage and food as much as a prize dahlia. A thin covering of topsoil over a heavy clay soil, rocks and building debris will never support a healthy lawn. It is real economy to spend time and money in the first instance to provide a proper growing medium.

If the job is too much for the home handyman, get help. Also read books on the subject. Have your soil tested. And make certain that there is grading and drainage.

After that, planting may come. The average family will find trees and shrubs costly. It is a better investment to spread out the purchases over a period of time than to try to do the whole job at once. This is not only easier on the budget, but ideas and tastes change from year to year.

Two warnings:

Everybody knows that little trees grow into big trees and that little shrubs grow into big shrubs. For some reason few gardeners can't seem to realize this when they are busily digging holes. The result is that in a few years the material becomes crowded and tangled and the gardener must start pulling out what he put in at so much expense.

Trees, shrubs and flower beds placed in the middle of grass areas are areas which must be worked around, and usually must have their edges trimmed. When planting a yard, think of your job as the man behind the mower and make the grass areas as clean and uncluttered as possible.